Teacher's Edition

Modern Biology
Laboratories

HOLT, RINEHART AND WINSTON

AUSTIN NEW YORK SAN DIEGO CHICAGO TORONTO MONTREAL

Developmental assistance by Ligature, Inc.

Special thanks to Carol Sue Davidson, New York City, and
Stephanie Shapiro, New York City.

Printed in the United States of America   89012 022 543

ISBN 0-03-013924-4

Table of Contents

The New Modern Biology Laboratories

This revision of *Modern Biology Laboratories* has been designed to teach students how to solve scientific problems in a critical and creative manner. To encourage student problem solving, *Laboratories* has been developed to be a more student-centered teaching tool. Both the visual design and the investigations themselves more vividly reflect the spirit of biological inquiry and encourage students to engage in it.

Students studying biology today do so at a time of significant change, not only in the science of biology but also in the society around them. As new technologies are developed, science presents as many new questions as it does answers. Our growing scientific knowledge must be applied within a modern social framework. Students, then, must develop the ability to identify the critical questions and evaluate their answers. Consequently, a basic understanding of scientific principles is becoming increasingly important if students are to become responsible adults and citizens.

Laboratories is an integrated approach to teaching these essential scientific thought processes as well as the concepts of biology. In this laboratory program, students are given the opportunity to develop hypotheses and draw conclusions, to generate their own questions, to design some of their own experiments. In short, they are given the opportunity to become independent and thoughtful learners of biology.

Primary to the development of these thinking skills are activities that spark and maintain students' interest. To this end, each investigation begins with an interesting yet challenging question which students should be able to answer after the investigation has been completed. Additionally, in the introductory sections of each investigation, students' past experiences and knowledge are related to new laboratory situations. Students are addressed in language that will help them feel comfortable in a laboratory environment. Procedures are clearly worded, explaining what to do, how, and why. Use of "you" and "we" and other vernacular allows students to overcome the confusion and uncertainty fostered by traditional scientific language and concentrate on the activity at hand. These features help to create a laboratory atmosphere that encourages an active and involving discovery of science, rather than a mere confirmation of text material. Elements of drama, mystery, and fun are not only permissable, but encouraged.

Inquiry Questions

Students are asked a variety of inquiry questions throughout the Procedure. The questions are designed to initiate student thought about not only the results, but the procedure steps themselves. The questions challenge the students and have been designed to be an integral part of the students' laboratory experience. Since questions are embedded directly into the Procedure, they will be answered as a part of the investigation process, not during study hall, after school, or during a late night telephone conversation.

Because these questions play such an integral role in the laboratory program, it is essential that students be made to feel that their answers are valid and significant. Students should be encouraged to report their own findings accurately and honestly . They should be discouraged from giving what they think are expected answers. Instead, they should be taught to carefully analyze their data. If data are inconsistent, they should be encouraged to examine it to pinpoint possible sources of error.

Process Skills

As students tackle each new investigation, they will continually be made aware of the skills they are developing as biologists. Their investigations will give them an appreciation of the scientific thought processes involved in the biological concepts they explore. These process skills are introduced and illustrated in Chapter 2 of the *Modern Biology* text and in the front matter of the pupil edition of this lab manual. The process skills are applied in *Process and Vocabulary Skills* also. Each investigation incorporates at least two process skills which are highlighted for both student and teacher. For example, the investigations on pH and calorimetry stress the process skill of measuring. The investigations on nucleic acid structure (the Velcro lab) and cell size (the agar cube lab) stress the process skill of modeling. In all labs, strategies for applying the process skills are given in the margin next to the step that focuses on that skill.

Variety Within Structure

In our efforts to promote a more student-centered learning enviroment, we have recognized the practical concerns of a modern high school laboratory environment. Throughout this revision, we have incorporated suggestions and field experience from both teachers and students concerning the content and the structure of the investigations. Our primary goal was to give flexibility and option to the teacher within the context of providing activities which would challenge and engage their students.

An interested and sound approach to science just does not happen in a student; it must be carefully and deliberately taught. To meet these teaching needs, the early labs in *Laboratories* feature more structured investigations (for example, the microscope lab and the cell structure lab) and builds toward student-initiated lab design by the end of the course (for example, the daphnia heartbeat and the interview lab on facts and fallacies in knowledge of infectious diseases). Throughout the program, students are encouraged to expand their application of the process skills, and teachers are given the support in the notes for each specific lab necessary to foster that growth.

Some investigations begin with a concrete procedure which teaches important scientific technique and then provides students with opportunites for branching activities in several Further Investigations. Other investigations allow students to design part of their own procedure with information given in teacher's edition notes to support this process. Still other investigations can be initial learning experiences that precede the reading of the text in order to develop student interest, especially those investigations where lab set up will be looked at over time (for example, the ecosystem-in-a-jar lab).

Time Required

An overriding concern of high school biology teachers is time, and usually the lack of it, for laboratory work. The laboratory program has been designed so that all investigations can be easily adapted for double lab periods. For those teachers not afforded this extended time, convenient stopping points have been highlighted for those labs that require more than the standard 40 minute period. Each investigation has been tested with students or reviewed by a practicing biology teacher in an effort to make these time allotments reasonable and accurate.

Teacher's Role

A student-centered laboratory program places additional responsibility on the teacher. Yet, the investment of time and effort put in by the teacher produces significant returns in the form of interested and thoughtful learners. As one teacher told us, "Sometimes it's difficult to make sure the students are seeing what they're supposed to be seeing without telling them what they're supposed to be seeing." Yet she, along with many other teachers heartily agree that a student-centered approach is "feasible, well worth the effort and infinitely better teaching."

Teachers are encouraged to see themselves as a coach or an aide during laboratory investigations. Students should feel as if they are being guided, rather than strictly evaluated during their lab period. It is suggested that teachers circulate among the students offering suggestions and tips and preventing unsafe practices.

Many of the investigations, especially those dealing with live animals, will produce a significant amount of the unexpected. Teachers must not be afraid to say "I don't know," and propose new questions. If the students' inquiry could lead to an answer they could discover for themselves, a teacher's reply of "I don't know" is suggested, even when the answer is obvious. Challenge students to discover science for themselves. Success is ridding classes of the "fill in the blank" mentality while preventing unsafe practices and answering answerable questions.

Cooperative Learning Atmosphere

Laboratories fosters an atmosphere of cooperative learning. Students are often encouraged to see themselves as part of a scientific team, to brainstorm ideas as a class, to discuss the labs with others as they perform it, and to pool the class data on a chalkboard. Yet, because this type of group learning is not traditional in other subject areas, it, like the scientific processes themselves, must be taught. Teachers are encouraged to help students grow as effective group members. Students should share tasks within the group. There should be no split between "doers" and "observers." Students of all ability levels should be expected to take an active role.

How to Use Modern Biology Laboratories

Modern Biology Laboratories is designed as a student-centered program, integrating *Modern Biology* text material into exciting and relevant laboratory experiences. Both the visual design and investigations themselves encourage students to be active learners of biology. While the investigations are numbered to correspond to a chapter in *Modern Biology* and there is at least one investigation for every *Modern Biology* chapter, this program seeks not only to confirm text material but also to extend and challenges students' knowledge of biology. *Laboratories* aims to involve students in the process of scientific discovery.

In many investigations, students are given the instructions necessary to develop hypotheses, to generate their own predictions, and to carry out experiments from which to draw conclusions. Yet, in other investigations, especially those with living things, students are asked simply to observe what is happening. Whichever scientific process they use, students are encouraged to be sensitive to the process and are given the opportunity to record their findings in a variety of ways. Moreover, they are given a number of formats in which to analyze their results and the results of their classmates.

The investigations in this program have been designed with three major components: Prelab Preparation (thinking about and planning what is to be done), Procedures (stepwise directions), and Postlab Analysis (interpreting data, pooling data, and error analysis). These three sections correspond to the thinking skills of planning, monitoring, and evaluating.

Investigation Question

Each investigation question seeks to capture student's interest in the activity. Curiosity is sparked and prior knowledge and experience are invoked in questions such as "How do you make sauerkraut?" and "In what ways do people and other mammals communicate non-verbally?" The investigation question gives the student an immediate purpose to do the lab, that is, to find the answer to the question. By the lab's end, they should be able to answer the question in their own words, through their own experience.

Materials

The materials that students need for the procedures are listed on the first page of each lab. Additional information about lab materials can be found in the frontmatter notes to the teacher for each lab. These notes offer practical and valuable information about preparation of solutions, maintenance of live animals and ordering or purchasing supplies. Moreover, tips and suggestions provide information about common pitfalls to be aware of, safety concerns and alternative materials. A complete list of the materials and supplies needed for the course begins on page T15.

Learning Objectives

The learning objectives are designed to highlight the correlation to text material and emphasize subject matter and content.

Process Objectives

The process skills introduced and illustrated in Chapter 2 of *Modern Biology* are integrated into each lab. A unifying theme throughout the lab manual is the emphasis on process skills. The process skills highlighted in each lab should serve as a focus for the students' development of scientific skills.

Strategies

To help students implement the process skills, strategies are located in the margin where the process skills are applied. The strategies include hints, guidelines, and ways to orient students' thinking about implementing the process skills as they work through the lab.

Introduction

An introduction attempts to relate the investigation with students' previous knowledge or life experiences. The introduction also provides the student with background information, especially when terminology is not presented in the *Modern Biology* chapter.

Prelab Preparation

The prelab sections of each investigation should be seen as an opportunity for students to become familiar with the activity before they enter the laboratory and to start thinking about what they will be doing and why. Prelab Preparation can be assigned as classwork the day before the investigation or at the beginning of a double lab period. It may also be assigned as homework the night before the investigation. For many investigations, you may want students to review the textbook chapter as a prelab activity, especially to emphasize specialized biological terminology. Other investigations may serve better as introductions to the textbook chapter. In all cases, however, students should be encouraged to review the whole investigation procedure during the prelab preparation.

The prelab questions have been designed to ensure that students have put a significant amount of thought into the activity and therefore are able to use their already limited laboratory time constructively. Prelab preparation time is also a good opportunity for you to demonstrate new techniques or to lead a class discussion about hypotheses and expectations, as well as preconceptions for the more open-ended labs. Prelab preparation time can also be a chance for students to design their own procedures with teacher and class support and input.

The teacher's notes in front matter provide suggestions for each lab for the most profitable ways to use this prelab time.

Procedure

As students perform the investigations, they are asked inquiry questions throughout the whole of the procedure. The questions are designed to initiate student thought about the results as well as the procedure steps themselves and to encourage them to think scientifically, rather than to follow seemingly arbitrary "recipes." The questions should be answered during the procedure with accurate and honest findings. Students should be encouraged to give their own findings, not what they think are expected "answers."

In many of the labs, students work in teams. Emphasis should be placed on cooperative learning. Students should be encouraged to discuss their work within the group, but at the same time to maintain their personal integrity. The group shouldn't copy answers, but should come to their conclusions through a well thought out process in which everyone actively participates.

Again, the teacher's notes provide you with common stumbling blocks to alert students to. The notes give hints necessary to get a certain step in the procedure to work, as well as special safety precautions to take. If a lab asks students to design their own procedure, one or more procedures that will work are given, as well as ways to support student creativity.

Postlab Analysis

Postlab Analysis can be completed at the end of the lab period or at the beginning of the next class. While some graphing and data analysis can be assigned as homework, the sharing process of class discussions is especially important. It is an integral process of the lab and should be given as much importance as doing the procedure itself. What matters more than when it is done is how it is done. Students should be encouraged to analyze their findings carefully. Individual data should be compared with pooled class data and known values when possible. Students should be asked to critique the procedure and to evaluate their performance of it. They

should also suggest changes to help deal with sources of error. Students should not be left with the feeling that because they didn't get the "right" answer that they failed to accomplish what they set out to do. One of the purposes of Postlab Analysis is to give validity to the experience of even finding "wrong answers" by emphasizing that the process of doing science is primary.

Postlab Analysis gives students the opportunity to review and revise hypotheses and predictions made earlier in the lab. It also allows them to think about what extension of their investigation might be of interest. The teacher's notes support this analysis.

Further Investigations

The Further Investigations, following the Postlab Analysis, offer alternative or more open-ended investigations. Most of these give directions for a particular exercise or field research. Students in an advanced course should be encouraged to develop their own laboratory reports using selected Further Investigations. The Further Investigations may also be used for independent research or extra-credit projects.

Safety in the Laboratory

A laboratory is a safe place for any one who knows what potential hazards exist in it. Carelessness and apathy toward safety are the major causes of accidents. Most accidents can be avoided with proper planning and with good safety habits on the part of both teacher and students. Sound safety habits and procedures, however, must be carefully and deliberately taught. Students must be taught specific techniques, proper handling of chemicals and apparatus, and emergency procedures, beginning with the introductory laboratory session and repeatedly throughout the year.

Introduction-to-Safety Lesson

A discussion of safety is necessary in the first laboratory session of the year. Point out the location and proper use of safety equipment and first aid kits. Demonstrate specific techniques, such as lighting Bunsen burners, disposal of broken glass, and mixing or heating solutions.

Review the safety symbols used throughout *Modern Biology Laboratories*. The safety symbols are clearly defined in the front matter of the pupil edition of this lab manual and safety rules and safety tips for each type of laboratory hazard are clearly presented also (pages xiii–xiv). Discuss these with your students and have them complete the Review of Laboratory Safety on pages xv–xvi.

You may want to have your students and their parents or guardians sign a Student Safety Contract. A copy of a standard safety contract is given in the *Teacher's Program Guide*.

Safety During the Year

For laboratory investigations that are new to you, try to find time to perform experiments yourself before assigning them to students. Demonstrate specific techniques and review safety procedures during the Prelab Preparation (for example, labs involving the transfer of microbial cultures, extraction of chlorophyll with alcohol, use of ultraviolet lamps). During Prelab Preparation highlight notations of CAUTION that appear in the Procedure section. Be consistent in insisting on the use of goggles, lab aprons, or gloves in all laboratory procedures where they are appropriate, whether students are performing techniques or observing.

If an accident does occur, be sure that broken glass and spilled chemicals are cleaned up and disposed of properly. Be sure you know how to use fire safety equipment when necessary. Give students first aid treatment for cuts, burns, or animal bites, and get immediate word to another adult if the situation warrants it.

Even if a student's injury seems minor, it is best to notify the school nurse and the student's parents. Your school system may require an accident report. A recommended form for Incident/Accident Reports has been given in the *Teacher's Program Guide*.

Students should never conduct investigations alone in the laboratory, nor should they perform unauthorized experiments.

Response to Infectious Disease Concerns

The likelihood of a student being contaminated with the AIDS virus or with any other infectious disease agent in a laboratory setting is very rare. We have, however, responded to the concerns of many science teachers, science department heads, and school administrators by modifying some of the investigations in this manual that involve body tissues and fluids. Students will examine only prepared slides of cheek cells and blood cells. Additionally, they will not be required to type any blood samples.

Maintaining a Safe Environment

You should inspect your classroom at the beginning of every school term and report anything that is unsafe or inoperative. You may want to use the School Safety Audit Form given in the *Teacher's Program Guide*. Enlist the support of other science teachers in insisting that your classrooms be safe for laboratory work. Frequent inspection of all equipment is essential throughout the year.

Make sure that all of the following are easily accessible and are in good working order:

first aid kit
multipurpose ABC-type dry-chemical fire extinguisher
fire blanket
sand bucket
eyewash facilities
emergency shower facilities
goggles, aprons, gloves (sufficient for all users)
specially marked containers for disposal of broken glass

Check the plugs and wiring of hot plates, incubators, water baths, and light sources prior to student use. Be sure you are familiar with the directions for operating a pressure cooker or autoclave. Sterilize materials in a classroom when students are not present or in another room. Students should never be allowed to operate a pressure cooker (autoclave).

Follow safe procedures yourself, not only when demonstrating techniques to your students but also in the preparation of solutions and other materials for each lab. Remember that you will often be handling acids and bases in a more concentrated form than will students, handling bulk quantities of toxic chemcials, and doing lab preparation in a room by yourself. Follow directions correctly, without shortcuts, for diluting acids and bases and for handling toxic or flammable materials.

Label all chemicals adequately (with name, concentration, date, and precautions) both on stockroom containers and especially on containers to which students will have access. Store chemicals by category (for example, flammable materials are ideally stored under a hood with adequate ventilation). Chemical storage areas should be secured at all times when not in use, with access to authorized personnel only.

Disposal of Chemicals

Disposal practices should be designed to avoid creating a safety hazard or damaging the environment. In most cases, exercising common sense should be sufficient. Keep in perspective that many substances (acetic acid, sodium bicarbonate, and sodium chloride, for example) are found in the average kitchen and many other substances (sodium phosphate salts in laundry detergents, strong acids and bases in household drain cleaners, and hypochlorites in household cleaning solutions, for example) routinely end up, in diluted form, in municipal sewage systems.

You should be aware of local, state, and federal regulations that govern which chemicals can be disposed of in municipal sewage systems, in sanitary landfills, or by commercial disposal companies, and also what quantities can be disposed. Usually the quantities of chemicals generated by lab experiments are relatively small, but if you are disposing of large quantities (cleaning out an obsolete store room, for example) you may want to check with governmental authorities or contact the manufacturer.

Before disposing of acids, neutralize the acid (to between pH 6 and 8) by adding an excess of sodium carbonate to the solution (verify with pH paper), and flush down the drain with large amounts of water. Similarly, basic solutions should be added carefully to a large container of water and neutralized by adding acid (verify with pH paper). Note that neutralization reactions generate heat, especially if the concentration of acid or base is greater than 1 N. The organic chemicals used in this lab manual, such as ethyl alcohol, isopropyl alcohol, acetone, or petroleum ether can be flushed down the drain using large amounts of water. Be care-ful to avoid splashing and continue to run tap water even after the chemical has cleared the sink.

Dispose of solids by placing them in a waste container designated for disposal in a local sanitary landfill. For your safety and that of your students, provide a separate, clearly labeled container for disposal of broken glass.

Disposal of Microbial Contamination

Contaminated growth medium, used plastic petri dishes, and contaminated cotton plugs should be autoclaved before disposal. Glass petri dishes, flasks, and test tubes should be soaked in a strong disinfectant (such as a 1:10 solution of liquid bleach) before washing and autoclaving.

Field Trip Safety

If you are planning to take your students out of the classroom, as you will, for example, in Investigation 52, you should visit the site before the class visits it. You should note during your preliminary visit any dangers, such as a steep dropoff, deep or swift water, plants to which students could react or be allergic, and animals (ticks, bees, snakes, etc.) Check with the school nurse for a record of students who have a history of allergies or special health problems. Bring a complete first aid kit and know how to use it. When containers for collecting are necessary, plastic jars and bottles are preferred over glass. Do not use mercury-filled thermometers in field situations. Upon returning from the trip, students should wash their hands, arms, and faces with soap and water. Report all accidents to the school administration, school nurse, and parents.

Poisonous Plants

Plant poisoning can involve allergies, dermatitis, and ingestion. Dermatitis is probably the most common to occur as a result of field work and can involve a number of irritating species:

Giant hogweed	Poison oak
Wood nettle	Poison sumac
Poisonwood	Stinging nettle
Poison ivy	

Allergic students will probably have the most trouble with field trips taken during the fall, especially during ragweed season. Students who suffer with allergic reactions, such as hay fever or asthma, should be identified.

Many plants or the parts of many plants (seeds, fruits, leaves, flowers, berries) can be highly toxic when ingested. The following plants are known to be poisonous:

House Plants
    Hyacinth, Narcissus, Daffodil (bulbs)
    Oleander (leaves, branches)
    Diffenbachia, Elephant's-ear (all parts)
    Rosary pea, Castor bean (seeds)
    Poinsettia
    Mistletoe (berries)

Flower Garden Plants
    Larkspur (young plant, seeds)
    Monkshood (fleshy roots)
    Autumn crocus, Star-of-Bethlehem (bulbs)
    Lily of the valley (leaves, flowers)
    Iris (underground stems)
    Foxglove (leaves)
    Bleeding heart, Dutchman's breeches (foliage, roots)
    Rhubarb (leaf blade)

Trees and Shrubs
    Wild and cultivated cherries (twigs, foliage)
    Oaks (foliage, acorns)
    Elderberry (shoots, leaves, bark)
    Black locust (bark, sprouts, foliage)

Wooded Area Plants
    Jack-in-the-pulpit (all parts, especially roots)
    Moonseed (berries)
    May apple (apple, foliage, roots)

Field Plants
    Buttercup (all parts)
    Nightshade (all parts, especially the unripe berry)
    Poison hemlock (all parts)
    Jimson weed (thorn apple) (all parts)

Poisonous Animals

On field trips, discourage students from catching and handling mammals. Bites can result in infections and there is the potential hazard of contracting rabies. Students should be discouraged from handling snakes unless under controlled and closely supervised circumstances. You should know if there is a serious problem with venomous snakes in your part of the country. You should warn your students to be extremely cautious when handling all organisms. Even seemingly harmless animals such as fish have spines that can cause very painful infections. Marine animals such as mollusks, sea urchins, and corals can also cause a dangerous wound. Other dangers in handling marine organisms include allergic reactions from things such as egg cases of whelks and the clusters of algae that may hide in many marine worms.

Probably the greatest danger during a field trip is from insect bites and stings; more people in the U.S. die each year of allergic reactions to insect bites and stings than from snake bite. In addition, mosquitoes and ticks are carriers of several infectious diseases. Long sleeve shirts and slacks that reach to the ankle should reduce the chance of bites and stings.

Ants	Jellyfish
Bedbugs	Millipede
Bees	Mosquitoes
Black Widow Spider	Mussels*
Blister Beetles	Nettling Caterpillar (Slug
Brown Recluse Spider	Caterpillar)
Centipedes	Oysters*
Chiggers	Pussmoth (Saddleback
Clams*	Caterpillar)
Copperhead Snake	Potato Beetles
Coral Snake	Rattlesnake
Cottonmouth Snake	Ticks (Dermacenter and
Fleas	Ixodidae)
Gnats	Wasps
Ioa Caterpillar	Yellow Jackets

*Toxic when living in polluted water or feeding on certain dinoflaggellates

Animals in the Classroom

From their earliest experiences with biology, students are taught the meaning of the word's roots—"bio" and "logy." For too many students, however, the "study of life" has been almost exclusively associated with lifeless animals in bottles and jars. This edition of *Modern Biology Laboratories* has been designed to be a more comprehensive and ultimately more meaningful tool for biology students. To this end, observations of live animals have been included in a number of the investigations. Our aim is to instill in students an interest in, understanding of, and respect for all living things within the context of their daily lives.

Since most high school students will not pursue science studies after graduation, it is increasingly important that their year spent in the biology laboratory give them a solid understanding of the natural world around them. As adults, some of the most pressing societal issues they will face will require sound biological knowledge. To confront these issues, young people need to understand and appreciate living organisms, relationships among organisms, and finally, the impact of humans on these organisms and the environment.

A major goal of biology teaching should be to lead students toward an actual encounter with nature. Isolated studies of tissues, organs and organ systems, without the experience of observing these living systems at work, denies students the excitement of the discovery of nature. While some things can be learned by passively poking at a preserved frog's hearing organs, other things can be learned by devising an experiment to determine just what a live frog can hear.

Animal Care

Having live animals in the classroom allows long-term observation (for several months, in some cases), an experience for students that extends beyond the confines of a one-period lab. It is also an added responsibility for you. As a biology teacher, you must create clear educational objectives for the use of live animals and assume a commitment for the responsible care of the animals. It is important that they not be mistreated. Prior to the arrival of the animals, provisions should be made for feeding and caring for them as well as for keeping them in a proper environment. Students should be instructed about how to handle the animals, for the protection of both the animal and the student.

Selection of proper animals

Before selection, it should be certain that a proper classroom environment can be created and maintained in the classroom for each species. Make sure that the proper temperature, space, and feed can be provided. Animals used in *Modern Biology Laboratories* can be obtained from biological supply houses or pet stores. Be sure that your sources have maintained the animals in good condition. If the animals you recieve are shipped in water-filled containers, place the shipping bag in or next to the holding tank until the temperature of the water in the shipping bag matches that for the water in the tank; this may take one to two hours. In addition, let animals become settled and acclimated to their cage or tank for several days or a week before they are handled in lab investigations.

Wild mammals and birds are generally not appropriate for use in the classroom. Their needs usually cannot be met and removing them from their habitat is inconsistent with the development of student respect and appreciation for living things and their environment.

Small, native animals (examples include frogs, toads, salamanders, insects, spiders, and earthworms) can sometimes be collected if their habitats can be simulated in the classroom. Arrangements, however, must be made to keep the animal in an area that resembles its natural environment, to ensure that the animal is handled infrequently, and to return the animal to its natural habitat after observation. Animals such as snapping turtles and poisonous snakes are not appropriate for the classroom. Also, be aware that many turtles carry *Salmonella*, and other animals carry external parasites. In addition, many local species may be on the list of "Federal Endangered/Threatened Species"; check with the U.S. Fish and Wildlife Service or the appropriate agency in your state before collecting species that may be under federal or state protection.

Providing the proper environment

Animals have special environmental requirements. Animal cages should be placed away from windows, radiators, and air conditioning ducts. Also, be aware that glass enclosures are susceptible to overheating and should be placed away from direct sunlight. Space allotment should be sufficient, and appropriate materials for behavioral needs, such as nesting, bedding or gnawing should be provided. Animals, especially small rodents, need a quiet place where they can rest or hide. Be sure to keep predator species at a distance from prey species. Even the sight or smell of one another can cause them great anxiety.

Fresh water aquaria

Aquarium water should be specially conditioned. Water from the tap should be dechlorinated by leaving it at room temperature for a day before it is used in a fresh-water aquarium. Aquarium plants may be especially appropriate for the habitat you are creating. Find out if the species you will keep need hiding places (rocks or even flower pots) and if they have special requirements for aeration, temperature, and so on. Once a tank is set up, it should never be moved or lifted for pouring; such stress will crack the seams.

Marine aquaria

A marine aquarium can be an exciting part of the classroom but generally will require more effort than a freshwater tank. For example, marine systems need a high rate of filtration (both mechanical and biological) and aeration, pH buffers, and, for northern species, a cool temperature. In addition, metal-bound aquaria are inappropriate because of corrosion. You must also decide if obtaining unpolluted natural seawater is practical or if use of synthetic salt mixtures will work best.

Terraria

A terrarium is a place for keeping and raising terrestrial animals. Terraria are usually made from a glass or plastic container, most often from an old glass aquarium tank that is no longer watertight. Small branches and rocks for climbing and hiding, a dish of water for swimming and drinking, and living plants should be chosen according to the size and species of the animals kept.

Feeding

Keep a record of who is feeding the animals and how much they are being fed. Also, be sure that the animals always have access to water and that the type and amount of food they receive is consistent with their nutritional needs.

Weekends and School Vacations

Arrangements must be made so that live animals are cared for over the weekends. If a proper level of care for animals cannot be maintained during periods when school is not in session, the animals should go home with you, other teachers, or with responsible students.

Humane experiments

All animals must be used in accordance with state and local laws. Experiments involving pathogenic, radioactive, carcinogenic, or toxic agents of any kind are prohibited to students.

Any experiment to be done with live animals should have clearly defined objectives and should be directly supervised by a qualified teacher. No experiment should cause pain or suffering.

Dissections

An increasing number of students have begun to object to dissecting animals on religious or moral grounds or for personal reasons. Moreover, many educational and scientific organizations are questioning the need to kill millions of animals for high school biology laboratories. Whether a student can learn better from doing a dissection or by working with models, diagrams, audiovisuals, and software has become a controversial topic. For *Modern Biology Laboratories,* we have chosen not to include dissections of animals such as earthworms, crayfish, starfish, and grasshoppers since for most students behavioral observations foster a greater respect for living organisms.

The frog lab contains the one dissection procedure in *Modern Biology Laboratories* and it is clearly marked as an optional exercise. Additionally, in a few animal behavioral labs, we have suggested dissection as an extension activity for only the very interested and capable student.

In the pupil's front matter for this lab manual, we have recommended that students research the issues of dissection and animal research by reading current literature and contacting individuals and groups with varying points of view on the subject. We suggest that you encourage your students to engage in this research and support their efforts to investigate all aspects of the issue.

Listed below are some resources for alternatives to dissection and for care for live animals. It is impossible to include a complete listing of all available products. Inclusion here is not intended to imply endorsement, nor does omission imply a negative evaluation. We also recommend that you send for catalogs from suppliers listed in the *Modern Biology Program Guide*. All audiovisual and computer software equipment should be previewed before you present it to the class to ensure that it is appropriate.

Resources

Print

Institute of Laboratory Animal Resources (ILAR) of National Resource Council, *Guide for the Care and Use of Laboratory Animals,* Animal Resources Program, Division of Research Resources, National Institutes of Health, (NIH), Bethesda, Maryland, 1985 (NIH Publication No. 85-23)

Daniel E. James, *Carolina Marine Aquaria,* Carolina Biological Supply Company, 1975.

Daniel E. James, *Carolina's Freshwater Aquarium Handbook*, Carolina Biological Supply Company, 1981.

Wynn Kapit and Lawrence M. Elson, *The Anatomy Coloring Book*, Barnes and Noble, 1977.

Lawrence M. Elson, *The Zoology Coloring Book*, Barnes and Noble, 1982.

Thomas Niesen, *The Marine Biology Coloring Book*, Barnes and Noble, 1982.

Guiding Principles for Use of Animals in Elementary and Secondary Education, Humane Society of the United States (HSUS), 2100 L Street, N.W., Washington, DC 20037

Endangered Species Handbook, Animal Welfare Institute, P.O. Box 3650, Washington, DC 2007.

Alternatives to Current Uses of Animals in Reserach, Safety Testing and Education, National Association for the Advancement of Humane Education, P.O. Box 362, East Haddam, CT 06423.

Software

"Operation: Frog", Scholastic, Inc., P.O. Box 7502, 2931 East McCarthy Street, Jefferson City, MO 65102. This program is for Apple II family and for Commodore 64. It simulates procedures of frog dissection and provides students with an opportunity to "reconstruct" the frog.

AV

"The Frog Inside-Out," Instructivision Inc., 3 Reagent Street, Livingston, NJ 07039. Videotape in VHS, Beta II or 3/4 U-matic format.

Models and Preserved Specimens for Demonstration

bio-LOGICAL™ Models, National Teaching Aids, Inc., 1845 Highland Avenue, New Hyde Park, NY 11040. Models can be dissected, assembled, and manipulated.

Biosmount, ™ Biocast ™ Bio-Dri Mounts, Bio-Plastic Mounts, and Bio-Mounts as well as many plastic models and similar products are available from Carolina Biological Supply Company, Connecticut Valley Biological Supply, Ward's Biology and other suppliers.

Charts

Denoyer-Geppert Science Company, 5711 North Milwaukee Avenue, Chicago, IL 60646.

Nystrom, Division of Herff Jones, Inc., 3333 Elston Avenue, Chicago, IL 60618. Also supplies models.

Equipment and Materials for the Year

Biological Materials, Living
Ameba culture, 21
Anguillula acetiglutinus (vinegar eels), 35
apple or carrot, fresh, 4
bivalves and bivalve extract, 34
brine shrimp eggs *(Artemia)*, 49
cabbage, shredded, 7.2
caterpillars (such as *Isia isabella* or *Gallevia mellonella)*, 33
chicken bones, 41
    meat, fresh, 4
Closterium (algae), 22
crayfish, 32
Daphnia culture, 43.1, 29
Dugesia (flatworms), 30, 35
eggs, raw, 3.1, 6
Elodea (Anacharis), 5, 24.2
Escherichia coli culture, 16, 20
Euglena culture, 21
fern fronds, 25
    gametophytes, 25
    spores, 25
    sporophytes (with sporangia), 25
flowers, such as gladioli, 27
food for insect larvae, 33
    for reptiles, 38
    pellets, 32
frogs, 37
fruits (including single-pit fleshy fruits, pomes, berries, wind-carried seeds, nuts, burrs, and coconuts), 27
goldfish, 36
Hydra culture, 29, 35
insects (crickets or mealworms), 37
leaves, fresh (such as *Coleus, Zebrina,* maple, birch), 7.1
liver, fresh, beef or chicken, 4
Lumbricus (earthworms), 31, 35
mealworms *(Tenebrio molitor)*, 33
Metridium (sea anemones), 35
moldy foods, 23
Mung beans, 15
onions, 5
Paramecium culture, 21
plants (such as geranium or ivy), 24.1, 26
plants and animals (such as earthworms, crickets, seeds, *Anacharis,* guppies, platies, mealworms, pond snails, *Daphnia, Chlamydemonas,* duckweed, isopods), 50
potatoes, 2, 4
reptiles (such as lizards, turtles, snakes), 38
rodents, small, 40
Sarcina subflava culture, 20

sea stars, 34
seeds, bean, pea or corn, 53
    oat (Avena sativa), 28
Serratia marcescens culture, biotype D1, 12
spinach leaves, fresh or frozen, 7.1
Tradescantia cuttings or fresh onion bulbs, 9.2
unknown specimens (such as unprocessed yogurt, blue cheese, bread mold, lichen), 1
yeast, dry baker's, 21, 49

Biological Materials, Preserved
birds, taxidermic specimens (or charts) including hawk, owl or similar bird of prey, loon or grebe, quail or partridge, wood thrush or robin, pelican, hummingbird, heron or egret, duck, kingfisher, and dove or pigeon, 39
frogs, 37
Nereis (marine worm), 35
owl pellets, 51
Romalea (grasshopper), 35
skeletons of fish, frog, pigeon, turtle, cat, and human (or charts), 41
mollusks (including clam, snail, squid), 35
unknown specimens (such as coral or shells, insects, tapeworm in acrylic block, fossil, loofah), 1

Chemicals, Stains, and Preparations
acetic acid, glacial, 9.2
acetone, 7.1
adrenaline or epinephrine, 43.1
agar, nonnutritive, 9.1
    nutrient, 12, 16, 20
    potato dextrose, 23
aspirin or other headache remedies, 43.1
bleach, household, 34, 53
bromophenol blue, 9.1
bromothymol blue, 3.1, 24.2
broth, beef, 29
    nutrient, 16
chloroform, 9.2
Clinistix or Testape, 45.2
coffee and/or cola drink, 43.1
common substances (such as vinegar, baking soda, shampoo, orange juice, tomato juice, grapefruit juice, soda water, cola, fabric softener, household ammonia, bleach liquid detergent, salt water, milk), 3.1
Congo red, 14, 21
diastase malt powder, 45.2
distilled water, 2, 3.1, 6, 9.2, 20, 21, 24.2, 26, 45.2, 49, 53
ethyl alcohol, 2, 7.1, 9.2, 20, 24.1, 43.1

Expendable Materials

toothpicks, flat-edged, 5, 7.2, 29
tubes, cardboard, 19
unknown specimens (such as cork, synthetic sponge, pumice or sand, powdered charcoal.), 1
Velcro strips, 3 colors, rough and smooth, 8
water, dechlorinated, 30, 32, 36, 50
wood or cork board (5 cm x 30 cm), 33
wooden splints, 24.2

Glassware
beakers, 50-mL, 45.2
   200–mL, 6
   250-mL, 24.1, 42, 53, 9.1
   400–mL, 6
   500–mL, 6, 36
   600-mL, 24.2
   1000-mL, 24.1
   1–L, 36
   Berzelius, tall form, 100-mL, 45.2
culture dishes, 49
culture tubes, 16, 20
cylinders, graduated, 10-mL, 3.2, 4, 7.1, 14, 22
   25-mL, 6, 45.2
   100-mL , 22, 24.1
dishes, 35
dropper bottles, 9.2
finger bowls, 31
flasks, Erlenmeyer, 1-L, 22
   125–mL , 3.1
   250-mL, 22
funnels, 6, 24.2
glass jars, large, 50
glass plates, 9.1, 25, 27
glass rods, 3.1
medicine droppers or pipettes, 2, 3.1, 9.2, 14, 21, 25, 26, 29, 30, 31, 34, 43.1, 45.2, 52, 49
petri dishes, 24.1, 27, 28, 30, 51, 52
petri dishes, 15-cm (or culture dishes), 29, 31
petri dishes, sterile, 12, 16, 20, 23
pipettes, 10-mL (with pipette bulb), 22
slides, depression, 29, 43.1
stirring rods, 4, 7.1, 14
stirring rods, plastic, 36
test tubes, 14
   15-cm length, 7.1
   25-mL, 3.2, 6, 45.2
   small, 3.1, 4, 24.2

Permanent Equipment
aquarium, 32, 36, 37
aquarium, marine, 34, 35
autoclave or pressure cooker, 12, 16, 20

baking pans, rectangular (glass, glazed ceramic, or enamel), 9.1, 49
balances, metric, 3.2, 6, 49
bar magnets, 30
bell jar, 37
blunt probes, 27, 32, 34, 37
books, 42
boxes, clear plastic, 32
burners, Bunsen (or alcohol lamps), 12, 16, 20, 23,
cages, small (or box) , 40
calculators, 11
casts or reproductions, life-size anthropoids, 17
compasses, magnetic, 30, 33
compasses, pencil, 19
container, plastic with tightly fitting cover, 7.2
dissecting needles, 2, 19, 29, 51
dissecting pans, 37
dowel, wood, 38
field and identification guides , 39, 51, 52
fish-feeding rings, 49
flashlights, 31, 32, 49
flowerpots, clay, 40
forceps, 2, 4, 5, 9.2, 23, 25, 26, 37, 51
   bent tips, 20
   fine, 29, 52
gloves, leather or thick, 40
goggles, ultraviolet, 16
hammer, 52
hole punch, 31
hot plates, 9.2, 24.1
incubators (2), 12, 16, 20
inoculating loops, 12, 16, 20
lab aprons, 3.1, 6, 14
laboratory tubing, rubber, 42
lamps, incandescent, 24.2
light source, ultraviolet, 16
light source, 24.1
metersticks, 33, 48, 52
microscopes, compound, 1, 2, 5, 7.2, 9.2, 14, 21, 22, 25, 26, 29, 34, 43.1, 47, 49
microscopes, dissecting or hand lenses,1, 23, 25, 27, 29, 30, 31, 34, 35, 49, 51, 52
microscopes, dissecting 30
mirror, hand-held, 11
mortar and pestle, 7.1
newspaper, 2
pots, plastic (2-inch diameter), 28
protractors, 17, 19, 28, 33
pumps, aeration, 49
razor blades, single-edged (or scalpels), 2, 4, 9.1, 9.2, 25, 26, 27, 28,
right triangles, 52

Prepared Microscope Slides

Preparing Solutions

Many of the stains, solutions, media, and reagents in *Modern Biology Laboratories* can be ordered from a supply house in a ready-made form. Ready-made preparations require less of your time but are much more expensive than powders, dehydrates, or concentrates that require weighing, reconstituting, or dilution. Detailed instructions for preparation of all materials are included as part of the notes to the teacher for each specific investigation in the front matter of this teacher's edition. Read the notes for each investigation well in advance for the lab day so you can allow sufficient time for preparatory work. Coordinate your scheduling of labs with other biology teachers in your department so you can share in preparing larger batches with less individual effort.

Labeling and Storage

Each stain, solution, growth medium, and reagent should be labeled with the name, concentration, the date, and any precautions needed in using the reagent, such as TOXIC, POISONOUS or FLAMMABLE. Follow directions on stock bottles or in the notes to the teacher for specific investigations for proper storage, for example: under refrigeration, in brown bottles, sterilized. Flammable or explosive materials (ethyl alcohol, isopropyl alcohol, formaldehyde, petroleum ether) should be stored separately, ideally under a hood with adequate ventilation. Make certain solutions and dilutions have been made fresh for the day of use, if directions call for this. Make a note of expiration dates for antibiotic disks or enzyme extracts.

Sterilizing Media

For certain labs, in particular Investigations 12, 16, 20, and 23, it is necessary to sterilize culture media (agars and broths), equipment, and glassware. Before sterilization, hydrated culture media are poured into test tubes or flasks closed with cotton plugs or loosely screwed-on plastic caps. Media can be sterilized in a pressure cooker or an autoclave at a pressure of 15–20 pounds of pressure for 15–20 minutes. Agar can be poured into sterile petri dishes by melting first by boiling and will solidify when cooled to about 40°C. While glass petri dishes can be reused with thorough washing and sterilizing, presterilized disposable plastic petri dishes are highly recommended for their convenience.

Acids and Bases

Particular caution should be exercised in preparing acid and base solutions. Never mouth pipette concentrated acids or bases. Never add water to concentrated acids or bases. Add the concentrated chemical slowly and carefully to water to avoid bubbling and splashing. Note that the dilution of acids and bases can be a highly exothermic reaction.

Student Assistants

The use of student assistants in the preparation of materials for laboratory investigations can prove very satisfactory, either with highly motivated students who may have an interest in science careers or with older students who have taken biology in years past. Student assistants should be well trained and well supervised at all times.

Precentage Solutions

In preparing solutions by weight, dissolve a number of grams of material equal to the desired percentage concentration in enough solvent to equal 100 mL. For example, a 10% sodium hydroxide solution is prepared by dissolving 10 g sodium hydroxide in 90 mL distilled water and mixing thoroughly and carefully.

In preparing concentrations of solutions that must be measured by volume, first measure out the number of milliliters of concentrated solution equal to the percent of the final dilution. Then add enough distilled water to bring the total volume in milliliters to a number equal to the percentage of the original solution. For example, to prepare a 60% solution of alcohol from a 95% solution, measure out 60 mL of the 95% alcohol and add sufficient distilled water to bring the volume to 95 mL.

Molar Solutions

To prepare a molar solution, dissolve the number of grams equal to the molecular mass of the substance in distilled water (or other solvent) and dilute to 1 liter. For example, the molecular mass of sodium chloride is 58.45; therefore 58.45 g NaCl dissolved in enough water to make a final volume of 1 liter would be a solution of 1 M NaCl. If a 0.1 M solution of NaCl is desired, 5.85 g NaCl is dissolved in water to a final volume of 1 liter (or 100 mL of 1 M NaCl is diluted to a final volume of 1 liter).

Directory of Suppliers

AO Reichert, Scientific Instruments
P.O. Box 123
Buffalo, NY 14215

Bausch & Lomb
Scientific Optical Products Division
1400 North Goodman St.
P.O. Box 450
Rochester, NY 14602

Carolina Biological Supply Co.
2700 York Rd.
Burlington, NC 27215

Connecticut Valley Biological Supply Co., Inc.
82 Valley Rd.
P.O. Box 326
Southhampton, MA 01073

Difco Laboratories
920 Henry St.
Detroit, MI 48232

Edmund Scientific Co.
101 East Glouster Pike
Barrington, NJ 08007

Fisher Scientific Co.
Education Materials Division
4901 West Lemoyne Ave.
Chicago, IL 60651

LaPine Scientific Co.
6001 South Knox Ave.
Chicago, IL 60629

Nasco
901 Janesville Ave.
Fort Atkinson, WI 55538
or
West P.O. Box 3837
Modesto, CA 95352

Sargent-Welch Scientific Co.
7300 North Linden Ave.
Skokie, IL 60076

Science Kit, Inc.
777 East Park Dr.
Tonawanda, NY 14150

Turtox, Inc.
500 West 128th Pl.
Alsip, IL 60658

Ward's Natural Science Establishment, Inc.
5100 West Henrietta Rd.
P.O. Box 92912
Rochester, NY 14692
or
11850 East Florence Dr.
Sante Fe Springs, CA 90670-4490

Notes and Answer Key

Investigation 1

Notes for the Teacher

Time
40 minutes. Advance preparation: 3 days before lab, soak mung beans; a few hours before lab, prepare the yeast and sucrose.

Materials
Specimens: Place 10 "unknown" specimens at different stations. You may want to provide descriptive labels providing their age, source, or location of discovery. Other information, such as that provided by a cross section on a microscope slide, may be included. Since students will not learn how to use a microscope until Investigation 2, you should provide them with the necessary information to use the microscopes correctly. Instruct them not to change any of the settings or to disturb the specimens.

1. unprocessed yogurt (living cultures of lactobacilli; prepared slide under high power of microscope)
2. blue cheese or mold on bread (living molds; wet mount under low power of microscope)
3. dried mung beans (dormant; optional, germinate with moisture)
4. dried yeast (dry powder, as well as activated culture; wet mount under high power of microscope, with Congo Red vital stain if desired.)
5. pumice or clean sand (prepare an aqueous preparation as for yeast; wet mount under high power of microscope)
6. powdered charcoal (prepare an aqueous suspension; wet mount under high power of microscope; this processed wood was once alive)
7. coral, seashells, sea sponge, or snail shell
8. dead insect
9. potted plant (especially cactus or stone plants)
10. plant membrane (on a slide, under low power of the microscope; use an iodine solution as stain)
11. lichen on a rock (a partnership of an alga and a fungus; include a prepared section on a slide under low power of the microscope)
12. cork (whole specimen; also thin section under low power of the microscope)
13. stained tapeworm specimen in acrylic block (use dissecting microscope to demonstrate the impressive cellular organization)
14. fossil (inanimate; all living parts were substituted by minerals)
15. loofah (*Luffa*) "sponge" (actually a dried plant stem, available in health stores and in department stores)
16. synthetic sponge (and/or natural sponge)

Three days before the lab, soak a tablespoon of mung beans in 125 mL of lukewarm tap water overnight. Drain and sow between wet paper towels in a plastic bag, place in a covered container, and keep in complete darkness until use.

A few hours before the lab, put a half tablespoon of dried yeast in 125 mL of lukewarm tap water with 2 tablespoons sucrose.

Randomly distribute the specimens among the stations so that living, nonliving, and inanimate objects are not grouped together.

Text and Lab Correlation
Chapter 1.

Prelab Preparation
For many students, this may be their first exposure to "life" in a biological sense, and they may have difficulty identifying what is or once was living. Some may classify only those organisms with fur as being animal. A general introduction to life forms that they may never have seen before can be very helpful.

It is not important to get the "right answer." Students should develop an appreciation of the common-sense skills that biologists use to test and measure their observations.

Procedure
Wherever possible, have duplicate stations to allow more opportunity for the students to review a difficult specimen. If the number of stations adds up to the total number of students in the class, you will be able to have one student at each station and do a timed rotation for each observation. Circulate among the students, particularly at the stations with microscopes, to be sure that they remain in focus. Encourage sharing information; teams of five students may work well together.

Postlab Analysis
Assist students in weighing the relative significance of their observations. The deductions of their classmates not only add insight but illustrate how different conclusions may be reached.

Further Investigations
2. Although the moon rocks are still being studied, scientists have yet to find any evidence of life in or on the rocks.
3. Viruses are essentially nucleic acids and protein and cannot live or reproduce on their own; they can thrive and reproduce only in cells. However, they also can avoid destruction in very inhospitable environments for very long periods.

Answer Key
1. Examples: Eat, move, see (respond to light), hear (respond to sound), feel (respond to pressure), smell (sense molecules dissolved in air or water), vocalize, respire, excrete wastes, heal wounds, reproduce, grow, produce energy.
2. Plants synthesize food directly from light energy. In general, they move only as a function of growth. Animals are unable to manufacture energy directly.

They are often capable of movement.

3. An organism generally grows by increasing the number of cells it contains. An icicle grows when more water molecules freeze on its surface.
4. See students' data charts.
5. Answers will vary.
6. Answers will vary. Library materials and more informed persons are sources of information.
7. Answers will vary.
8.–10. Answers according to conclusions in data chart.
11. Answers will vary but, in general, will be the converse of traits of living organisms.
12. Answers will vary, but could include culturing for men almost never reveals all of the traits even if they were all present.
13. Answers will vary but could include culturing for further growth, observation over time, or microscopic observation under higher power.

Investigation 2

Notes for the Teacher

Time
Three 40-minute lab periods

Materials
Microscopes: Most standard student compound light microscopes are 100× on low power and 400×–440× on high power. If your microscopes have a scanning power or an oil immersion objective, tell students not to use them in this lab.

To save time, prepare a small stock of the letter "R," cork slices, and potato shavings in advance.

Text and Lab Correlation
Chapter 2 in the text.

Prelab Preparation
Explain what the microscope is—how it works and what it is used for. It will take the students a while to get a sense of the size of the objects they will be observing. Provide examples to aid in their understanding of magnitude.

Procedure
The students may be confused about the relationship between the area visible under the low power objective as compared to the high power if they try to compare the diameters of each. The areas of each are related via $A = \pi r^2$ (the equation for the area of a circle). This relationship is needed to answer question #18.

When the students measure the diameter of a cork cell, you may wish to have them do it as described in the manual, using low power, or you can have them make their observations under high power. If you want them to use high power, they will have to cut the cork to near paper thinness (which may take several attempts). They should then use a drop of water and a cover slip to prepare their mounts.

Answer Key
1. 400×
2. 400 times larger.

3. Usually 100× under low power and 400×–440× under high power.
4. Low power.
5. Upside down and backward.
6. Right then left (opposite the actual direction of movement).
7. Away from, then toward, the student (opposite the actual movement).
8. It moves in the direction opposite that of the actual movement.
9. Less.
10. Different parts of the thread are in different planes of focus.
11. Answers will vary; for instance, the colored spots of pigment in the thread will become darker or lighter at different diaphragm settings.
12. Answers will vary, but they should be more than 1 mm and less than 2 mm.
13. Answers will vary, but the area of the low power field is about 16 times larger than area of the high power field (see Procedure). The diameter of the low power field is about 4 times larger than the high power field diameter.
14. Answers will vary; about 0.4 to 0.5 mm.
15. Answers will vary, but the low power field should be about 1,500 to 2,000 microns wide and the high power field should be about 400 to 500 microns.
16. Answers will vary.
17. Answers will vary, but be approximately equal to 1/16th of the answer to question #16.
18. Answers will vary (see the answer to question #13).
19. See students' drawings.

20. Answers will vary.
21 Answers will vary, but they should generally be less than 1,500 microns.
22. So somebody else can understand what was seen.
23. To indicate the size of the original specimen.
24. Via photographs or drawings (visually).

Investigation 3.1

Notes for the Teacher

Time
40 minutes for each of 2 parts. Part I is easily accomplished and understood. Part II is more challenging and can be done with one or more of the further investigations.

Materials

Bromothymol blue indicator: Add 0.1g bromothymol blue to 16.0 mL 0.1 N NaOH and 234 mL distilled water.

0.1 N HCl: 8.6 mL concentrated hydrochloric acid to a final volume of 1000 mL with distilled water.

Egg white solution: Make a 1:4 dilution of raw egg white with distilled water.

Dilute basic solution: Distilled water, brought to the same pH as the egg white solution (approximately pH 8) with dilute sodium hydroxide added dropwise. In order to appreciate the buffering action of the egg white as requiring more HCl to create a color change (Step G), the initial solution (Step E) must be of the same pH as the egg white solution.

Alternative: Bromothymol blue as an indicator (color change in pH range of 6.0–8.0) will function to show increased acidity by turning blue to yellow. It will not show a color change from a basic solution (such as the egg white solution) to a solution of greater alkalinity (such as by adding NaOH). If you want to show the buffering capacity of egg white in increasing alkalinity, you can use the indicator phenol red (color change in pH range of 6.8–8.4) and NaOH. For the phenol red solution, add 100 mg of solid phenol red to 2.85 mL of 0.1 N NaOH. Then, dilute this to 100 mL with distilled water. For 0.1 N NaOH, dissolve 4 g sodium hydroxide pellets in 1000 mL distilled water. Use the procedure described in Procedure Steps E, F, and G, substituting NaOH for HCl and phenol red for bromothymol blue.

The water to be used for rinsing and for solutions should be neutral (pH 7). Adjust pH if necessary.

Text and Lab Correlation

Chapter 3. Labs 4, 7.2, 9.1, 14, 21, 24.2, 43.1, and 53 all use pH or discuss acids and bases.

Prelab Preparation

This is a very important lab because the concept of pH appears again and again in subsequent labs. You may want to do the red cabbage indicator (Further Investigations) as a demonstration the day before.

Procedure

In Part I make sure every student tests vinegar, baking soda, and distilled water to see a spread of pH. Emphasize that too much handling of pH paper will give erroneous results. Students can also use spot plates (glass or porcelain) for testing a few drops.

In Part II, make sure students swirl their flask after adding each drop of acid.

Answer Key

1. Excess of hydronium ions. Sour taste (as in throwing up, the sourness of vinegar or lemons), burning sensation.
2. Excess of hydroxide ions. Bitter taste (of soap), feels slippery.
3. Equal numbers of hydronium and hydroxide ions. pH 7.
4. An indicator's color depends on pH and does not give a permanent color to the material it contacts. A dye colors the material it contacts permanently, and a dye does not change as pH changes.
5. Answers may vary. Possible answers: inside the cells, digestive tract, kidneys.
6. solution A—few hydronium ions, base; solution B—equal hydronium and hydroxide ions, neutral; Solution C—many hydronium ions, acid.
7.–9., 11. See students' data charts for Part I.
10. If the color of the litmus is already the color that indicates an acid, adding an acid would not change its color. The converse is also true.
12.–13. See students' data charts for Part II.
14. Egg white should have a pH of approximately 8. The number of drops of HCl should be significantly greater than in question 13. See students' data chart.
15. Egg white solution; dilute basic solution.
16. Acids have a pH below 7 and turn the blue form of litmus to red. Bases have a pH above 7 and turn the red form of litmus blue.
17. Answers will vary; increasing pH corresponds to decreasing acidity and so on.
18. The egg white.
19. Buffering stabilizes the internal chemistry of an organism. For instance, a buffering agent in the bloodstream would modify the unbalancing effects of acids from nutrients or from cellular wastes.

Investigation 3.2

Notes for the Teacher

Time

Two 40-minute labs. In advance, ask students to bring in several food samples. More samples can be tested if fresh or moist items are eliminated.

Materials

Test Foods: Include some items that are easy to burn, such as dry cereals, marshmallows, and pasta.

Text and Lab Correlation

Chapters 4, 5, 7, and 45 and Lab 45.1 on nutrition.

Procedure

Matches have a cooler flame than Bunsen burners and alcohol lamps and are less likely to cause the food items to flame quickly and give off heat away from the test tube of water.

Suggest that students subdivide tasks, such as food testing, ash weighing, preparing apparatus between tests, and data tabulation. Students can weigh non perishable food in advance and store in vials or snap-cap tubes.

Review the use of the metric balance for weighing food samples.

Should a mercury thermometer break, students should not touch it or try to clean it up. The mercury should be drawn up with a pasteur pipette and rubber bulb. Mercury must be disposed of as a hazardous chemical according to local regulations.

If the samples do not burn completely on the first attempt, have students relight their samples. Make sure they measure the difference in water temperature from the first burning before proceeding to the second burning. This step should be repeated until the sample will no longer burn. Each measurable increase in water temperature should then be added to obtain the final count.

Explain that in the metric system the mass of water is the same as its volume if measured on earth. Thus, for water, 1 g of water has a volume of 1 cc = 1 mL. A sample problem using the 2 equations is important for showing how the units of measurement cancel out.

Postlab Analysis

The point of the lab is not to strive for the caloric values of published tables, but to appreciate the process of calorimetry. If students are interested in the sources of error of their setup, have them improve on the calorimeter design by, for example, adding insulation.

Since dry materials burn better than those that contain water, students may conclude (incorrectly) that shredded wheat has more calories than chocolate. Ask your students to consider why foods known to be fattening did not measure up to their predictions.

Answer Key

1. The heat should be in contact with a measured mass of water in the test tube. Calories can be calculated.when the water temperature rises
2. No. The calorie is a measure of any increase in water temperature.
3. The drier items and those with more fats and carbohydrates should heat water more.
4.–7. See students' data charts.
8. See students' data charts (food mass – ash weight = weight change).
9. See students' data charts (final temp. – beg. temp. = temp. change).
10.–11. See students' data charts.
12. The apparatus is not shielded, which allows some of the heat to escape. You are not only raising the temperature of the water but also of the test tube. Foods may burn incompletely, producing smoke which escapes. Combustion is not started with the apparatus in place, allowing further heat loss.
13. Answers will vary.
14. Fat-containing foods give off the most (~9 Kcal/g), then carbohydrates and proteins (~4 Kcal/g).
15. Answers will vary.

Investigation 4

Notes for the Teacher

Time
Parts I and II require 40 minutes each.

Materials
Hydrogen peroxide (H_2O_2): Hydrogen peroxide can be obtained from a pharmacy or supermarket. If purchased as 30%, dilute with water to achieve a 3% solution. The lab requires about 40 mL per group.
Liver: One pound of fresh beef or chicken liver per class. Each group will need 12 pea-sized pieces.
Hydroxylamine (NH_2OH): Available from biological supply houses. Dilute to a concentration of 5%.

1 N HCl: Add 8.6 mL concentrated hydrochloric acid to 100 mL distilled water. Handle concentrated acids cautiously. Do not mouth pipette.
1 N NaOH: Dissolve 4.0 g sodium hydroxide in 100 mL distilled water.

Text and Lab Correlation
Chapter 4 on carbohydrates, proteins, and lipids and Investigation 3.1 on pH testing.

Procedure
You do not have to do all 6 parts in the order given. Occurence of Catalase could work well if you assign students to bring in different fruits, vegetables, and meats.

The hardest part is the students understanding that the dead liver contains a functioning enzyme.

Students are asked to make two graphs: reaction rate as a function of temperature (Part K) and reaction rate as a function of pH (Part M). You may want to spend some time on assigning independent and dependent variables to the x and y axes, labeling axes, using available space, plotting points, and drawing a "best-fit" line.

Answer Key

1. Catalase. Hydrogen peroxide. Water and oxygen.
2. No.
3. Oxygen.
4. See students' data charts.
5. Warmer; exothermic.
6. Bubbling.
7. Water. Nothing. No hydrogen peroxide to break down.
8. Yes. It would bubble again.
9. Yes.
10. See students' data tables.
11. Answers will vary.
12. Denature it.
13. Nothing. The enzyme in the liver was destroyed by boiling. See students' data tables.
14. See students' data tables.
15. See students' graphs.
16. Answers will vary. At approximately 37°C.
17. Because the energy level is so low, molecules move very slowly and do not make contact.
18. Because the enzyme was denatured.
19.–20. See students' data tables.
21. See students' graphs.
22. Answers will vary but should be near 7.
23. It denatures the enzyme.
24. See students' data tables.
25. Hydroxylamine competed with hydrogen peroxide for the active sites of the enzyme.
26. It contributes to normal body temperature.
27. Boiling, high and low pH, high and low temperature, hydroxylamine.
28. The optimal temperature and pH were at normal physiologic conditions, where the enzyme normally would be functioning.
29. Yes; because hydrogen peroxide is made in many kinds of tissues.

Investigation 5

Notes for the Teacher

Time
Part 1 and Part II each take 40 minutes.

Materials
Prepared slides: Order from biological supply houses.
Oral smear (Stratified squamous epithelium): Carolina #H6002, Ward's #93 W 6003
Blood smear (Wright stained): Carolina #H6455, Ward's #93 W 6541

Slides of unknowns could include wet mounts of any organisms you may have, or you can order specific slides. Three or more unknowns should be examined per student. Some of the unknowns should exhibit characteristics that would make the classification somewhat challenging. Unknowns should be prepared in advance and any labels on the slides should be covered with tape. *Elodea:* Can be purchased at pet/aquarium stores as a bunch of sprigs. You may find them under the name *Anacharis.* Illumination for 12 hours before the lab brings out the cytoplasmic streaming.
Iodine solution: Use Gram's iodine or dissolve 3 g potassium iodide in 25 mL distilled water. Add 0.6 g iodine to KI solution and stir to dissolve. Add distilled water to final volume of 200 mL. Store in dark bottle.

Text and Lab Correlation
Chapter 5 on cell structure. You can refer back to this lab when teaching Chapter 43 on the circulatory system. Students will observe a variety of protist cells in Lab 21.

Prelab Preparation
Students should review their text and become familiar with the organelles that animal and plant cells contain. For this lab, they do not need to know the function of each organelle but they should understand what cellular structures can be found in animal and plant cells.

Procedure
Demonstrate how to stain a specimen on a slide. In order to remove excess dye, add water to one side of the coverslip and use a paper towel to draw off dye from the other side of the coverslip.

For the unknowns, you should set up several stations with one slide and microscope at each station. Students should rotate from station to station, spending about 2 minutes at each microscope.

Postlab Analysis
Students should be aware that unobservable differences between plant cells and animal cells will not aid them in determining whether a cell is of animal or plant origin.

Answer Key
1. An organized nucleus.
2. The greater the specialization, the more different cell types exist.
3. They are individual cells that move. They comprise a tissue because they are an aggregate of different cell types that work together to perform specific functions.
4. No, erythrocytes are called corpuscles; they do not contain a nucleus or organelles and cannot divide or repair themselves.
5. One.
6. Shape is square or rectangular. Size should be estimated based on how many cells fit across the diameter of the field of view and have the units of micrometers (μm).
7. A brown oval.
8. Near the outer surface. They are displaced by the central vacuole.
9. See student drawings with labels.
10. A bricklike arrangement of rectangular cells along the length of the filament.
11. A green oval.
12. They move. They are transported by the liquid cytoplasm (cytoplasmic streaming).
13. Answers may vary but both have a cell wall, a nucleus, and a central vacuole. Onion bulbs lack obvious chloroplasts (since they are under the ground); *Elodea's* green chloroplasts mask the nucleus.
14. Presence of chloroplasts, a cell wall, and a central vacuole.
15. They were scraped from the surface of the oral cavity where they only loosely adhere to one another.
16. See student drawings with labels.
17. Answers may vary. At least 3 types should be present, but more may be reported if students can discern subtypes of leucocytes.
18. They are small, individual cells of various shapes and organelles that are too small to discern. They differ in size, shape, and color of stain.
19. No. None can be seen with only 400–430× magnification.
20. Answers will vary. See students' data tables.
21. Answers may vary but should include a cell membrane not confined by a cell wall.
22. Both have cytoplasm and a nucleus.
23. Only plant cells have chloroplasts, a cell wall, and a central vacuole. Plant cells are generally of a more consistent shape (less round or irregular than an animal cell).
24. Answers will vary. The chloroplasts allow plant cells to make their own food. The cell wall makes plant cells more rigid and gives them greater protection. The central vacuole contains water and cellular wastes.
25. In the leaves that extend out of the ground.

Investigation 6

Notes for the Teacher

Time
60–80 minutes for Part I. 60–80 minutes for Part II, plus 20–25 minutes on the next day for final weighing (Step K) and Postlab Analysis. If you want to complete either Part I or Part II in one 40-minute period, you can do a demonstration with the class participating in data collection. You may wish to do all of this lab, Part I only, or Part II only; they are independent but complementary exercises.

Materials

Dialysis tubing: Presoak 24 hours in distilled water.
1 M HCl: 80 mL concentrated hydrochloric acid in 1,000 mL water.
Karo Syrup: from grocery store
50% Glucose solution: 50 g glucose in distilled water, to make 100 mL final volume.
Eggs: Bring in additional eggs to allow for breakage.

Text and Lab Correlation

Chapter 6 on homeostasis and transport. The technique of modeling with dialysis tubing can be related to Lab 45.2 on starch digestion.

Prelab Preparation

If possible, spend one class period before this investigation reviewing osmosis and diffusion. Students should also be able to distinguish passive transport from active transport. It would be helpful if you could demonstrate the technique for preparing dialysis bags in this period.

Discuss with your students their ideas for the setup of Beaker 4, the experimental setup. You can suggest variables such as different glucose concentrations, temperature, or disturbance of the solution. Provide additional solutions for them to choose, such as a 1% starch solution or 1.0% egg albumen solution.

Procedure

For Part I, show your students how to fill and tie dialysis tubing. Make sure they know how to seal the ends, fill the tubes, and especially how to eliminate air and leave enough slack in the tubing for water to enter. Explain to the class that if this is not done, the pressure inside the tubing might affect their results.

In Part II the HCl will dissolve the shell but not interfere with the semipermeable membrane inside. It is through the inner membrane that osmosis will occur. You may use vinegar instead of HCl to dissolve the egg shell; you will have to let the eggs soak overnight. You may allow the students to leave the eggs in Karo syrup more than 24 hours before the final weighing.

Further Investigations

1. Use 1% soluble starch solution and 1% egg albumin solution for this investigation. These tests should be conducted as described in the Tests Table following.
   Have just enough water in a 100-mL beaker to cover the dialysis bags.

Answer Key

1. Diffusion is the movement of molecules from an area of greater concentration to one of lower concentration. Osmosis is the diffusion of water molecules through a membrane.
2. When concentration is the same on either side.
3. Answers will vary.
4. See students' data charts.
5. See students' data charts. Bag 2 should swell and Bag 3 shrink.

Tests Table

Test for	Reagent	Procedure	Result
Starch	Iodine	1. Place 5 mL unknown in test tube 2. Add 3 drops iodine	deep blue black color
Glucose	Benedict's	1. Set up hot water bath 2. Add 5 mL unknown to test tube 3. Add 30 drops Benedict's 4. Place tube in water bath for 5 minutes	reddish-brown color
Protein	Biuret	1. Place 5 mL unknown in test tube 2. Add 10 drops of Biuret	purple color

6. The dialysis bag containing water that is placed in a beaker of water is an experimental control.
7. See students' data charts.
8. Osmosis should occur in Beakers 2 and 3.
9. Answers will vary, but in Beaker 2, water should move into the dialysis bag, and in Beaker 3, water should move out of the dialysis bag.
10. See students' data charts.
11.–12. See students' data charts. Should see the egg shell dissolving.
13.–14. See students' data charts. Egg should get larger as it fills with water.
15.–16. See students' data charts. The egg should have become smaller as water moved out.
17.–18. Answers will vary.
19. A selectively permeable cell membrane. Both will not allow large molecules to cross. However, the dialysis tubing does not have special pumps and is only capable of passive molecular movement.
20. Osmosis begins rapidly and then slows as equilibrium is approached.
21. The acid removed the egg shell, exposing the egg's selectively permeable membrane.
22. Osmosis. Egg is in a hypotonic solution when it is put into water beaker. The egg is in a hypertonic solution when put into the syrup.

Investigation 7.1

Notes for the Teacher

Time

Making the two chromatograms takes no more than 40 minutes. The lab may be done during 2 class periods: 1 period to introduce chromatography and prepare materials (Steps A and B), and a 2nd period to do Steps C to G.

Materials

Spinach: 20 fresh leaves; frozen will work if thawed.
Ethyl alcohol: 300 mL (Acetone may be used for extraction, but it is more hazardous.)
Chromatography solvent: 200 mL (92 parts petroleum ether to 8 parts acetone).

Leaves may be stored in alcohol for future use; the pigments will dissolve in the alcohol over time. Thus, containers of leaves sorted either by species or by color may be "put up," and the pigments readied for future chromatography.

Text and Lab Correlation
Chapter 7 on photosynthesis, especially the section on the light reactions. Investigations 24.1 on starch production in plants and 24.2 on oxygen production in plants.

Prelab Preparation
This lab is most useful in autumn, when a variety of leaves/colors are available. Encourage students to collect a large number of different species and colors.

Procedure
After the leaves have been grouped by the class, assign one type and color of leaf to each team.

Pigment bands in the finished chromatogram will be most distinct if the extract is very concentrated at the initial spot on the paper. A way to increase pigment concentration in the alcohol is to evaporate as much alcohol as possible from the extract. However, heating alcohol, even in a water bath, is hazardous; an alcohol fire is a real possibility. For this reason, the heating step has been skipped, and students are directed to apply many small drops to the paper instead. If you feel that heating the extract is worth the risk, it should be done in test tubes in a water bath over a hot plate, away from students and under careful observation.

The chromatography solvent is both flammable and poisonous; students should neither breathe its fumes nor let it come in contact with their skin. You may wish to set up the chromatography tubes in advance, with solvent and corks.

For Step D, it is important that each chromatogram be taken out of the solvent the instant the solvent front reaches the end of the paper strip. This is crucial to the valid determination of R_f values; pigments must not be allowed to migrate farther up the strip after the pure solvent reaches the end. Although the R_f values *should* be constant, expect lots of variability in the values students actually get.

Postlab Analysis
In general, the leaf's overall color is reflected in the chromatogram's dominant pigments. Minor pigments in the chromatogram are most likely hidden in the overall color. Discuss why a green leaf contains red and yellow pigments.

The R_f value for each pigment tends to be constant among different leaves, chromatograms, and species (if the same solvent and paper are used). Discuss with students factors that might account for variability of R_f values among teams.

Answer Key
1. Answers will vary. Species and color will depend upon geographic location and time of year.
2. Examples: xanthophyll, yellow; carotene, orange; chlorophyll, green; anthocyanin, red.
3. The grana, which contain the chlorophyll, are stacks of pigment-containing membrane disks in which the light reactions occur.
4. Answers should include leaves, green stems, fruits, and flower parts, especially petals.
5. Pigments in leaves probably function in photosyn-

thesis; pigments in flowers function in attracting pollinators; pigments in fruits may function in food production or attraction of seed dispersers.
6. The solvent will spread the pigment mixture into separate bands of pigment up the paper strip. Adequately dark bands require a heavy starting concentration of the pigment mix.
7. See students' data charts.
8. See students' data charts. Answers will reflect relative pigment migration in decimal form.
9. Yes, but students' data may not show this, because of large experimental error.
10. Spinach will show all the pigments described. Other leaves may also show them.
11. Answers may show hidden pigments.

Investigation 7.2

Notes for the Teacher
Time
Total duration 2 weeks (10 school days): 40 minutes each on the first and last days; 1–2 minutes at the beginning of class on the days in between.

Materials
Cabbage: Shred a few heads of cabbage in advance.
Salt solution: To make 2.5% salt solution, add 2.5 g NaCl to 97.5 mL water. You will need several liters, depending on the number and size of containers.
Containers: One-pine sherbet containers or 1/2-pint food-storage containers are large enough. Covers on containers should fit securely so that liquid does not escape as fermentation progresses. If sanitary conditions are maintained with materials, students may be allowed to taste the sauerkraut.

Text and Lab Correlation
Section 7.1 on the need for energy, and Section 7.3 on respiration, and Chapter 3 on basic chemistry.

Prelab Preparation
Fermentation proceeds best between 65–75°F at 2.5% NaCl. Neither temperature nor salt concentration is so critical that special care is needed. Have students review what they learned about pH in Lab 3.1.

Procedure
This experiment may be performed as a teacher demonstration. If done as a demonstration, it can be quickly set up one day and just passed around the room every day until fermentation is complete. Students will be surprised at the production of gas (the gas bubbles produced at the top of the container are impressive) and the change in color and odor. Sometimes it is sufficient for the teacher to explain that these changes are caused by fermenting bacteria. When bubbling ceases, the sauerkraut is finished. You may want to use an oil immersion lens if the bacteria are difficult to see.

Postlab Analysis
At optimum conditions, the primary organism is *Leuconistic mesenteroides,* which converts sugar to lactic acid, carbon dioxide, acetic acid, ethanol, mannitol,

dextran, and esters, all of which combine to produce the characteristic odor and flavor of sauerkraut. As fermentation progresses and lactic acid rises above 1%, *Leuconistic mesenteroides* is replaced by *Lactobacillus plantarum* and *Lactobacillus brevis,* which, if sugar content is high enough, may produce a lactic acid content to 2.4%. The final level of lactic acid is usually 1.7%.

At 90°F or 3.5% salt, *Pediociccus cerevisiae* is favored and a more acid sauerkraut results. If oxygen reaches the sauerkraut in a high salt environment, a red yeast will grow and the sauerkraut will be pink. Black or brown sauerkraut results when there are high temperatures and oxygen. All of them are edible but not highly regarded by sauerkraut fanciers.

Further Investigations

An excellent source for information and recipes is ISIS, *Food and Microorganisms,* Ginn and Company, 1976. The 76-page minicourse is out of print, but teachers in your school or district who have used the ISIS program will probably still have copies for you to use. Included are classroom recipes for cucumber pickles, pickled spareribs, yogurt, and pancakes made with yeast.

Answer Key

1. The human body produces lactic acid when cells do not received enough oxygen for aerobic respiration during a long session of strenuous exercise.
2. The CO_2 produces the small holes in bread and makes the bread rise.
3. See students' data charts.
4. 2.5% NaCl is hypertonic to cells, so water will leave the cells causing the separation.
5. Yes.
6. It goes from almost odorless to having a strong odor, becomes paler, goes from being crisp to being limp, and the pH goes down.
7. Sketches and estimates will vary.
8. See students' graphs of pH changes.
9. The cells plasmolyze (the cell membrane separates from the cell wall) because they have lost water; 2.5% NaCl is hypertonic to the cells.
10. Possible answer: Because brine creates conditions favorable for the right kind of bacteria to grow. Different bacteria would have grown in water, and you would not have gotten sauerkraut.
11. Carbon dioxide is a product of fermentation.
12. It went down due to production of lactic acid and other acids.
13. Production of acetic acid: $CH_3COCOOH + 2H \longrightarrow CH_3COOH + CO_2$.
14. So that conditions will remain anaerobic.

Investigation 8

Notes for the Teacher

Time
40 minutes for Parts I–III and 20 minutes for Part IV.

Materials
Amino acid pairs: Tape two $3'' \times 5''$ index cards next to each other. See the teacher notes on procedure below.
Velcro and ribbon: Purchase at a sewing supply shop. Grosgrain ribbon is recommended because it is especially sturdy. You will need the following amounts for each team (any color ribbon or Velcro may be substituted for those listed)—24 cm yellow ribbon (mRNA), 96 cm blue ribbon (DNA); 20 cm black Velcro (both a rough strip and a smooth strip), 20 cm beige velcro (rough and smooth strips), and 10 cm white velcro (rough and smooth strips). Give students Velcro strips and instruct them to cut 2-cm × 3-cm pieces. Have a staple remover available so that students may correct mistakes if necessary.
Marking pen: The Sharpie laundry marker, made by Sanford, is suitable for the purpose of this lab.

Text and Lab Correlation
Chapter 8 on nucleic acids. This lab can serve as an introduction to Investigation 13.

Procedure
It will help the students if you prepare a model in advance to show them what it will look like when they finish. Give each group a pair of amino acids on $3'' \times 5''$ cards that are different from each other team's. One group should get a pair in which the amino acid is repeated (e.g., Tyr, Tyr), one group the start codon (AUG), and another group a stop codon.

For Step I, you may want to have worked out a specific sequence for the model "gene" in advance or perhaps a specific amino acid sequence. Be sure the students are aware of the directionality of each strand when they assemble the whole model and they understand the structures and processes that these models represent.

Further Investigations
1. There are 720 permutations or ways to arrange 6 items in a sequence of 6.

Answer Key
1. Their backbones are different; RNA contains ribose sugars and DNA contains deoxyribose sugars.
2. Possible answers: RNA can leave the nucleus and DNA cannot; DNA is double-stranded, RNA is not; the ribose sugars of RNA have an extra hydroxyl group (OH) on them.
3. RNA is made in the nucleus (by RNA polymerase) during transcription.
4.–7. Answers will vary.
8. In the nucleus.
9.–10. Answers will vary.
11. Hydrogen bonds
12. The 2 pieces of DNA will be exact copies of the double-stranded DNA made in Step G.
13. For a class of 5 teams: 30 nucleotide base pairs, 10 codons, 10 amino acids.
14. Transcription
15. Translation
16. Replication
17. G C T C C C T A T
    C G A G G G A T A

Investigation 9.1

Notes for the Teacher

Time
40 minutes for the lab itself. Postlab Analysis can be done on another day. Preparing materials for this lab takes approximately one hour.

Materials
0.2 N HCl: Prepare approximately 100 mL per team. Bring 16.6 mL of concentrated (12 N) hydrochloric acid to 1 liter with water.

Agar blocks with bromophenol blue: Measure the flat sides of a rectangular baking pan (glass, glazed ceramic, or enamel) to estimate how many $6 \times 3 \times 3$-cm blocks you may cut out of each pan. Fill it with known values of water (e.g., 1000 mL at a time) to a depth of 3 cm, then calculate the total required volumes for all classes.

Prepare a 3% solution of agar powder (nonnutritive) in boiling tap water (30 g agar/1000 mL H_2O). Add 2 mL 1N NaOH (40 g NaOH/liter) per liter. Once the agar has cooled enough to pour, measure the pH for your records. Add bromophenol blue to a concentration of 250 mg/liter (first dissolve the powder in 5 mL distilled water to ease mixing and to compensate for evaporation during boiling). Pour into the pan to a depth of 3 cm. After the agar has cooled and hardened, keep the pan covered with plastic wrap until use to prevent dehydration.

Slice away the corners of the agar slab with a scalpel or single-edged razor blade, and use a metric ruler to measure rows 6 cm wide. Cut these into 3-cm lengths. Place the agar blocks directly onto a glassplate, using a thin plastic pancake turner. Be careful not to gouge the agar surface; lift rather than push. The cuts may be made in advance, but do not remove individual blocks until immediately before use.

Do a test yourself in advance of the lab with a large block (at least 4 cm) to find a crude rate of diffusion. This will help to determine at what time students should remove each block. (The average rate of diffusion is approximately 0.5 mm/minute.)

Text and Lab Correlation
Chapter 9 on chromosones, Chapter 6 on homeostasis, Chapter 5 on cell structure, and Lab 6 on osmosis.

Prelab Preparation
Copy the Class Data Chart on the chalkboard so that students can pool their data at the end of the investigation.

Procedure
Students may work in groups of 2 or 3. You may have different teams cut their blocks at 2-minute intervals in order to show that the rate of diffusion is constant over time. Use information from your test block to cover all points, from no diffusion to complete diffusion. Stress that all blocks in a set should be examined as closely together in time as possible.

When students answer Question 5, they may find that distances vary. It may be that all surfaces were not evenly exposed to the acid solution; that one side is the surface that polymerized in contact with air and may have hardened slightly; that some of the agar was incompletely dissolved or started to harden earlier than the rest (inhomogeneity in the gel); or that dehydration has occurred.

Have the students pick up any agar drops on the floor. They can be the source of treacherous falls.

Postlab Analysis
These calculations can make the correlation of mathematics with biology come alive for the students.

Further Investigations
1. Folding is used wherever adsorption is required. Surfaces of high transport capability, such as the intestinal lumen, kidney, and egg cell, use a combination of folds, loops, and microvilli to maximize the available surface area.
2. Students should describe the limits of specialization when intercellular dependence (tissues) does not occur. Encourage a far-reaching approach.

Answer Key
1. 10–20 microns.
2. Answers will vary. Possible answer: Model does not take into account active transport.
3.–4. See students' records kept on separate pieces of paper.
5.–7. See student teams' tables.
8. Answers should be in mm/minute (average of those blocks in which diffusion was not total).
9. See students' data charts.

Surface Area/Volume Ratio

Block No.	L	W	H	Area $A = 6 \times l \times w$	Surface Volume $V = l \times w \times h$	Ratio A /V
1	30	30	30	5400	27,000	0.2
2	20	20	20	2400	8,000	0.3
3	10	10	10	600	1,000	0.6
4	5	5	5	150	125	1.2

10. As a cell's size increases, its ratio of surface area to volume decreases. Since the rate of diffusion is constant, a greater internal volume will not be reached.
11. Answers will vary.
12. Diffusion will be poor. Oxygen and glucose will be transported in more slowly; excretion of ammonia and other wastes will also be inhibited.

Investigation 9.2

Notes for the Teacher

Time
For Part I, 40 minutes plus extra time if some students need to repeat the procedures. For Part II, 40 minutes. Part II may also be done the day before Part I. Allow one to two weeks before the lab to grow the roots.

Materials

Tradescantia **root tips:** One to two weeks before the lab, prepare cuttings from *Tradescantia*. Make clean cuts with a single-edged razor blade and place the cut stems into fine, moist sand to root. Keep the bed moist at all times. Darkness promotes root growth.
Alternative: Use root tips from fresh onion bulbs or scallions. Start growing onion roots about one week before the lab by propping an onion bulb at the top of a bottle, with its roots in water. Have available prepared slides of onion and/or hyacinth root mitosis.

Clean the microscope slides, even if new, by rinsing with hot soapy water, clean water, and ethanol or acetone, and then polishing with a clean, dry cloth. Use 22 mm square plastic coverslips.
Fixative: Mix 30 mL glacial acetic acid with 90 mL 95% ethanol. Each class will need 50 mL.
Pectin solvent: Add 50 mL concentrated hydrochloric acid to 50 mL 95% ethanol. Each class will need 50 mL. Caution: Pectin solvent is corrosive. Wear safety goggles while preparing it. If it gets on your skin, flush it immediately with water.
Carnoy's solution: Prepare in a 100-mL graduated cylinder. Add 10 mL glacial acetic acid to 50 mL 95% ethanol. Add 30 mL chloroform. Bring the volume up to 95 mL with 95% ethanol. Caution: Work in a hood and avoid breathing vapors. The storage container should be dark and resistant to the volatile chloroform.
Toluidine blue O stain: Dissolve 0.5 g stain in 99.5 mL distilled water.

Text and Lab Correlation

Chapter 9 on mitosis. Lab 9.1 on cell size.

Procedure

For Step E, the hot plate should be set at low. If the hot plate is too hot, the slides may crack. Caution students about picking up the hot slides. For Step F, forceps may squash the root tips. Camel-hair brushes on the edges of coverslips may also be used to transfer the root tips. For Step G, not all students may see the chromosomes. In Step H, not all students will see a variety of mitotic stages. Failure to succeed at these steps may lead to frustration. For these students, supply prepared slides that clearly show chromosomes and mitotic stages.

Postlab Analysis

Discuss the significance of the mitotic events the students have just modeled. Encourage students to think about what might go wrong during mitosis from what they have observed.

Further Investigations

1. Fruit fly colonies are available from local universities and/or biological supply houses. Detailed procedures for performing this procedure are available in *Carolina Drosophila Manual*, Carolina Biological Supply Company, or other appropriate manuals. Alternatively, these slides may be purchased already prepared from biological supply houses.
2. The students may cut out the chromosomes and position the centromeres on a line drawn on a card. This will help to show similarities and differences

that are not noticeable at first. Then they can assign pair numbers. The karyotype should show 8 pairs of chromosomes in descending order of size, although it is more important to match the shapes rather than the absolute lengths. Chromosome pair number 3 is satellited, with small arms attached to the main body by tenuous stalks.

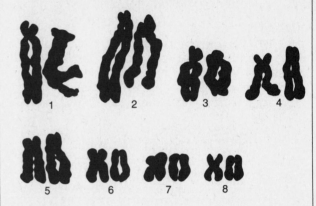

Answer Key

1. DNA is replicated during the S (synthesis) in interphase.
2. Prophase: (early)—Chromatin condenses to form chromosomes; nuclear membrane breaks down; and centrioles appear; (late)—Spindle fibers form. Metaphase—Chromosomes line up along the cell equator. Anaphase—Centromere divides, and chromatids separate. Telophase: (early)—Chromatids polarize, and centrioles and spindle fibers disappear; (late)—Chromatids decondense, and nuclear membranes form.
3. Cytokinesis is division of the cytoplasm. Mitosis is division of the nuclear material.
4. Chromatin contains thin, uncoiled strands of DNA. Chromosomes are chromatin in which the DNA has coiled up to look like rod-shaped structures.
5. The specimen must permit light to pass through it.
6. See students' drawings.

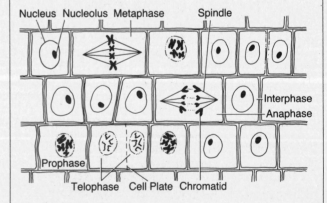

7. There are 16 chromosomes.
8. The mitotic stage is prophase.
9. The 2 colors represent the DNA of each parent.
10. The notched line represents the doubling of the DNA, which occurred earlier during S in interphase.

11. The mitotic stage is metaphase.
12. The mitotic stage is anaphase.
13. They are modeling the pulling apart of the chromatids by the spindle fibers.
14. Answers will vary, but the chromosomes may make V shapes.
15. Central centromeres make a diamond; terminal centromeres make a V or seagull shape; and eccentric centromeres make a kite shape.
16. The mitotic stage is telophase.
17. Cytokinesis normally occurs next. It is the division of the cytoplasm.
18. Answers may vary, but probably perpendicular to and away from the center of the chromosomes (towards which the spatulas are drawn).
19. The nuclear membrane disappears in middle prophase and two new ones (around each group of chromosomes) appear during late telophase.
20. No. It is the splitting of the centromere that decides how sorting occurs.
21. No, unless there is an abnormal division. Mitosis maintains genetic stability.

Investigation 10

Notes for the Teacher

Time
40–50 minutes. Allow time for the students to pool data.

Materials
Beans: Navy and black beans can be substituted for white and colored beans, as can any materials that differ only in color. Two colors of poker chips, buttons, and paper clips are other possibilities. Obtain at least 300 of each color for a class of 24. If you want to allow students to eat their experiment at the end, you can use jelly beans or other candies. For jelly beans, bring in clean yogurt containers so that students do not eat out of lab equipment. Also, warn students not to eat the materials until the lab is finished because that will throw off ratios.
Containers: Small paper bags, boxes, or plastic containers (anything opaque) can be used to hold the beans. Obtain at least 12 containers for a class of 24.

Text and Lab Correlation
Chapter 9 on meiosis and Chapter 10 on genetic probability. Investigations 11 and 15.

Prelab Preparation
You may wish to clarify the distinction made between characteristic and trait in Chapter 10. **Characteristic** is used for a particular feature of an organism (e.g., seed color); **trait** is used for one of the forms in which a characteristic occurs (e.g., yellow seeds or green seeds).

Also, review the rules of probability as they pertain to the Punnett square. Emphasize that each pairing of gametes to form an offspring is considered a random event—that is, the result of one pairing does not affect the outcomes of other pairings of gametes from the same parents. You may wish to have students demonstrate this rule with a coin toss activity.

Procedure
In Step C, explain that the beans are mixed in each container to ensure a random contribution of an allele by each parent during the pairing of gametes. Students should realize that their predicted ratios will be the probabilities indicated by the Punnett square.

In Step E, each bean (allele) is returned to its respective container after each trial pairing so that the numbers of available alleles remain constant at each pairing. In discussing the Strategy for Modeling, students should realize that they can ensure random pairings by returning the beans to their respective containers, mixing the beans, and closing their eyes before they remove a bean from a container.

Postlab Analysis
Allow time for data pooling within the teams and by the whole class. Construct team and class data tables and display on an overhead projector or the board.

Further Investigations
1. For a heterozygous individual, students should use 25 beans of each color in the container. For a homozygous individual, they should use 50 beans of the color representing the dominant allele or the recessive allele. They should choose a large number of pairings in order to produce ratios close to the predictions of a Punnett square.
2. A dihybrid cross of 2 heterozygous individuals involves 4 kinds of gametes (4 possible allele combinations within the gametes), for example: DT, dT, Dt, and dt. The Punnett square will have 16 offspring blocks. Students should place in each of the 4 "parental containers" 25 beans of each of 2 colors. For each offspring, they should draw one bean from each container for a total of 4 beans per offspring.

Answer Key
1. An allele is one form of a gene. A dominant allele masks the effect of a recessive allele, therefore a recessive allele is not expressed when paired with a dominant allele. An organism is homozygous for a characteristic when the 2 alleles in a pair are the same, and is heterozygous when the 2 alleles in a pair are different. Genotype is the combination of alleles within an organism; phenotype is the physical feature produced by an organism's alleles.
2. Female = DD; male = dd; possible offspring = Dd, Dd, Dd, and Dd.
3. Heterozygous.
4. Female.
5. Yes, because each possible offspring has a dominant allele for dimples.
6. See Punnett square 1 on the next page
7. DD = dimples; Dd = dimples; dd = no dimples.
8. 1 out of 4, or 25%.
9. Both parents are Tt = tall. Offspring are TT = tall, Tt = tall, and tt = short.
10. See Punnett square 2 on the next page.
11. Genotypic ratio of 1 TT : 2 Tt : 1 tt; phenotypic ratio of 3 tall : 1 short.

Punnett Square 1

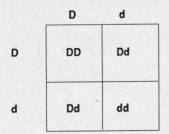

	D	d
D	DD	Dd
d	Dd	dd

Punnett Square 2

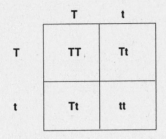

	T	t
T	TT	Tt
t	Tt	tt

12. Answers will vary because results will be random. Students should record one of the following combinations in each row of their data tables: TT—tall, Tt—tall, or tt—short.
13. See second data table. Answers may be close to or far from 1 : 2 : 1 (TT : Tt : tt) and 3 : 1 (tall : short).
14.–17. Answers will vary; but teams' ratios should be closer to predicted ratios than are individuals' ratios, and the class ratios should be closer to predicted ratios than are teams' ratios.
18. When the number of offspring is large.

Investigation 11

Notes for the Teacher

Time
Allow 40 minutes each for Parts I and II. At least one evening must elapse between parts to permit data collection.

Materials
Make sure calculators have a square root function and that students know how to use it.

Text and Lab Correlation:
Chapter 11 and Lab 15 on patterns or variations.

Prelab Preparation
For Part I, review Mendel's principles and the concepts of dominant and recessive gene inheritance and expression. For Part II, discuss the importance of gene frequency. Geneticists use population sampling to study gene distributions and their variations over time. Point out that some dominant genes, such as the one that causes Huntington's Chorea, are very rare.

You may wish to expand this lab by having students investigate other single-allele traits, such as left-over-right thumb crossing, cleft chin, straight thumb, straight little finger, and free ear lobe.

Procedure
In Part I, have students pool their data as efficiently as possible. Display a large table on a chalkboard where students may fill in their own phenotype. Make sure students work together to calculate the percentages.

At the end of Part 1, assign the collection of tongue-rolling data for the remaining subjects as homework for Part 2 of the lab. Remind students to record the name and ability of each subject tested. If possible, share data with another lab section to decrease the number of additional ("outside") subjects be to tested. To ensure that you will have a sample of 100 to work with, ask groups to interview some extra people (but use data for only 100).

For answering Question 13 with 100 individuals you should be fairly close to the national average. However, there could be differences due to a small population size or to your students being from a particular ethnic background that has different gene frequencies.

For answering Questions 14 and 15, students may need help with the calculations. First work through the problem with the example data. Students first must finish calculating q (r) from Question 14 by taking the square root of the r^2 decimal percentage value and then calculating p (R).

Further Investigation
1. Recall the textbook example of the Amish manic-depression study. List the Hardy-Weinberg characteristics of this population. Work with your school librarian to develop a bibliography for students to consult.

Answer Key
1. Answers will vary.
2.- 6. See students' data charts.
7. No. Each trait is found in different frequencies in a population.
8.–9. Answers will vary.
10. The recessive gene may be more common than the dominant gene.
11. To make sure you sample 100 *different* people.
12. See students' data charts.
13. Answers will vary, based on data.
14.–20. Answers will vary
21. Values may approximate the national average more closely.
22. Answers will vary. Students who interviewed only close relatives probably biased the data.

Investigation 12

Notes for the Teacher

Time
For Part I, 40 minutes, and for Part II, 20 minutes. Allow 2 days to elapse between Parts I and II. Prepare agar plates 3 days before Part I. Inoculate stock cultures 2 days before Part I.

Materials
Serratia marcescens **culture, biotype D1:** Carolina Biological #15-5452, Ward's #85-W-0997, or other biological supply houses.

Nutrient agar plates: Pour agar plates 3 days before class. To make nutrient agar you can use Bacto nutrient agar or tryptic soy agar and follow the accompanying directions for preparation. To make 1 liter of nutrient agar from scratch, mix 3 g beef extract, 5 g peptone, and 15 g agar in 1,000 mL distilled or deionized water.

Autoclave the solution (or use a pressure cooker) at 15 pounds pressure for 15 minutes in a flask stoppered with a cotton or gauze plug and covered loosely with foil. When the flask is cool enough to handle, swab a clean bench surface with 70% ethanol and spread out the culture dishes. Open the lids as little as possible and pour enough molten agar directly into the dishes to cover the bottoms. Let them sit undisturbed until hardened. Pour several extras to allow for contamination and for growing the stock cultures.

Incubate the plates overnight at 37°C to ensure that no contamination has occurred. Check all plates, discarding those showing growth. Invert and refrigerate in a covered container for the class, removing them about one hour before use.

Sleeves of sterile disposable culture dishes (110 mm in diameter) are available through supply houses. Alternatively, clean glass petri dishes may be autoclaved.

Two days before Part I, streak stock culture plates and incubate 2–4 at 27°C and 2–4 at 37°C.

Prelab Preparation
There will be more opportunity to practice aseptic technique in the laboratory exercises on bacteristasis (Investigation 20) and the inheritance of resistance (Investigation 16).

It is essential that the students understand that the red pigment, prodigiosin, is not made directly by the expression of a gene. It is the product of a metabolic pathway in which the final, regulatory step is a condensation reaction of several amino acids. The enzyme that catalyzes this reaction is induced by temperature. In essence then, what the students will be observing is a bioassay for gene expression of the rate-limiting enzyme. This is a valid assay for enzyme levels because all conditions will be identical except for temperature.

Procedure
Explain that the sterile plates should not be opened except for inoculation, and that hands, air, and all materials should be considered a source of contamination. Demonstrate how to inoculate an agar plate with a wire loop. (This procedure is described in more detail in Investigations 16 and 20.)

Further Investigations
1. Prodigiosin is a good example of a naturally occurring compound with possible therapeutic activity. Several pharmaceutical companies are currently looking at it for this purpose.

Answer Key
1. Lactose levels in medium.
2. Several amino acids are enzymatically joined.
3. Temperature.
4. Appearance of red bacterial colonies.
5. Bright red at 27°C; white-pink at 37°C.
6.–7. See student answers in data table.
8. See student data charts. Expected results: (1) red; (2) red, but delayed in appearance; (3) intermediate pink; (4) faint pink-white.
9. Predictions will vary, but pink is most likely.
10.–11. Predictions will vary.
12. Average room temperature is 22°C; normal (human) body temperature is 37°C.
13. Answers will vary. Example: to compete successfully with other microorganisms (at 27°C), which is not necessary at 37°C.
14. Answers will vary. Example: to kill other microorganisms such as ameba or fungi.

Investigation 13

Notes for the Teacher

Time
Parts I and II will each require about 40 minutes. If this material is new to you, allow yourself some time in advance to work through the lab yourself. You may want to read the references suggested in the notes on the procedure.

Text and Lab Correlation
Chapter 13. Review Sections 8.3 and 11.3 and Lab 8 on modeling DNA.

Prelab Preparation
This information is designed to get students thinking in terms of base pairs of DNA. You should review this material by doing blackboard exercises on matching bases. Some students might find this abstract and difficult. Try explaining by analogy if students experience trouble comprehending. However, you should be aware that the concepts of recombinant DNA technology may be too difficult for some students.

Make sure the students understand that DNA bases can be in any sequence but that the sequence is specific to that DNA which defines the gene. The minor differences in DNA from person to person are part of what makes each of us unique.

Procedure
Part I is designed to familiarize students with the concepts of recognition sites, the base sequences that restriction enzymes identify, and restriction sites, the places within a recognition site that the enzyme cuts.

The restriction enzymes cut only bases defined by the recognition site even though the same two bases might exist together at other locations along the DNA strand.

A thorough understanding of the use of restriction enzymes in Part I will prepare students to understand how DNA fragments are analyzed in Part II. Remind the students that scientists cannot actually see the individual molecular fragments but they can see thousands of these fragments, sequestered in a gel, stained with ethidium bromide, and illuminated with UV light. The scientists must analyze the fragments by the pattern they leave on the gel. Stress that DNA patterns are read not only by measurement but also by comparison with a known sequence of DNA.

For more information, see J.D. Watson, J. Tooze, and D.T. Kurtz, *Recombinant DNA: A Short Course,* Scientific American Books, 1983 and *Science,* June 5, 1987, special edition on "Frontiers in Recombinant DNA."

Postlab Analysis

Have students compare their gel electrophoresis patterns to make sure they all comprehend the migraton patterns: the fewer the numbers of base pairs the farther the DNA fragments travel in the gel.

Further Investigation

1. Instead of using the DNA sequences in the lab, you can have the students make up their own DNA chains. Several common restriction enzymes and their recognition sites are:

Bam HI
```
      ↓
G  G  A  T  C  C
C  C  T  A  G  G
               ↑
```

Bgl II
```
      ↓
A  G  A  T  C  T
T  C  T  A  G  A
               ↑
```

Eco RI
```
      ↓
G  A  A  T  T  C
C  T  T  A  A  G
               ↑
```

Hind III
```
      ↓
A  A  G  C  T  T
T  T  C  G  A  A
            ↑
```

Pst I
```
               ↓
C  T  G  C  A  G
G  A  C  G  T  C
   ↑
```

Sma I
```
         ↓
C  C  C  G  G  G
G  G  G  C  C  C
         ↑
```

DNA IA
```
                        ↓
T T G C A A G T C A G A A G A A T T C A A C C T A G G A A T T C T A A G C G C
A A C G T T C A G T C T T C T T A A G T T G G A T C C T T A A G A T T C G C G
                        ↑                   ↑
```

DNA IB
```
                                                    ↓
T T G C A A G T C A G A A G A A G T C A A C C T A G G A A T T C T A A G C G C
A A C G T T C A G T C T T C T T C A G T T G G A T C C G G A A G A T T C G C G
                                                    ↑
```

Answer Key

1. ATTCGGCATCCAACCTTGAGG
2. Answers will vary but all pairs should be A-T's or C-G's.
3. TAAGCC**GTCGAC**TCGAACTCC
4. ATTCGG**CAGCTG**AGCTTGAGG
5. GCCTCTA A **G*AATT** CAGTTGG
   CGGAGAT T C **TTAA*G**TCAAGC
6. Two fragments. Sticky ends. Nine and eleven.
7. A single base pair mutation in the recognition site. T—A became G—C.
8. See the bottom of this page.
9. Three from DNA IA and two from DNA IB.
10. From DNA IA: 14 bp, 13 bp, and 12 bp; from DNA IB 27 bp and 12 bp.
11. Students should make a different DNA molecule from DNA IA than from DNA IB. An almost unlimited number of possibilities are feasible.
12. Three fragments from DNA X and two fragments from DNA Y.
13. DNA X: 100, 300 and 600 base pairs. DNA Y: 400 and 600 base pairs.
14. See students' simulated gels.

Simulated Gel

15. The restriction enzyme will not recognize the site if even one of the base pairs changes and therefore cleavage will not occur at that site.
16. Scientists cut the DNA to be studied and a known, standard DNA with the same restriction enzymes, then place the different DNAs next to each other on the same gel electrophoresis and separate the fragments. They then compare the gel patterns.

Investigation 14

Notes for the Teacher

Time
40–60 minutes, plus time for postlab analysis. This lab may be completed in 40 minutes if one partner from each team moves on to Step F as soon as the mixture becomes cloudy (after Step C) and the 2 students work simultaneously. Some monitoring of time and progress will be helpful. The prelab activity of generating a list of life characteristics may be assigned as homework on the previous day.

Materials
Solutions: These quantities are sufficient for 30 students.

1% gelatin: 2 g dry gelatin powder; add hot water (not boiling, 80°–90°C) to 200 mL.

1% gum arabic: 2 g dry gum arabic powder; add hot water (not boiling 80°–90°C) to 200 mL.

1% HCl: 5.2 mL of concentrated (38%) hydrochloric acid diluted with distilled water to 200 mL.

Water-soluble dyes: A variety of dyes will work. Congo red usually works best and living ameba will take it better. To prepare Congo red, dissolve 0.5 g in 100 mL distilled water. Stir to dissolve. Filter to remove any large particles.

Alternatives: Coacervates may form in a variety of protein/carbohydrate combinations. In addition to gelatin/gum arabic, these substitutes may work:

    alcohol added to gelatin in water,
    dilute sodium sulfate added to gelatin in water,
    gum arabic, sucrose, or alcohol added to egg
      albumin in water,
    sugar added to egg yolk in water.

It appears necessary to have protein in water first, and then add carbohydrate to it.

Text and Lab Correlation
Chapter 14 on the origin of life and Lab 1 on describing life.

Prelab Preparation
These life activities include: getting energy, reproduction, getting oxygen, protection against enemies, maintaining water balance, dispersal/movement, waste removal, perception/reaction to environmental stimuli, adaptation to the environment. Students will generate some of these on their own, and more of them in the context of class discussion.

Refer to Investigation 3.1. pH is defined as $1/\log_{10}$ of the hydrogen-ion concentration. The lower the pH number, the more acidic the solution. The pH range in this experiment is acidic (less than 7); narrow range pH papers in this range give the most sensitive results. However, wide range papers (1–11) may be used. This investigation requires quantifiable measuring.

Procedure
Students should use 4 test tubes: one for creating the mixture, and the other three for collecting the lab ingredients. *Do not let students shake test tubes at any time;*

this inhibits coacervate formation. Stress the meaningful use of time, and keep students moving.

It may take a few trials to see coacervates. As acid is added, coacervates may change in size, in shape, and in aggregation (clumping). They may show movement. If pH is about 6.8–7.1, they should take up the dye that is added.

Further Investigations
1. Amebas should take up the dye and become easier to observe. (Too high a concentration of dye may kill the amebas.) Life activities of amebas may be observed or inferred. Let students compare coacervates and amebas.

Answer Key
1. Cellular organization, movement, response, need for energy, production of waste, presence of water, presence of organic chemicals, presence of cells, inherit characteristics, metabolism, growth, reproduction.
2.–5. See students' data charts. pH is about 7.1–7.4. Most solutions will be cloudy. There may be hundreds of cell-like coacervates.
6. See students' drawings.
7.–8. See students' data charts. pH is about 6. The mixture is cloudy. (The interface rearranges.)
9. See students' drawings.

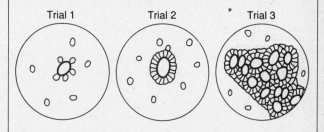

Trial 1   Trial 2   Trial 3

10. Coacervates should concentrate the dye and become darker than the surrounding field.
11. Answers will vary; the behavior could be taken as evidence of "life"; ability to selectively take in and concentrate materials.
12. They appear to move, grow, even reproduce, associate with one another, and ingest and concentrate material (dye). Students may choose to interpret some of these observations differently.
13. Answers may vary. Coacervates lack nuclei, do not carry on metabolism, and do not produce wastes.
14. Possible answer: size and number increased between pH 3 and 5.
15. Coacervate structure is affected by pH.

Investigation 15

Notes for the Teacher

Time
40 minutes. You will need to prepare the mung beans 4–5 days in advance of the lab.

Materials
Mung beans: You can order these from seed catalogues (Parks, Burpee, etc.). Sometimes, they are available in supermarkets and health food stores. Use only seed that is certified edible and not chemically treated, so that the students can eat any leftover sprouts. Seeds that are pink or green are treated with fungicides and are poisonous!

Screen: If you sprout the mung beans on a screen, you can rinse them easily. The beans should be rinsed daily to prevent bacterial and fungal contamination.

Paper towels: Use a layer of 2 or 3 wet towels. Drain off excess water.

To prepare the bean sprouts, presoak the mung beans in cold water overnight. Pick out any poorly developed beans and spread the rest out on a screen. Place the screen on moist paper toweling. If you have no screen, grow the beans on paper towels in a shallow baking pan. You will have to respread them when you rinse them. Keep beans separated to allow space for the sprouts to grow. Keep moist at all times. Grow the beans in a dark, warm place for 3 or 4 days, until the sprouts are about 1–3 cm long.

Text and Lab Correlation
Chapter 15.3 on natural selection and adaptation and Chapter 16.1 on variation. Lab 16 demonstrates how natural genetic variation in a population can be advantageous to a species.

Prelab Preparation
Statistical analysis is important to population studies. When dealing with relatively small sample sizes, the averages or means can be misleading when examining characteristics of a population, so it is not included in this study.

Procedure
The graphs for population variability should show bell-shaped curves or, at least, a double bell-shaped curve, if done correctly.

Groups can take turns measuring each others' earlobes until each group has measured 10 different people. You can also have them look at the way the earlobes are attached. An attachment such that the earlobe does not hang free is caused by a recessive gene. You can work out the ratio of free to attached ear lobes in the class to determine what the frequencies are in the class population. If the students are interested, they can find out how their parents' earlobes are attached (as well as those of their siblings) and hypothesize on the nature of inheritance of this trait.

Postlab Analysis
Have the groups compare their graphs to see if there is a large difference in the shape of the distribution curves and in the modes. If so, they should determine if these differences are due to an error in graphing, in sample size, or in sample selection.

Further Investigations
1. The relationship of second-finger length to fourth-finger length is a sex-linked trait. Students should be able to determine from the pattern that emerges which finger is shorter for females and which for males, if their samples are large enough.
2. Stress that the amount of light and temperature must be the same for all the individuals in the population; they should measure the amount of water and any fertilizer given.

Answer Key
1. 5' 7".
2. Probably not. Factors such as age, sex, or nutrition level of the people will bias the results.
3. Usually, the greatest number of individuals falls close to the midpoint of the range.
4. See students' data charts.
5. Answers will vary.
6. If measured a day earlier, the sprouts would have been shorter and would have shown less relative variation. A day later they will be longer and will show a higher degree of variation.
7. See students' graphs.
8. Answers will vary.
9. The longest sprouts might have broken through the soil sooner and reached the sunlight faster to increase growth. On the other hand, they may use up the stored nutrients in the seed earlier, before they become able to do photosynthesis.
10. Before it was picked, the bean may not have been well nourished as it was forming in the plant; the reason could be genetic; the bean may represent a shorter variety within the population.
11. See students' data tables.
12. Answers will vary.
13. Most of the individual measurements cluster around a modal or middle measurement; therefore the curve should resemble a bell.
14. With increasing sample size, the curves will assume a better or smoother bell shape centered around the true mode for the population.
15. One could expect to see 2 overlapping bell-shaped curves with different values for the mode.
16. Some examples are disease resistance, increased speed to escape predation, thicker fur to protect against cold, better camouflage.
17. Examples for plants are resistance to diseases (viruses or microbial) or insect predators, salt tolerance, shade tolerance, cold tolerance.

Investigation 16

Notes for the Teacher

Time

Because this investigation requires the incubation of bacteria, it must take place over 4 days. Each day's procedure will not take an entire class period; Part I, 30 minutes; Part II, 10 minutes; Part III, 15 minutes; Part IV, 15–20 minutes. Advance preparation of glassware and media will be necessary.

Materials

Stock Culture: *Escherichia coli* may be obtained from biological supply companies.

Nutrient broth media: Dissolve 5 g peptone and 3 g beef extract in 1 liter distilled water. Heat gently until all solids are dissolved. Fill culture tubes with 10 mL solution, insert cotton plug, and sterilize at 120°C, 15 lbs pressure, for 15 minutes.

Nutrient agar media: Add 5 g peptone, 3 g beef extract, and 15 g agar to 1 liter distilled water in an Erlenmeyer flask and mix well. Heat gently and stir until solids are dissolved. Boil dissolved mixture for one minute. Plug flask with cotton and sterilize at 120°C, 15 lbs pressure, for 15 minutes. While solution is still hot, pour plates with sterile petri dishes. Note: Prepared sterile petri plates may be obtained from biological supply companies.

Ultraviolet light source: One source is a General Electric 15-watt germicidal light bulb in a fluorescent desk light. Use in an enclosed area so that students do not accidently look into the ultraviolet light.

Text and Lab Correlation

Chapter 16, Section 16.2, on disruption of genetic equilibrium. Lab 12 on gene regulation and Lab 20 on bactericide effectiveness both emphasize sterile technique.

Prelab Preparation

Make sure that the students understand the terms *sensitive* and *resistant*. These are important in understanding bacteriological concepts, as well as comprehending adaptations to stressful environmental conditions.

Explain to your students that even though they will be working with nonpathogenic *E. coli*, it is important to maintain sterile technique. Not only does it preserve the validity of an experiment, but the sterile technique protects against accidental contamination with pathogens. Sterile technique is also used in Investigations 12 and 20.

Procedure

Demonstrate the sterile technique before the students try it. Sterilize all materials contaminated by bacteria before disposal or cleansing, either by autoclave or in 95% ethanol.

If teams work in groups, they may irradiate their plates together, saving time. Monitor students when they are using the ultraviolet light to be sure that they are observing safety precautions.

Further Investigation

1. If students perform this investigation, caution them not to let any streptomycin-resistant organisms escape. Use stock *Bacillus mycoides* as the experimental bacteria. Gradient streptomycin culture plates may be made. This reduces the numbers of plates necessary for different dilutions.

Answer Key

1. To prevent contamination of the stock culture and to kill any microscopic drops that escape while transferring the culture.
2. For 2 reasons: you may contaminate the stopper or cover with organisms on the work table, or you may contaminate the work table with organisms from the stopper or cover.
3. To show that the sterile technique is adequate.
4. To confirm that the nutrient agar is capable of sustaining normal growth.
5. Only ultraviolet-resistant *E. coli* should grow after irradiation. Other strains should be killed.
6. You can hypothesize that either no colonies will grow because you have no mutant UV-resistant organisms on the plate, or that there are UV-resistant individuals that will produce colonies.
7. Answers will vary.
8. Because all cells in a colony are progeny of the same original cell, they should have the same genetic makeup. Thus, they should all be UV resistant. However, there is a very slight possibility that there are individuals in which the UV-resistant gene has mutated.
9. When exposed to sunlight.
10. There should be no growth in the media control (MC), much growth in the culture control (CC), and decreasing amounts of growth for increased exposure time to ultraviolet irradiation (Plates 1, 2, 3).

11. The plate that was exposed for 3 minutes. If any cells survive the more stringent test, they are more likely to have the mutation for resistance.
12.–13. See students' data charts.
14. If all were negative, media and plates were properly sterile and sterile technique was used.
15. If any colony control (CC) plate did not have growth, it means that either there was no growth in the nutrient broth or all the cells were killed during the transfer.
16. Most likely that the original colony transferred was not actually ultraviolet-resistant.
17. Natural selection occurs only when the mutation is already present in the gene pool and when conditions favor survival for organisms with that trait. The speed of natural selection is affected by the length of time that the organisms take to produce new generations.
18. Possible answer: Environmental conditions (UV exposure) favored survival of only those organisms (bacteria) with the UV-resistant trait present.

Investigation 17

Notes for the Teacher

Time
40–50 minutes

Materials
Life-size reproductions or casts of hominoid forms are optional.

Text and Lab Correlation
Chapter 17 on Human Evolution, and Chapter 40, Section 40.1 on the Evolution of Mammals.

Prelab Preparation
Encourage students to explore the topic and approach the topic openly. Acknowledge that your students may have preset notions about the specimens they are studying. The purpose of the investigation is not to prove that evolution has occurred. It is hoped that students will relate to the specimens and gain an appreciation of how anthropologists evaluate fossil evidence.

Procedure
Since the drawings of human, ape, and fossil hominid skulls on page 102 are approximately 1/6–1/7 of life size, the checklist of features in Step B has incorporated conversion factors of 40 for square area and 1000 for cubic volume to approximate answers in the order of magnitude of "life size." (The factor of 1000, in calculating brain capacity, also includes $4/3 \times \pi$) If you have life size casts of skulls available for your students to measure, eliminate the factor of 40 in calculating square area and substitute 4.2 (or $4/3 \times \pi$) for the factor of 1000 in calculating cubic volume.

In the Neanderthal specimen the braincase is longer and flatter than that of modern humans; but it is still seen as more human-like. *Homo erectus* is more human-like than ape-like because of its larger brain size, and its front teeth are smaller than those of apes. Perplexing are the australopithecines, with their ape-size brains, large jaws, and prominent jaw muscle attachments. However, they also possessed the more human-like reduced front teeth and erect bipedalism.

Note that in question 4, students will correctly discover that they lack sufficient information for estimating brain capacity (first column in data sheet).

Postlab Analysis
Discuss in class the issue of the "common ancestor"; when did the "split" occur?

Further Investigations
Natural History, *National Geographic*, and *Smithsonian* are excellent sources for articles on archaeological or anthropological study.

Answer Key
1. Answers will vary. Students may suggest that early humans did not bury their dead; consequently, the dead bodies on the ground may have decayed or have been used as food by scavengers.

2. See students' data charts. Humans have long slender fingers, while gorillas' are short and full. Both have five digits. Both have an opposable thumb; however, the human thumb is larger in proportion to the other fingers.
3. See students' data charts.

Comparison of Human and Ape Skulls

	Human	Ape
Brain Capacity (cu. cm)	1950 cm³	422 cm³
Lower Face Area (sq. cm)	95.2 cm²	345.6 cm²
Brain Area (sq. cm)	297.6 cm²	230.0 cm²
Jaw Angle (degrees)	90°	130°
Sagittal Crest (absent/present)	absent	present
Brow Ridge (absent/present)	absent	present
Teeth (one jaw)	16	16
molars	6	6
premolars	4	4
canines	2	2
incisors	4	4

4.–5 See students' data charts.

Comparison of Fossil Hominids

Fossil Hominid	Brain Cap. (cu. cm)	Lower Face Area (sq. cm)	Brain Area (sq. cm)	Jaw Angle (deg.)	Sagittal Crest (yes/no)	Brow Ridge (yes/no)	Teeth (no. of each)
Neanderthal	H	H	H	H	No	Yes	H
Homo erectus	I	H	H	I	No	Yes	I
Australopithecus africanus	A	A	A	A	Yes	Yes	I
Australopithecus robustus	A	A	A	A	Yes	Yes	I

H=human-like
A=ape-like
I=intermediate

6. The chimpanzee has sharp teeth in the front of its jaw used for biting into fruits, and large flat teeth in back that are used for grinding food. Human teeth are smaller and are adapted for a more varied diet. The u-shaped jaw of the chimp, and the spacing of its teeth, enable the animal to bite off leaves and pull branches through the spaces, retaining the

leaves in the mouth for chewing. The round jaw of humans is related to our particular method of chewing and is shaped for receiving food from the hands.

7. In comparing the skulls of the human and another primate, the ape, the human has a high forehead that provides more space for the development of the front part of the brain. Also the ratio between the size of the braincase and the size of the whole skull is larger for humans.

8. Some students may be surprised about the same number and kind of teeth.

9. Some of the hominids show both human and ape-like characteristics. It suggests a stage in evolution. Refer to the information on australopithecines in teacher notes for "Procedure."

Investigation 18

Notes for the Teacher

Time
You will need 20–40 minutes for Prelab Preparation, depending upon the ability level of the students. 40 minutes for Part I and 40 minutes for Part II.

Materials
Geometric shapes: Make 2 photocopies for each pair of students.
Alternative: You can substitute other objects for the geometric figures and make slight modifications in the wording of the procedure accordingly. You should include 12–20 different items in each set of objects. Possible sets of objects: nuts and bolts, kitchen utensils, deck of playing cards, writing utensils, cosmetics, books, musical instruments. Have students participate in bringing objects.

Prelab Preparation
Spend some time explaining dichotomous keys. Give students an opportunity to discuss and ask questions. If you wish to give your students some hands-on experience with established dichotomous keys before they create their own key, you can schedule the Further Investigation as a Prelab Preparation.

Procedure
Some students might need assistance in getting started. They need to understand that each division of groups does not result in equal numbers of geometric figures in each group. The ultimate objective is to get each figure into a separate group.

Postlab Analysis
Students should end this activity with the understanding that using a dichotomous key improves the efficiency and effectiveness of human communications regarding biological diversity. It is important that students understand that the geometric figures in the individual groups in this exercise would represent a group of closely related organisms and not individual organisms. Further, students should understand that the skill of classifying has important application to their daily lives.

Answer Key
1. Fungi; Protista; Plantae.
2. Categories are mutually exclusive (either/or).
3. Answers should include creating mutually exclusive categories at each step.
4.–5. Answers will vary; see students' diagrams.
6. Possible answer: We looked for more specific characteristics as each group was split into two parts.
7. Answers will vary.
8. If no, must specify where there were problems.
9. Less diverse in general characteristics after the third division.
10. This process allows one to sort objects more accurately, to apply the same procedures for grouping, and to facilitate communications.
11. Yes. It depends upon what the observer sees as special characteristics. This fact is illustrated by the different grouping systems of the different laboratory teams.

Investigation 19

Notes for the Teacher

Time
40 minutes

Materials
Have a circle picture of a small animal cell drawn in scale since students will find it impractical to do themselves. Have students bring text to class.

Text and Lab Correlation
Chapter 9 for viruses and Chapter 44 for discussion of antigens and host immune system.

Procedure
Try out these models before class to determine any needed modifications. Divide class into 3 groups. One group should work on each model. Students who find geometry difficult should work on the influenza model. Leave time before and after lab to answer questions in lab. Have templates prepared so students can spend time cutting and putting together. Make sure angles for icosahedron are carefully measured.

Postlab Analysis
Give students chance to compare models and discuss the problems they had building and visualizing their models.

Further Investigations
Direct students to *Scientific American* for more detailed articles on specific viruses and to *The Molecular Biology of the Gene* by James Watson for a more detailed treatment of viruses in general.

Answer Key:
1. Polio—icosahedron, nerve cells; common cold—icosahedron, cells in the nasal passages; rabies—helical, nerve cells; mumps—helical, cells in salivary glands; AIDS—complex, cells of the immune system.

2. See students' drawings.

Icosahedron   Helix

3. There are 20 faces (triangles), 60 edges, and 8 vertices (8 axes of symmetry starting at each vertix).
4. A solid figure having 20 faces. The icosahedron model also has 20 faces (triangles).
5. Capsid contains and protects the nucleic acid.
6. Answer should be based on the actual length of the ribbon used for one turn in the model.
7. Answer should be approximately 17 times the diameter of the whole model.
8. A corkscrew movement by the capsid peeling off of the RNA.
9. Molecules jutting out on spikes will interact easily with the surfaces of host cells.
10. Helping penetration by the virus by digesting the host's cell membrane.
11. The outer capsule would be most recognized and is probably most subject to selctive pressure. The viral nucleic acid could change the capsid so that the host's immune system would not recognize it.
12. If the models were made carefully, they should be identical. Models are limited to the templates and do not show the diversity of the actual organisms being modeled. Models do not show the nucleic acid component of the virus or how the virus infects the cell.
13. The nucleic acid must code for the capsid components, be packaged inside the capsid, and be replicated by host cell.
14. Studying the virus directly would require an electron microscope as well as protection from infection by actual viruses.

Investigation 20

Notes for the Teacher

Time
40 minutes each for Parts I and II and 20 minutes for Part III (24–48 hrs after Part II). Note: Parts I and II can be done in one day if a double lab period is provided.

Materials
Bacteria: *Escherischia coli* and *Sarcina subflava* may be obtained from biological supply houses. Stock tubes should be subcultured onto sterile test tube slants of nutrient agar to make at least 3 tubes of each species per class of 24 (plus a new stock tube which can be stored under refrigeration until next year). Do all subcultures 2–3 days ahead of lab to obtain good growth (or earlier and refrigerate after growth).

Medium: Prepare 350 mL nutrient agar per class of 24 (including stock subcultures) Mix 1 g beef extract,

1.6 g peptone, 5 g agar and 350 mL distilled or deionized water in a 500-mL, or larger, Erlenmeyer flask. Heat until the agar dissolves and pour 20 mL aliquots into clean, loosely capped culture tubes. Sterilize at 15 p.s.i. for 20 minutes in an autoclave or pressure cooker. Also sterilize petri dishes, or purchase sterile disposable ones. Remove all items from the sterilizer while still warm and slant the tubes needed for the stock subcultures. Tighten caps on all tubes and refrigerate.
Bactericides: Remind students to check expiration dates to make sure the bactericides are fresh.

Penicillin and streptomycin disks in sterile vials are available from supply houses. Keep these refrigerated and do not use them beyond their expiration dates.

Text and Lab Correlation
Chapter 20. Labs 12 on gene expression in *Serratia* and 16 on developing bacterial resistance. (Aseptic technique may be taught once for these three labs.)

Prelab Preparation
Assign students to bring in various antiseptic, disinfectant, and antibiotic substances used in their homes. Supply small, preferably sterile, jars or vials if possible. Students in different groups can share the same bactericides. You may wish to assign or coordinate which students will bring in what bactericides and you might want to bring in several test substances yourself. Ask for volunteer groups to try each bactericide available so all are used. Students should work together in groups of two pairs.

Procedure
The hot water baths and bacteria stock cultures should be shared by the class. Bring the water baths to a boil before the Part I lab starts. Students should use cooperative working habits since time will be an important limiting factor. The melted agar medium might solidify if there is a delay. If that happens, students may return tubes to the water bath for a few seconds—being careful not to overheat the medium after it is inoculated with living bacteria. Note: Make certain that the medium is not too hot when it is inoculated or the bacteria will die.

Clean contaminated glassware in chlorine bleach.

Postlab Analysis
Students should understand that disinfectants, antiseptics, and antibiotics are not all equal in their effectiveness as bactericides because: different bacterial species have genetically different sensitivities; chemicals with bactericidal properties are packaged in different concentrations; chemical diffusion rates into the medium may vary; chemicals break down or deteriorate variably. Therefore, the presence of a zone of inhibition is more important for this lab than the absolute size of the zone.

Further Investigations
1. Most of the original antibiotics (penicillin, tetracycline, etc.) were obtained from yeasts, fungi, and molds. More modern antibiotics are frequently semisynthetic derivatives of these. Naturally made antibiotics may enable organisms to compete successfully with bacteria for nutrients.

Answer Key

1. Sterilize equipment.
2. Tie back long hair. Keep the flame at a safe distance from skin, clothing, and flammable liquids.
3. The glassware and culture medium were pressure heated (autoclaved).
4. Saprophytic bacteria need an organic energy source (carbohydrate, fat, or protein), minerals, and water.
5. Nutrient agar contains peptone (amino acid source), beef extract, agar, and water.
6. Agar is added to solidify media. It is a protein extracted from the cell walls of certain red algae.
7. Spores in the air or on surfaces, cells in water, microbes on skin, hair, etc.
8. To prevent contaminating spores from the air from getting into the tube.
9. Autoclaving glassware and flaming and/or alcohol soaking of instruments.
10. See data table.
11. Many bactericides should produce a zone of inhibition. If not, they may be too dilute or lack effective ingredients.
12. Results should be similar if procedures and species used were similar.
13. Answers will vary.
14. The bactericide diffuses into the medium. Sensitive cells will have some essential metabolic process prevented and will cease growth.
15. The substance that produces larger zones may have a greater bactericide action or it might diffuse better in the medium.
16. It depends on concentration and toxicity to particular species.
17. Different bacteria have different vulnerabilities to particular bactericides.
18. Disks with higher concentrations of bactericides will give larger zones of inhibition. This could be controlled (in theory) by adding the same amount of bactericide to each disk.
19. Many advertisers exaggerate the effectiveness of their products, but products sold as disinfectants or antiseptics are usually effective.

Investigation 21

Notes for the Teacher

Time
40 minutes for each species of protist. Assign the Prelab Preparation as homework, and allow 30 minutes for Postlab Analysis and discussion.

Materials
Cultures of Protista: Order from biological supply houses to arrive the week your lab is scheduled. Recommended species are as follows: *Euglena gracilis* (small, 30–40 μ, but easy to grow) or *Euglena oxyuris* (larger and easier to observe) are best. *Ameba proteus* is large (up to 600 μ) and easy to observe. *Paramecium* species range from 120–180 μ. *P. caudatum* (fair) and *P. aurelia* (smaller) are colorless and are best for the investigation.

Microscopes: Microscopes need lamps that provide adequate light at high power. Also, a stage mechanism for efficient movement of the slide while viewing is helpful, though not essential.

Yeast solution stained with Congo red: Prepare yeast suspension one day in advance (1/2 package dry yeast in 50 mL water). The day of the lab, heat the yeast to boiling and add 0.1 g Congo red powder. Continue to heat for 5–10 minutes. Congo red turns blue under strongly acidic condition (pH 3).

Methyl cellulose: Dissolve 10 g powdered methyl cellulose in 90 mL distilled water. (A ready-made solution may also be purchased.) A ring of methyl cellulose will help slow down *Paramecium* and *Euglena*.

Text and Lab Correlation
Chapter 21 on protozoa and Chapter 22 on algae. Lab 1 on defining life, Lab 14 on coacervates, and Lab 22 on algal blooms.

Prelab Preparation
Be sure to clarify the difference between the euglenoids (a division of algae) and the 2 phyla of protozoans. Behaviors common to all organisms may be reviewed using this lab, relating the observed characteristics of protists to those of multicellular plants and animals.

Procedure
In the interest of saving time, you may want students to do Steps D–F with only *Ameba* or *Paramecium* and then do Step G with *Paramecium* only.

In observing protists under the microscope, several points of technique must be stressed. As students try to observe the active organisms, they must remember that if they move the slide up, the image moves down. Also, readjustment of the condenser/iris diaphragm is necessary to keep the light level optimal. With microscopes having a multiple hold (rotating wheel) condenser, set the light aperture a little to the side. Such a setting darkens the background and illuminates the organism from the side, giving a glassy, three-dimensional appearance.

Students may check the correct depth of focus with the strands of cotton. Methyl cellulose will slow down the vigorous movement of *Euglena* and *Paramecium*. Apply a dime-sized ring of methyl cellulose to the slide, put in strands of cottons, and then the drop of culture. Air bubbles in the wet mount slides are very confusing to students; be sure they know the difference between bubbles and organisms.

Occasionally, have students turn the microscope lamps off to give the organisms on the slides a rest. It may also be desirable to refresh the slide with an additional drop of culture at the edge of the coverslip. This will add moisture if the slide begins to dry out. After the addition of distilled water (hypotonic) in Step E, add another culture drop to bring osmotic concentration back to near the original value. Modify the number of drops of distilled water, Congo red yeast, and vinegar as needed depending on the results of the initial experiment. If it turns out that the progression of treatments attempted on one slide (distilled water, Congo red yeast, vinegar) is too much for one group of organisms, encourage students to make more than one slide of each kind of organism.

Answer Key

1. Answers should include a number of the following: the ability to get energy, maintain water balance, remove wastes, adapt to environment, reproduce, protect themselves from enemies, disperse and move, and perceive and respond to environmental stimuli.
2. See students' labeled drawings.
3. Answers will vary. Different sizes would suggest ongoing individual growth and reproduction.
4. Organisms of the same species that are attached or side-by-side are probably reproducing. All 3 species undergo asexual (mitotic) fission.
5.–11. See students' data charts.

Observations for Questions 5–11

	Ameba	Paramecium	Euglena
General Shape	changing, blob-like	torpedo, slipper, shape defined by pellicle	like a long pancake
Overall Color	gray-clear	clear	greenish
Color of Parts	shadowy nucleus	darker nucleus and food spots	chloroplasts are green, eyespot is reddish
Style of Movement	slow, oozing (ameboid)	rapid, spiraling	quick, jerky, or wriggles
Structures for Movement	cytoplasm flows into projecting pseudopod	rapidly pulsing cilia	whiplike flagellum
Number of Pulses of Contractile Vacuole			
Vacuole Reaction to Distilled Water		(should show increased pulses)	

12. Fresh water is hypotonic to a protist, and water will enter the cell by osmosis. The contractile vacuole excretes excess water.
13. *Ameba* and *Paramecium* (heterotrophic) might be expected to eat the yeast (endocytosis), which then should be visible inside them. *Euglena* (autotrophic under these conditions) would not be expected to eat the yeast.
14. As yeast are digested within the food vacuoles of ameba and paramecia, acidic digestive juices may change the Congo red dye to blue.
15. Answers will vary.
16. *Paramecium* should become covered with a bristly layer of trichocysts. These would probably make the paramecium more difficult for a predator to eat.
17. Protozoa are one-celled, like early organisms, and may exhibit primitive animal behaviors. However, modern protozoa have evolved far beyond the ancestral condition; some are extremely complex.
18. Ameboid movement uses unspecialized structures.

Flagellum movement (*Euglena*) uses structures common to most animals and many plants (sperm cells have flagella) and must have evolved very early. Movement by cilia seems most specialized because of the great number of cilia and the coordinated rhythm of their movement.

19.–20. Answers will vary.
21. Answers will vary. *Euglena* does not eat when light conditions allow it to make its own food by photosynthesis.
22. Answers may vary. Amebas can change shape to avoid capture, the paramecium has a pellicle covering and also has the ability to back up, and the euglena has a pellicle covering.
23. *Euglena* seems most primitive because it has aspects of both plant and animal—although some students may perceive this as a highly evolved characteristic. *Ameba* shows strictly animal (heterotrophic) character, but it is very simple with few specialized characteristics. *Paramecium*, with its complex structure and behavior, seems most specialized.
24. Answers will vary. The following may be observed: getting energy, reproduction, maintaining water balance, dispersal/movement, protection against enemies, adaptation to the environment, perception/response to stimuli. The following are not obvious but could be inferred: getting oxygen, waste removal.

Investigation 22

Notes for the Teacher

Time
Pollutant solutions must be prepared one week ahead of Part I. Allow 20 minutes for Part I and 40 minutes for Part II, a week apart to allow the algae to grow. Set aside a few minutes before each class during that interval for your students to observe the cultures.

Materials
Pollutant Solutions: These must be prepared one week in advance.
Powdered laundry detergent: Make sure the label says that it contains phosphates and is biodegradable.
Fertilizer—plant food: Use either liquid or solid with nitrates.
Preparation of pollutants: If time permits, have students prepare the stock solutions of pollutants. Prepare 1% solutions of the pollutants by adding 10 g (e.g. powdered detergent) or 10 mL (e.g. liquid fertilizer) to make 1 L with algal growth medium (freshwater) in an Erlenmeyer flask. Break down the pollutants so the nutrients are available to the algae: add about 5 g dirt that is *not* sterile and has had no insecticides or herbicides added to it; stopper the flask with cotton and leave in a dark place for one week. Filter before using.
Algal growth medium—freshwater: May be obtained from biological supply houses in ready-to-use or concentrated form. Adjust the pH to 7.8 by adding 1N po-

tassium hydroxide dropwise and checking the pH with pH papers. It is helpful to sterilize the medium if possible. If you cannot, make it up fresh before the lab begins.

Algae: Use unicultures of freshwater green algae in liquid medium. *Closterium,* a desmid, is recommended because it is pollution-tolerant, hardy, and single-celled. Cells are large and visible under compound or dissecting microscopes.

Erlenmeyer flasks, 250 mL: If enough flasks are not available, equal-sized jars or similar containers may be used as substitutes. All glassware used should be free of any chemicals and well rinsed.

Microscopes: Make sure your students use the same power of similar microscopes for all counts, so the numbers will be comparable.

Text and Lab Correlation
Chapter 22 on algae and Chapter 53 on water pollution.

Prelab Preparation
This lab provides a good opportunity to experience collective learning. Each team of 5 students can test one kind of pollutant. Several teams can use the same solutions unless you wish to introduce other substances like those suggested in Further Investigations 2 and 3.

This lab deals with how a simple life form has a very significant, contemporary impact on the environment and combines the topics of algae growth, nutrients, ecosystems, and pollution.

Procedure
Remind students to mix solutions well both before and after adding algae. If local temperature fluctuates or if the laboratory does not have constant temperatures, plan the experiment for an appropriate time of year to minimize extreme temperature changes.

Postlab Analysis
In examining and comparing graphs from different students, take note of how your students calibrated the y-axis for the Number of Cells. Variations in the scale among the graphs will make it hard to visually compare the data.

It is essential to this study that students have time at the end of Part II to pool data from all teams. This will give them a chance to compare the effects of the different pollutants. Draw a large set of graph axes on the chalkboard similar to the model graph on page 133. Each team of students may use a different color chalk to graph their results on the board.

Further Investigations
2. Zinc is a micronutrient but is toxic at higher concentrations. If students test zinc salts they should use a very broad range of concentrations. Ideally, they will see a growth curve that rises, then drops at higher concentrations.
3. You can include these toxic substances with the

original experiment. Students can then compare the effects of nutrient pollutants with the effects of toxic pollutants.

Answer Key
1. Light, water, carbon dioxide, nutrients (e.g. nitrates and phosphates), and simple minerals.
2. Sunlight.
3. Air, light, water, warmth, nutrient source.
4. Set up one flask with algae, medium, but no pollutant and incubate it just like the other cultures.
5. The more complex chemicals will be broken down into simpler forms that the growing algae can use.
6. Answers will vary.
7. A *Closterium* cell:

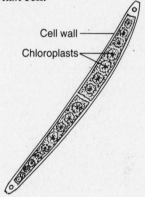

Cell wall
Chloroplasts

8.–9. See students' data tables.
10. Controls should have the lowest counts unless your variable is a toxic pollutant.
11. See students' graphs on separate paper.
12. Answers will vary but rates of growth should increase.
13. If the growth decreases at high concentrations, the pollutant could be toxic at higher concentrations.
14. If growth rate exceeds that of the control then the pollutant solution contains a nutrient.
15. Answers will vary.
16. It may be a nutrient, toxin, or have no effect.
17. Possible answers: Sample pond water weekly and incubate algal cultures with your water samples. Look for changes in algal populations over time.

Investigation 23

Notes for the Teacher
Time
Part I: 20–30 minutes. Part II: 40 minutes. One week must elapse between Parts I and II.

Materials
Mold Cultures: You may want to furnish some mold cultures as a backup or substitute for samples students bring in from home. Put a damp piece of filter paper on the bottom of each of several finger bowls. In each bowl, put a food item likely to grow mold, such as a slice of homemade bread or natural bread (no preservatives), a piece of orange (with rind intact), and a piece of

cheddar or longhorn cheese. Stack the bowls, cover the top one with plastic wrap, and store them for a week in an incubator at room temperature or in a styrofoam box. Also place a dish of warm water in the storage container to provide moisture. Add drops of water to the filter paper in each bowl as needed during storage to prevent drying. If mold forms too quickly, refrigerate cultures.

Medium: Prepare 250 mL of potato dextrose agar for each class of 24 according to the directions on the container. Sterilize at 15 p.s.i. for 20 minutes along with 24 petri dishes per class (or buy sterile, disposable dishes). While the medium is still warm, pour it into the petri dishes, covering the bottoms generously. Allow the agar to harden, and refrigerate the dishes until the lab.

Solution: Make a 5% (w/v) solution of sodium propionate in water. For a class of 24, make 500 mL by dissolving 25 grams of sodium propionate in about 400 mL of water, and then adjusting the volume to 500 mL. Sodium propionate is a more effective fungistat at acidic pH, so stir 25 drops of concentrated HCl (37%) into the 500 mL solution. The pH will be about 6. The sodium salt is cheaper than the free acid (which is a liquid) and should have a much longer shelf-life.

Text and Lab Correlation
Chapter 23. Lab 20 on bactericide effectiveness.

Prelab Preparation
Have the students develop a list of suggested environmental factors that affect mold growth (e.g., nutrients, moisture, warmth, light, absence of inhibitory or toxic chemicals). Help students focus on factors that can be readily tested in the lab.

Procedure
It is helpful for students to work cooperatively. For example, suggest that one student lift the cover of the petri dish slightly while another inoculates the agar with the fungal sample. In Step C, caution the students to be careful not to dig into the agar; they should gently flick the mold sample evenly over the agar. Discuss why the forceps should be sterilized and why the lids of the petri dishes should be lifted as little as possible (to avoid contamination by other organisms). Fungal spores can be highly allergenic. Caution the students about dispensing the spores into the air, and inquire if any of them have severe pollen allergies or asthma.

In Step D, students should apply the propionic acid solution sparingly and carefully to only one half of the agar in each dish. Discuss the purpose of the untreated half (to act as a control). In Step E, be sure each group places one dish in one environment and the other in an opposite environment. In Steps F and G, you may wish to have students sketch the fungal colonies as they appear in the petri dishes.

Postlab Analysis
Allow time for data sharing, perhaps by means of a class data table on an overhead projector or chalkboard. In discussing the students' conclusions regarding the growth requirements of molds, explain that only a certain degree of warmth encourages mold growth, because extreme heat is lethal to most life. Light has little effect on mold growth unless the cultures dry out in direct sunlight. In that case, loss of moisture is the actual growth inhibitor.

For Question 9, ask students to explain why raisins keep for a long time (lack of moisture).

Further Investigations
1. Some groups may observe green or blue-green colored molds, species of *Penicillium* and *Aspergillus*. The pigments are in the spores and are not chlorophylls. Students should realize that fungi are heterotrophic. Attempting to grow fungi on a medium devoid of a carbon source in the presence and absence of light will readily show that the fungi cannot obtain carbon from the CO_2 in air.
2. Mold would be likely to grow sooner on additive-free bread (no preservatives), whole-grain bread (more complete nutrients), natural cheese (no preservatives), and low-sugar spread. (Excess sugar plasmolyzes cells and is, in a sense, a preservative. A high concentration of salt is a preservative for the same reason.)
3. Sodium propionate and calcium propionate are common fungistats that have been ruled safe for human consumption in small amounts. Students may find labels showing other additives that are not necessarily fungistats. These may include: nitrites and nitrates (preserve the red color in cured meats); butylated hydroxy toluene or anisole (BHT or BHA) (preserves crispness in cereals); citrates and gluconates (buffers and antioxidants). The Federal Food and Drug Administration allows the use of these additives based on the assumption that the risk of harm from their use in small amounts is probably lower than the risk of harm to public health from microorganism contamination or unpalatable food.

Answer Key
1. Possible answers include old gravies and sauces, homemade or additive-free whole-grain breads, old fruits and vegetables, and cheeses.
2. The effect of propionic acid also is being tested at each location.
3. Some students may be likely to hypothesize that warmth and moisture favor the growth of mold, whereas cold and dryness inhibit it. Students should include a hypothesis regarding the effect of propionic acid.
4.–5. Answers shown on the next page are the expected range of results.
6.–7. The propionic acid-treated sides should show growth inhibition or lack of mold growth. Most fungi should show similar results.

Condition 1		Condition 2	
Room Temperature		**Refrigerator Temperature**	
Normal	Prop. Acid	Normal	Prop. Acid
++, +++	0, +	0, +	0
Light		**Dark**	
Normal	Prop. Acid	Normal	Prop. Acid
+, ++, +++	0, +	++, +++	0, +
Moisture		**Dryness**	
Normal	Prop. Acid	Normal	Prop. Acid
+++	0, +	+	0

Key: 0 = no growth
        + = poor growth
        ++ = good growth
        +++ = excellent growth; signs of reproduction (pigmentation)

8. Students may conclude that warmth, moisture, darkness (possibly), and absence of propionic acid favor fungal growth, while cold, dryness, light (possibly), and propionic acid inhibit it.
9. Possible answers: drying the product, refrigerating it, salting it, and using chemical mold inhibitors.

Investigation 24.1

Notes for the Teacher

Time
40 minutes. Get the plants 4–5 days before lab begins. Plan class time the day before this lab to demonstrate how the plants have been kept in the dark and then set up with the light screens.

Materials
Geranium plants: Two or more, depending on the number of students in your class. Each pair of students should have both a light-screened leaf and an unscreened leaf to test as a control. If you want the students to complete the Further Investigations, additional plants will be needed.

Light screens: Prepare these by cutting out a bold alphabet letter from a 4-cm square of black construction paper. Use a single-edged razor blade rather than scissors. Discard the letter and use the cutout block.

Alternative: Light screens can also be prepared as photonegatives. Use black-and-white film in your camera.

Photograph a complete roll of exposures of a solid dark object with thick lines, such as block letters, against a light background. For a control light screen, photograph a solid dark background (the photonegative will be clear), or cut out a piece of clear plastic the same thickness as a negative. Be sure to allow at least a week to have your film developed. Also, increase the time that the plants with light screens are exposed to light for 2–3 days.

Iodine solution: Lugol's iodine or Gram's iodine is available from biological supply companies. To make it yourself, dissolve 10 g potassium iodide and 5 g iodine in 500 mL distilled water. Store in a brown bottle or in a dark room.

Text and Lab Correlation
Chapter 24.1 on plants as food. Lab 7.1 on leaf pigments and Lab 50 on pyramids of energy in ecosystems.

Prelab Preparation
First, place the plants in darkness for 3 or 4 days in order to deplete the leaves' store of starch. Second, attach the letter cutout squares as screens, one per leaf. Students may help you attach the letter screens. This must be done before the plants are exposed to light. Clip a letter screen to the upper surface of each leaf and a solid piece of black construction paper to the under surface. Third, expose the plants to light for 24 hours or more. You should use artificial light in addition to sunlight to speed up the process.

Procedure
Review the procedure with the students and remind them to observe caution in handling hot alcohol. Keep a fire blanket or fire extinguisher handy. Make sure each student heats the beaker of alcohol in the water, not directly on the hot plate. Safety goggles should be worn. Let the beakers cool before moving. Optional: You may want to set up 2 or 3 hot plates at the front of the room for all students to share under your close supervision.

Postlab Analysis
If students understand the relationship between light and photosynthesis, they will realize that the carbohydrates and, indeed, all food we eat, come from the process of starch production observed in this lab.

Further Investigations
1. Cover a lamp with cellophane paper. (Make sure the hood of the lamp remains cool so that the cellophane does not melt.) You may use red, blue, and/or red and blue cellphane together. Use clear cellophane as a control. Expose the starch-depleted plants to this filtered light for 24 hours or more, and then test the leaves for starch.
2. Expose the starch-depleted, light-screened plants at different distances from one light source in a dark room. For example, place one plant 1 m away and another plant 3 m away from the light source.

Answer Key
1. After the plant has used up available sugar and stored starch, it will die if unable to photosynthesize.
2. The part of the green leaf not exposed to light should not manufacture starch.
3. There probably will not be a dramatic color change in a healthy, grown plant.
4. Probably not.
5. There should not be much difference before the chlorophyll is extracted and the iodine added.
6. The area exposed to light.
7. The color of the chlorophyll will hide the reaction of iodine with starch.
8. See students' notes.
9. Chlorophyll is not water-soluble.

10. Possible answer: The parts blocked from light remain whitish or the color of the iodine solution. The parts exposed to light turn blue-black.
11. Photosynthesis only takes place in the presence of light.
12. Possible answer: The leaf part covered by the screen was whitish; the part exposed to light was blue-black. The entire control leaf was blue-black.
13. The covered part of the leaf did not turn blue-black because there was no starch there.
14. Plants are important because they supply directly or indirectly all of our food. The earth's food web depends upon the plants' ability to photosynthesize.

Investigation 24.2

Notes for the Teacher
Time
Allow 20 minutes for the Prelab and 40 minutes for Part I. A 24-hour period should elapse between Part I and Part II. Allow 40 minutes for Part II and Postlab Analysis.

Materials
Elodea is readily available at most local pet stores that carry tropical fish. Pet stores use the name *Anacharis*.
Bromothymol Blue Indicator: Mix 0.5 g bromothymol blue powder with 500 mL of distilled water. If the solution appears green, add 0.1N sodium hydroxide drop by drop until it turns blue. The color changes quickly; precision work is important here.
Sodium Bicarbonate Solution: A large amount of sodium bicarbonate solution is needed for each class, approximately one liter for each team. To be most effective, the solution should be *freshly prepared* just before the class. Place it in a *tightly capped* bottle until you are ready to use it. A 0.5% solution should be sufficient: 5 g sodium bicarbonate per liter of distilled or deonized water. Make sure before the lab that the bromothymol blue will be yellowish-green in the sodium bicarbonate (Steps B, F, and G) If not, adjust the sodium bicarbonate to 1%.
Glassware: The glassware used should be clean and free of any possible acid or base residue. The size of the beaker should be coordinated to the size of the available funnels.
Lamps: A 75-watt incandescent bulb will be sufficient. A film strip or slide projector are other possible sources. The distance between the set up and the light source should be as small as possible without touching the beaker. Use fluorescent or plant lights for Further Investigations.

Text and Lab Correlation
Chapters 24 and 25 on plants and Investigation 24.1 on the role of plants in converting light energy into chemical energy.

Prelab Preparation
Allow adequate time for the completion of the prelab activities. The concepts of controlled experiment and double control can be clarified.

Procedure
If materials or time are in short supply, the control set-ups can be performed by different teams of students who will share data. Have each team set up one gas-collecting apparatus; keep half in the light and the other half in the dark.

In Step B, the blue indicator should turn greenish or yellowish. If the indicator is added too rapidly, the color may be missed. If no color change is observed, check the concentration of sodium bicarbonate.

Further Investigations
1.–3. Students should create an experimental design similar to that used for collecting oxygen gas. The rate of bubble release would indicate the rate of photosynthesis. A red filter would produce a higher rate than a green filter. Students should be aware of controls, variable influences, and recording of data.

Answer Key
1. The substances that enter a factory are an indication of the raw materials required to carry on the process performed in the factory. The substances that leave the factory are an indication of the products and wastes produced by the processes.
2.

$$6\,CO_2 + 6\,H_2O \xrightarrow{\substack{\text{Light Energy} \\ \\ \text{Chloroplasts} \\ \text{Enzymes}}} C_6H_{12}O_6 + 6\,O_2$$

3. The reactants: CO_2 and H_2O.
4. Light.
5. By adding or eliminating the light.
6. Glucose and oxygen. Because the purpose of the process is to produce glucose for plant growth, oxygen might be considered a waste product.
7.–9. The bromothymol blue turns yellowish. Carbon dioxide in your breath lowers the pH of the water.
10. Light.
11. Answers should describe how the experimental design allows for the comparison of two groups that differ only in one variable.
12. If carbon dioxide is consumed the bromothymol blue will turn from yellowish to blue-green.
13. The bromothymol blue will remain yellowish if the hypothesis is false.
14. The solution should appear yellowish (or yellow green), indicating acidity or low pH.
15. Bubbles appear to be rising from the plant and gas is collecting at the top of the tube.
16. If the hypothesis is correct, the gas is oxygen released by the plant during photosynthesis.
17. The effect of light on bromothymol blue.
18. Answers should describe a color change from yellowish to blue-green in the tube in the beaker kept

in the light. The tube in the beaker placed in the dark may be a lighter yellow. No change should be observed in the control tubes of only sodium bicarbonate and bromothymol blue.

19. Gas should be observed in the tube for the experimental set up kept in the light. A smaller amount of gas should be observed in the tube kept in the dark.
20. Answers should describe the re-ignition of the glowing splint.
21. The gas in the tube is oxygen.
22. The glowing splint does not re-ignite.
23. The tests performed on each tube yielded different results indicating that the gases are different. In the dark, a plant uses oxygen and produces CO_2 as a waste product, as part of aerobic respiration.
24. The color change from yellow to blue-green supports the hypothesis.
25. No changes in the tubes containing only bromothymol blue and sodium bicarbonate even though one was kept in the light and the other in the dark.
26. The tubes without Elodea.
27. Yes, the results of burning splint test support the hypothesis that plants produce oxygen.
28. Answers will vary.
29. Yes. Animals require both the food and oxygen produced by plants during photosynthesis. Plants can exist with a closed cycle, using the wastes produced by one process as ingredients for the next function.
30. A plant kept in the dark produces carbon dioxide.

Investigation 25

Notes for the Teacher

Time
Part I: 40 minutes. If you decide to prepare your own fern spore culture, allow 4–6 weeks for the gametophytes to grow. Part II: 40 minutes, including class discussion.

Text and Lab Correlation
Chapter 25 on plant diversity, Chapter 26 on plant structure, Chapter 27 on plant reproduction, Chapter 9 on mitosis and meiosis. Lab 27 on flowers and fruits.

Materials
Living fern sporophytes, gametophytes, and prepared slides of gametophytes and spore cultures can be ordered from biological supply houses. Fern spores can also be collected by placing mature fern fronds on white paper with their sporangial surfaces facing the paper. In a few days, the spores on the paper can be transferred to a clean vial and refrigerated until needed. Be sure to keep the fronds in a clean place (such as a cardboard box) while spores are being shed. If you live in a rural area or have access to the country, you can collect fern sporophytes from the wild. Many species of fern are common in places where there is adequate moisture. Good places to look are in northern pine forests, in shaded areas by streams and lakes, and in rock crevices near water. Collect only common species for lab use. Sporophytes with rhizomes are ideal for making cross sections of the stem. Ferns are also readily available from stores and greenhouses that deal in ornamental houseplants.

To make a spore culture, sterilize a supply of dechlorinated tap water, broken bits of unglazed crockery (old clay pots), wet sphagnum moss, and 400–600 mL beakers (to fit over the plastic pots as humidifiers). Soak 3–4" plastic pots in a 5–10% liquid bleach solution and rinse thoroughly with clean water. Air dry. Fill the plastic pots with the wet moss and cover with the pottery shards. Using a new cotton swab, flick fern spores generously over the shards. Set the plastic pot in a dish of water and cover it with a beaker. Place it in good light but away from direct sun. Add water to the dish each day or as needed and check with a good hand lens every few days for signs of growth. The young protonemata will be filamentous (a chain of cells). In time, they will develop into flattened and somewhat heart-shaped prothalli. Have patience, growth is slow (4–6 weeks).

Prelab Preparation
Assign the Parts of a Fern table as a homework assignment.

Procedure
Gametophytes—Have students observe antheridia and archegonia in either fresh wet mounts or in prepared slides. Remind students that protection of the sex cells (with sterile jacket cells) was a key adaptation to life on land. The restricting archegonial jacket prevented zygotes from escaping and immediately dividing into spores. Instead, they were constrained, developing into sporophytes.
Sporophytes—If a variety of fern leaves is available, some students may find it interesting to compare the patterns of venation and sporangial distribution. Sporangial dehiscence and spore shedding may be simulated by applying a drop of glycerine to a mature dry sporangium on a glass slide.

Answer Key
1. See the table of Parts of a Fern on the next page.
2. Possible answers: germinating spores, small groups of cells, heart-shaped prothalli.
3. Damp soil and bright but not full light.
4. Small, heart-shaped, green.
5. Separate. Rhizoids must grow down to anchor the plant; the cells in the apical notch grow out and up.
6. To make sperm.
7. To make and store eggs.
8. By swimming through water.
9. Yes. Eggs and sperm are not always produced on the same gametophyte.
10. Diploid.
11. See students' drawings on separate sheet of paper.
12. In a solid cylinder in the center (primitive ferns), a

Parts of a Fern

Part	Chromosome Number (1N or 2N)	Definition and Description	Where Found
Spore	1N	Single cell for asexual reproduction. Grows into a gametophyte.	Inside sporangia on leaves or on ground under or near ferns.
Gametophyte (prothallus)	1N	Small, haploid, sexual plant. Grows from a spore	On ground in damp, shady habitats.
Zygote	2N	Fertilized egg.	Inside the archegonium
Embryo	2N	Developing sporophyte plant, growing from a zygote by mitosis.	Inside the archegonium.
Young Sporophyte	2N	The sporophyte plant, still very small, but with developing shoot and root meristems & tiny leaves.	Bulging from the archegonium of the gametophyte.
Mature Sporophyte	2N	The adult sprophyte with typical stem, root & leaf structure; the latter with sporangia.	Woods, rocks, aquatic places; usually shaded or protected habitats.
Rhizoid	1N	Hairlike outgrowth of epidermal cells for anchoring the gametophyte & absorbing water.	Lower surface of gametophyte.
Antheridium	1N	Sterile, protective jacket cells making a round-shaped enclosure for sperm cells.	Lower surface of gametophyte among rhizoids.
Archegonium	1N	Sterile, protective jacket cells making a vase-shaped enclosure for the egg cell.	Lower surface of gametophyte, near growth notch.
Stem	2N	Upright or horizontal organ with meristem, branches, vascular tissue.	Soil, rock crevices, etc.
Root	2N	Downward growing, modified stem with meristem & vascular tissues.	In soil.
Frond (leaf)	2N	Green, photosynthetic and/or spore-bearing organ; may be simple or complex (with pinnae etc.)	Produced in a pattern on the fern stem.
Sporangium	2N – 1N	Capsule or sac consisting of a sterile jacket & a group of spore mother cells or spores.	Fern leaves (underside).
Sorus	2N	A cluster group of sporangia.	Fern leaves (underside).
Fiddlehead	2N	The young leaf as it emerges from the shoot meristem in a tightly coiled shape.	Tip (meristem area) of the fern stem.

hollow, C-shaped cylinder (intermediate ferns), or separate bundles arrayed in a ring (advanced ferns).

13. They supply water and nutrients to all parts of the plant.
14. Roots serve for absorption and anchoring; rhizoids are primarily anchors.
15. Roots.
16.–17. See students' drawings on separate sheet of paper.
18. Increases the chances of a spore landing in a suitable place to grow.
19. When the sporangium opens, spores will drop out.
20. See students' table begun in Prelab Preparation.
21. Plants that have free-living gametophytes (ferns and mosses) live in damp places; seed plants exist in many types of habitats.
22. By branching from the rhizome.
23. Sporophyte. Gametophyte.
24. Possible answers include: reproduction by seed; a greater degree of vascularization; more efficient roots; larger and more efficient leaves.

Investigation 26

Notes for the Teacher

Time
Allow 40 minutes.

Materials
Plants: Obtain one plant each of at least 2 different species from which epidermis is easily removed, such as geranium, most ivies, many types of tree leaves, snapdragons, Shasta daisy, iris, and daffodil. Test your chosen species of plants for ease of epidermis removal.

Use well-watered plants that are kept out of direct sunlight but in a well-lighted area; this ensures open stomata. Direct sunlight or heat might result in leaf dessication and the closing of the stomata. All the plants must have at least 3 hours of light on the day of the lab. Use artificial lighting to prepare for morning classes.

Students should remove only one leaf at a time from a plant, as new leaves are needed for the lab, so that leaves will be as fresh as possible. Leaves removed before they are needed might wilt.

Solutions: Dissolve 50g of NaCl in 1 L of distilled water.

Text and Lab Correlation
Chapter 26.4 on leaves. Lab 24.2 on plants and air.

Prelab Preparation
When stomata are open, oxygen (a product of photosynthesis) moves from the region of high concentration inside the leaf to the region of relatively low concentration outside the leaf (the atmosphere). The reverse is true for carbon dioxide, which is used in photosynthesis. Whenever stomata are open, water vapor is lost to the atmosphere. The only exception is in foggy weather, when the air contains more water than a leaf does. Stomata are not always completely open or closed. There are many intermediate positions of the guard cells.

Procedure
As an alternative to cutting or pulling the epidermis off the leaf in Step B, your students may find it easier to leave it hanging from the torn edge and to make a wet mount of the entire leaf section. They then should focus just on the area where the epidermis is hanging over the edge.

In discussing Questions 13–15, point out that when guard cells are under great turgor pressure, the thinner outer walls balloon out and carry the thicker inner walls along with them, causing the stoma to open. When the guard cells are flaccid (as when a leaf loses water rapidly on a hot day), the thick, elastic inner walls pull the rest of the cells inward toward the stoma, thus closing it. This helps a partly wilted plant prevent excessive loss of water vapor.

In Step H, replacing the water that bathes the epidermis with 5% NaCl solution causes turgor pressure to decrease in the guard cells. The students can watch this happen if they handle the fluid exchange carefully while the slide is still on the microscope stage.

Postlab Analysis
Stomata closure, under environmental conditions such as a hot, dry day, results in reduced absorption of carbon dioxide from the air, thus interfering with photosynthesis. Closure of the stomata would also prevent the removal of oxygen produced during photosynthesis. Excess oxygen trapped inside the leaf would be used in respiration, which in turn would produce carbon dioxide that could be used in photosynthesis. When stomata are closed, the rate of photosynthesis (which can be 10–20 times the maximal rate of respiration) is limited to the rate of respiration within the leaf.

Further Investigation
1. Stomata on leaves from plants kept in the dark should be closed, whereas stomata on leaves from plants kept in light should be open (unless excessive heat causes water loss, resulting in closed stomata). Students should avoid exposing the plants to direct sunlight or to heat.

Answer Key
1. Photosynthetic activity is greatest in daylight, so gas exchange would also be greatest.
2. Under conditions of heat and dryness, the plant would dry out if water loss were not reduced.
3.–4. Possible answers include holes or tubes in the leaf and removable covers or a valve system.
5. See diagram. Students' drawings should show open stomata.

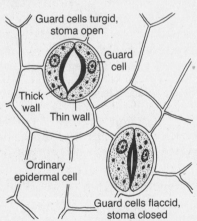

Guard cells turgid, stoma open
Guard cell
Thick wall
Thin wall
Ordinary epidermal cell
Guard cells flaccid, stoma closed

6. The inner walls are thicker than the outer walls. Students' drawings should indicate this characteristic.
7.–9. Answers will vary. The density of stomata on a leaf varies with the species of plant.
10. Most students should find more stomata on the lower surface than on the upper surface of a leaf. Advantages include stomates being shaded from the direct sunlight.
11. On the upper surface, because it is the surface that is in contact with the atmosphere.

12. Should hypothesize that if guard cells lose water, they will collapse, causing the stoma to close.
13. The guard cells collapsed, closing the stoma.
14. The difference in inner and outer wall thickness.
15. Answers will vary depending on the hypothesis.
16. Gas exchange would be severely reduced if not prevented.
17. Guard cells would lose water and stomata would close. Photosynthetic activity would be inhibited due to the reduction of carbon dioxide available to the leaves.

Investigation 27

Notes for the Teacher
Time
You may have to order flower blossoms several days ahead. Allow 40 minutes each for Part I and Part II.

Materials
Flowers: Gladioli are excellent flowers for dissection; they may be purchased at any florist, but may have to be specially ordered. You could ask for older flowers that would normally be discarded. Be sure to get some leaves with the flowers. If gladioli are not available, you could substitute lilies, tulips, or crocuses (all are monocots and all have colored sepals to match the petals). Make certain before the lab that the flowers have readily distinguished parts. Roses, carnations, daisies, and chrysanthemums are difficult for beginners.
Fruits: At Station 1, any single-seeded fleshy fruit (olives, cherries, or avocados) could be used. For Station 2, pears or squash would also work. Any wind-carried seeds, such as ailanthus, ash, or thistle could be substituted at Station 3. If burrs are unavailable for station 4, substitute seeds from beggar ticks, Queen Anne's Lace, or tick trefoil. You may want to have students use a hand lens or a dissecting microscope to observe "hitchhiker" type seeds. Opening the burr fruit will reveal the seeds.

Text and Lab Correlation
Chapter 27 on flowers, fruits, and seeds. Lab 25 on plant diversity.

Prelab Preparation
You may want to explain that cultivated flowers may not show clear adaptations to the behavior of pollinators, since human selection may be producing "artificial" flowers that exist only through propagation by cuttings. The same situation applies to fruit (e.g., seedless grapefruit, bananas, etc).

Procedure
Caution the students in using the razor blade while dissecting the flower ovary.

In Part II, the fruits at Stations 1 and 2 could be cut in sections along a variety of axes. This would help student to get a three-dimensional idea of the shape of the original ovary. Encourage students to orient the fruit in various ways and to imagine the fruit as a giant ovary within giant petals. You may wish to have 2 coconuts, one for display in a whole state, one cut to show the endosperm "meat." Several interested students may want to try to find the small embryo in one of the 3 "soft spots."

Further Investigations

1. Staining the dissected seeds with iodine will indicate the location of stored starch. Soaking dried seeds makes dissection easier.
2. Flowers containing only female or only male reproductive parts are found on ashes, maples, willows, poplars, date palm, and holly. In some cases, plants have separate male and female flowers on the same individual. Maple, oak, hickory, ash, birch, corn, and squash are among plants with flowers of a single sex on an individual plant.

Answer Key:

1. Monocots: one cotyledon, principal veins in leaves parallel, flower structures arranged in threes. Dicots: two cotyledons, principal veins branching from midrib or base, flower structures arranged in twos, fours, or fives.
2. Pistil—female reproductive structure of flower. Stamen—male reproductive structure of flower.
3. By animals that carry the pollen from plant to plant as they feed. By the wind.
4. Animals disperse the seeds of fleshy fruits. Wind disperses fruits with feathery or winglike structures.
5. The ovule becomes the seed, the ovary becomes the fruit.
6. Gladioli are monocots (parallel-veined leaves, petals in multiples of 3).
7. To attract pollinators with color and scent and to help funnel the pollinator towards the central sexual parts of the flower.
8. Six.
9. See students' drawings on separate piece of paper.

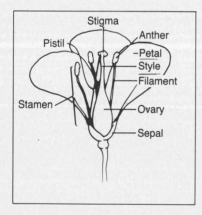

10. See students' drawings on separate piece of paper.

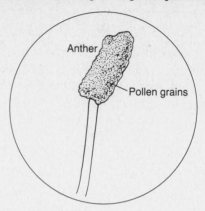

11. Hundreds. Each pollen grain has only a small chance of reaching a stigma of another gladiolus flower.
12. One.
13. Pollen sticks more easily to the stigma; 3 long "prongs" provide a lot of surface area for pollen to attach to.
14. See drawings for Answer 9.
15. See students' drawings on separate piece of paper.

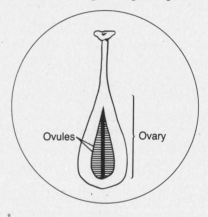

16. Many seeds, one for each ovule. They will be arranged in stacks like the ovules.
17. One seed is inside pit.
18. The fleshy part protects the seed and attracts animals who eat the fruit and disperse the seeds.
19. Many seeds, arranged radially in compartmentalized areas.
20. The fruits listed in this lab will have 3 to 5 carpels.
21. By wind. Maple seeds have stiff, winglike structures. Dandelion and milkweed seeds have light, spreading structures that act as parachutes.
22. Lightness, some means of keeping aloft, access to breezes (held high on the plant).
23. By attaching to the fur of animals.
24. Attributes should include stickiness or entangling hooks.
25. By water, it floats.
26. Only on land will the coconut come in contact with fresh water. When the fresh water passes into the 3 softer spots at the end of the coconut, the embryo in one of the soft spots begins to grow.

27. Color attracts and directs the pollinator. Flower shape tends to aid the pollinator in landing. Stigma and anthers are often situated so that incoming pollen is captured by the stigma and the pollinator picks up pollen on its way out, avoiding the stigma and the possibility of self-pollination. Nectar is secreted at the base of the petals, so the pollinator must pass by stigma and anthers to get to the food.
28. Seeds may have structures that facilitate their dispersal by being eaten, carried on the wind or water, or attached to the fur of animals.
29. Hypothesis should be stated using *"if/then"* phraseology and include testable statements.
30. Answer should include information based on visual experiences (number and arrangement of ovules in the blooming flower are similar to those of the seed in the fruit) as well as text information (haploid sexual cells in the flower have become diploid structures in the fruit).

Investigation 28

Notes for the Teacher

Time
The oat seeds must be soaked overnight. To save lab time, you may want to assign Part I (the shelling of the oat seeds) to 1 or 2 students or do it yourself. This lab continues over a 5-day period. Beware of weekends interrupting the investigation. Part I—10–15 minutes (day 1); Part II—15–20 minutes (day 2); Part III—40 minutes (day 4); Part IV—40 minutes (day 5)

Materials
Seeds: Oat (*Avena sativa*) seeds are available from biological supply houses or feed stores. Purchase about 4 oz. each year and refrigerate them until needed (they are better when fresh).

Text and Lab Correlation
Chapter 28 on plant growth and Chapter 26 on plant structure.

Prelab Preparation
Divide the class into teams of 4. Two members of each team will perform the phototropism experiment and the other two will do the gravitropism experiment. Make sure that all students understand that they are responsible for understanding *all* the material and answering *all* the questions.

Procedure
Designate a growth and storage area in cupboards or closets and a greenhouse area or illuminated plant bench. If only a window sill is available, the plants that require uniform illumination must be turned 90° every hour during the day. All manipulations of seeds and seedlings should be done in a semi-darkened room. The less light that reaches the plants, the more accurate will be the results.

Demonstrate how to use a protractor for measuring angles of curvature.

It is essential that all seeds and seedlings be kept moist at all times. Have the students check them regularly. The time schedule may need some modification based on germination rates of your seeds and your ambient room temperatures. Keeping the seeds at a warm room temperature, at least 25°C, will help maintain the schedule given in this lab. If germination is slow, allow 3 or 4 days between Part II and Part III (doing Part II on day 5 or 6 and Part IV on day 6 or 7).

Postlab Analysis
Make sure that enough time is set aside on day 5 (Part IV) for pooling data, analyzing possible sources of error, and analyzing results.

Answer Key
1. The root, because water is needed to enable sprouting and growth.
2. Yes, pot 1.
3. Stems (meristems) distribute auxin preferentially toward less illuminated parts of the plant. (This causes the darker sides to grow or elongate faster than the lighter sides, resulting in the plant's bending toward the source of light.)
4. See students' data tables.
5. See students' drawings of plant curvatures.

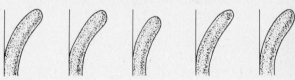

6. See students' drawings of roots and shoots, day 4.

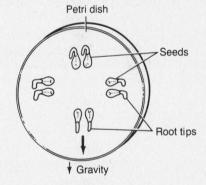

7. See students' drawings of roots and shoots, day 5.

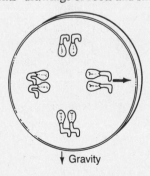

8. Gravity causes auxin to accumulate in the lower sides of the root where root growth is then inhib-

ited, causing the root to curve downward as it grows.

9. The plants were grown in the dark and handled under very dim light.
10. The shoots grow toward the light.
11. Yes. Phototropism was clearly evident in pot 4. The plants all grew in the direction of the light.
12. Foil-covered-tip and dark-grown plants (should) both grow straight but the ones covered with foil might be shorter.
13. If all seeds are not genetically identical, growth responses may vary. Prolonged or incorrect handling of plants during the experiment may also cause variation.
14. The roots grow downward, pulled by gravity.
15. Shoots grow in the direction opposite the pull of gravity.
16. (See Question 8.) Although auxin inhibits cell growth in roots, it enhances growth in stems, causing them to grow upward.
17. The roots and shoots reorient so that their root tips always face toward the force of gravity and shoots always face away from the force of gravity.

Investigation 29

Notes for the Teacher

Time
40 minutes for Part I. Part II, watching a hydra capture a daphnia, may not fill a 40-minute period.

Materials
Hydras: These may be obtained from biological supply houses or from lakes and ponds. If you choose to catch them, they are usually found attached to pond weeds and bottom materials. These organsims and others you might use during the year can be maintained in a freshwater aquarium. Note: Hydras are very sensitive to metals and chlorine, so do *not* use tap water unless it has been dechlorinated by letting it stand for 24 hours. You may also use filtered, sterile pond water or bottled spring water.

Hydras can be fed daily or every other day. Do not feed hydras for 1–2 days before the students are to perform this lab. Hydras that are not hungry will not discharge nematocysts.

Hydras will not survive if their water becomes too dirty, so the aquarium or culture dish should be cleaned and the water changed. The hydra can be moved while cleaning. For more information on culturing hydras, see F. Barbara Orlans, *Animal Care from Protozoa to Small Animals*, Addison-Wesley, 1977.

Daphnia cultures: You may wish to get enough to use in Investigation 43.1.

Beef broth: Use the specially prepared beef extract used in making bacteriological media. However, you can try bouillon cubes (preferably unsalted), or try salted and unsalted and compare them.

Filter paper: Any highly absorbent paper that will not break up on contact with water may be used. Students can tear off tiny pieces with the forceps.

Text and Lab Correlation
Chapter 29 on sponges and cnidarians and Lab 43.1 on daphnia.

Prelab Preparation
You may wish to tell your students about some of the basic principles of sensory organs, neurons, and effector cells. Although these concepts are treated in detail in Chapter 46 of the text, it might help to explain briefly that sensory cells detect chemical (e.g. amino acids) and physical (e.g. touch) stimuli. The sensory cells transmit information to the nerve cells, which, in hydras, form a simple neural network. The neurons, in turn, communicate with the effector cells which create the observable response—the firing of nematocysts and the drawing in of the food.

Procedure
Demonstrate to students how to pick up organisms with droppers. If the tips of the droppers are too small, the droppers can be inverted. Refer to Lab 43.1 for an illustration of this technique.

If your students cannot see the nematocysts under the dissecting microscope, they should be able to see them under $100\times$ or $400\times$ magnification with the compound microscope. Demonstrate how to place a drop of water with a hydra on a coverslip. The students should then add a drop each of methylene blue and vinegar, place a depression slide over the coverslip, and invert it with the coverslip in place so the hydra is suspended into the hole. The vinegar will make the nematocysts eject and the methylene blue will make the filaments more visible. This happens very quickly and may be difficult to observe.

Postlab Analysis
Some students may have recalcitrant hydras that do not demonstrate these effects. Encourage sharing of information and looking at each others' specimens.

Further Investigations
1. Hydras will move around if they are not getting enough food, sometimes like a cartwheel.
2. Each bud is counted as a prospective individual. Healthy, well-fed hydras should grow logarithmically. A number of different growth-rate experiments can be tried by counting buds on control and experimental groups.

Answer Key
1. They eat animals only. Their receptors on sensory cells would not respond to plants.
2. They must live in water with a very slow current. Their preference for sitting on plant material brings them close to locations where prey are feeding.
3. Enzymes are secreted in the gastrovascular cavity that break down food into simpler forms, which then enter the animal's cells through endocytosis.

4. See students' drawings.

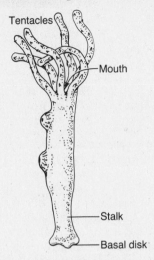

Tentacles

Mouth

Stalk

Basal disk

5. Protective response by withdrawing tentacle.
6. Yes, the hydra should send out nematocysts and reach for the filter paper in response to beef broth.
7. The hydra should attempt to grab the probe with its tentacles in response to touch, but may contract.
8. The hydra contracts when it is threatened.
9. Feeding responses are induced by both chemical and tactile stimuli, but answers will depend on observations and animals.
10. The hydra shoots a dart at the prey, the nematocyst filament then retracts, and the tentacles grasp the prey, bringing it into the opened mouth.
11. The hydra releases any undigested material from its mouth.
12. The cnidarians can remain fixed in one position but still capture moving prey.
13. Porifera are sessile and filter feed. Cnidarians have some mobility and have special adaptations—the nematocysts—that allow them to feed on small, motile animals.
14. Hydras are sessile polyps and jellyfish are mobile medusae.

Investigation 30

Notes for the Teacher

Time
40 minutes

Materials
Magnets: Bar magnets can be obtained from the physics department or a general science teacher. The stronger the magnet, the more effective the lab is. "Recharge" magnets before the lab with a magnetizer.
Flatworms: If you specify brown Planaria when ordering from a biological supply house, they often send *Dugesia* sp.
Dechlorinated water: Let tap water sit for 24 hours in an open container. You may also use pond or aquarium water.

Text and Lab Correlation
Chapter 30 on flatworms, including an in-text lab. Lab 35 on comparing invertebrates and Lab 46 on vision and hearing.

Prelab Preparation
Encourage students to contemplate the many sensory stimuli that may exist beyond human consciousness and what scientific instruments exist or would be needed to study the stimuli themselves. Students should understand that what they are observing is how the animal *responds* to the stimuli, not how it receives or perceives the stimuli.

As part of Prelab Preparation, you may want to give your students a chance just to observe planaria the day before they do the Procedure (perhaps by using the in-text lab for Chapter 30).

Procedure
Teach your students how a compass works, emphasizing the time required for the needle to point to true north. Students need to be shown how to line up the readings so that "N" is pointed to by the red bar. Encourage students to be patient and to give flatworms enough time to move in a clear direction. Students should get clear results for Steps C and D as long as they do not put the north pole of their magnet at N and the south pole of their magnet at S. Usually the worms are strongly attracted to the south pole; they either avoid the north pole or show no response.

At the end of the experiments, add the flatworms to your freshwater aquarium if you maintain one.

Postlab Analysis
Have the students share their results on the chalkboard.

The hows and whys of magnetotaxis in flatworms (and in many other species as well) are still an unknown for research scientists. The element of mystery in this lab may intrigue some students but frustrate others. Encourage speculation and an open-ended class discussion. Flatworms do orient themselves against currents (to face upstream). You could introduce this information in postlab discussion.

Further Investigations
1. Ward's Natural Science Establishment has a "Train-a-Tray" in its biology catalog that might be suitable for phototaxis experiments, or you could design your own.
2. To separate the influences of gravity and light, try lighting from below.

Answer Key
1. You would expect attraction toward or avoidance of one pole or the other.
2. Step A (with only the earth's weak magnetic field) is a control for Steps C and D (with the stronger magnetic field of the bar magnet).
3. Observing the worm immediately might reveal

only its response to being placed in the dish.
4. See students' data charts.
5. Most students will observe random movement.
6. No. In the earth's weak magnetic field, no particular direction is preferred.
7. Clarifies whether the response is an individual peculiarity or a response attributable to all flatworms.
8. Possible answer: The bar magnet will create a stronger magnetic field.
9. See students' data charts.
10. Ambigious results are possible, but, in many cases, most worms avoid the north pole.
11. See students' data charts.
12. Yes. Usually most worms are attracted to the south pole, but results may vary.
13. Possible answer: Yes, the south pole in Step D.
14. Some worms may have stronger responses than others.
15. The magnetic field of the earth's pole might be causing an interaction in that there will be 2 magnetic fields instead of one. This can be controlled for by orienting the bar magnet in all compass directions and comparing the responses at north, south, east, and west.
16. Iron.
17. Answers will vary. Encourage creative thought.
18. Possible answer: Not all mutations necessarily have an evolutionary advantage and no one know why flatworms have this capability.

Investigation 31

Notes for the Teacher

Time
40 minutes for Part I and 40 minutes for Part II.

Materials
Earthworms: Live earthworms may be obtained from biological supply houses or from local soil. Get the largest available. They may be stored in a refrigerator if they are kept moist and covered. After the lab, the earthworms may be returned to the soil.
Ice cubes: Make these ahead of time.

Text and Lab Correlation
Chapter 31 on annelids. Lab 35: Comparing Invertebrates.

Prelab Preparation
Discuss experiences students have had with earthworms (used for fishing bait, seen on the sidewalk after a heavy rain, etc.).

Procedure
In Part I, Step E, encourage the students to wet their fingers with water and feel along the ventral surface sides of the earthworms. The setae very noticeably feel like stiff bristles.

In Part II, Step M, assign these 3 temperature points to different groups: 10°C, 20°C, 30°C (no higher). If worms are still cold from being stored in the refrigerator, the first temperature point may be when they are first taken out of the refrigerator.

Postlab Analysis
Make sure that students know how to graph properly. Tell them to circle their data points and to draw the best possible straight line through the data. Emphasize that the line does not have to touch every point. If one point is way off the graph, discuss why (maybe the students made an error in counting, the worm was sickly or too cold, etc.). This is an excellent exercise for students to see how conclusions can be drawn from good, not perfect, numbers.

Further Investigations
1. Controversy over dissection has increased in recent years. In this context, we've chosen not to require an earthworm dissection. For most students, behavioral observation is a more effective learning activity than dissection. Nevertheless, for the interested and capable student, a dissection may be an excellent extension activity. We recommend "Investigation 30: Two Eucoelomate Groups" Part II C, *Biology Investigations*, Holt, Rinehart and Winston, 1985, pp. 185–186, 189–190. For others, films, models, etc., can provide much better insight into internal anatomy.
3. Have the worm crawl over paper towels soaked with increasing amounts of water.
4. In the "robin response," the earthworm will quickly retract in escape behavior if its anterior end is grabbed. It will pull back and greatly flatten, which forces the setae into the ground, making it very difficult to pull the worm from its burrow.

Answer Key
1. Segmented worms with complete digestive, circulatory, excretory, and nervous systems.
2. Earthworms are invertebrates.
3. Starve, because there would be no organic matter in the sand.
4. The earthworms turn and aerate the soil as they eat.
5. No.
6. The terms are more specific and correspond more closely to the animal's internal organs. For example, the "top" of a human is the head (anterior), whereas the "top" of a dog is its back (dorsal).
7. It rolls over.
8. Since they are all hermaphrodites, they all lay eggs and must secrete the mucous sac.
9.–10. Answers will vary, but there probably will be little response.
11.–12. The anterior end is usually most sensitive. It helps keep the worm under the soil, away from the drying sun.
13.–14. Answers will vary, but the earthworms will probably prefer wet towels. The earthworm's natural environment is moist soil.
15.–16. Answers will vary, but the earthworm will probably prefer the sandpaper. The dry paper towel absorbs water from the worm's body. Since a worm is hydrotactic, it shuns the drying environment of the towel. Also, the sandpaper may feel more like the texture of dirt, the worm's natural habitat.

17.–18. Answers will depend on the temperature assigned.
19. Pooled class data on blackboard.
20. See students' graphs.
21. As the temperature decreases, the heartbeat rate slows down.
22. They prevent it from leaving the soil and drying out.
23. Gas exchange directly through the skin is possible because the skin is moist and it has two long blood vessels (dorsal and ventral) into which oxygen can easily diffuse.
24. Reactions occur more slowly at cold temperatures. The earthworm can conserve its energy when it is cold.

Investigation 32

Notes for the Teacher

Time
40 minutes each for Parts 1 and 2.

Materials
Crayfish: Can be obtained from biological supply houses. They should be maintained in 5- or 10-gallon aquariums for the duration of the labs. Note: Metal containers are toxic. Several crayfish may be kept in a single tank as long as enough space and shelter are provided for each crayfish. Note: Crayfish are very territorial and will tend to fight, especially if they are overcrowded. The bottom of the tanks should be covered with gravel. Add dechlorinated tap, pond or spring water to a level about 2.5 cm above the gravel. Include one or two flat stones that extend slightly above the water level. Each crayfish must have its own shelter—a small clay pot placed on its side works nicely. They should not be kept near the radiator or in the sun. Room temperature should be between 18 and 24°C (64 to 75°F). Water in small containers should be changed every other day. Note: Avoid significant temperature changes when replacing water. Crayfish may remain outside of the water for 5 to 10 minutes without harm. They can be fed crumbled cat food pellets, or similar food, but they should not be fed for at least 24 hours prior to Part II.
Saturated salt solution: Dissolve 3.5 gm NaCl in 10 mL water.

Text and Lab Correlation
Chapter 32 on the Class Crustacea. This lab can be related to Investigation 33 on caterpillar behavior.

Prelab Preparation
Discuss the Class Crustacea and go over the other classes of the Phylum Arthropoda so that the students will understand how the crayfish compares with other members of the phylum.

Procedure
Have rubber gloves handy if any students feel uneasy about picking up the crayfish. They should be cautioned but not unduly frightened about avoiding the chelipeds, which can pinch fairly hard.

Students should be advised not to fill their tanks up with water. This is harmful to the crayfish in addition to making the tanks or boxes hard to handle.

Postlab Analysis
Encourage the students to think creatively about the reasons *why* the crayfish behaves as it does. Remind them that virtually every reaction has a specific survival advantage related to it.

Further Investigations
1. The students should obtain an understanding of the nocturnal and somewhat sedentary lifestyle of the crayfish.
2. Some species of crayfish have the capacity to alter the color of their shells to that of their surroundings. In their natural habitat this would help protect them from predators.
3. If some of your students would like to learn about the internal anatomy of the crayfish, you might want to encourage them to perform a dissection of preserved specimens, create a model, make drawings, etc. Procedures for dissecting a crayfish may be found in "Laboratory 46: Dissecting a Crustacean—The Crayfish," Procedures B and C, *Biology Laboratory Manual*, Scott, Foresman, pp. 163-164, 1985.

Answer Key
1. Jointed appendages and hard external skeletons.
2. Invertebrates, because they do not have a backbone.
3. Head, thorax, abdomen.
4. Anterior, dorsal surface.
5. Possible answer: The crayfish lies motionless.
6. Six or seven (depending on the species).
7. Its body cannot be too big because there would be nothing to support the internal tissues.
8. See students' drawings.

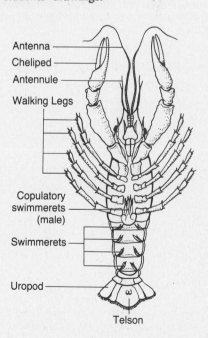

Antenna

Cheliped

Antennule

Walking Legs

Copulatory swimmerets (male)

Swimmerets

Uropod

Telson

9. See students' crayfish.
10. Breathing.
11. They move and come into contact with more oxygen.

12. It maximizes the amount of light that can be sensed.
13. Wastes are released away from the mouth and sense organs.
14. First avoids the rock but later hides behind it or under it.
15. The crayfish is nocturnal and hides behind rocks during the day.
16. Usually forward, raised on walking legs with abdomen extended.
17. Moves its antennae and backs away.
18.–19. Crayfish moved away from light on its anterior end. It may show a similar movement or no response to light on the posterior end. Light receptors are on the anterior end.
20. The crayfish, being nocturnal, seeks out dark places. This protects it from predators during the day.

21. Possible answer: Uses maxillae and maxillipeds to guide food into its mouth.
22. It becomes very excited and moves away from the salt.
23. They detect physical and chemical stimuli.
24. Possible answer: Crayfish shuns light, hides behind a rock, propels water through its gill chambers, has compound eyes and antennae.

Investigation 33

Notes for the Teacher

Time
Parts I and Part II: 40 minutes for each. Students may need additional time to run caterpillar trials under different conditions of temperature and light. Part I must be done outside of the school building, although Step C may be done inside.

Materials
Insect Larvae: You can collect Woolly Bear caterpillars (larvae of the moth *Isia isabella*) in the fall and early spring; store them in the refrigerator for several days, making sure that you allow them to warm up for about an hour before the lab begins. Tent caterpillars also make excellent subjects. They are negatively geotactic in the morning and positively geotactic in the evening. The widespread Eastern species (*Mala cosoma americanum*) is abundant in the early spring on crabapple or cherry trees. The Western Tent Caterpillar prefers oaks. Any caterpillars found away from an obvious food source can be presumed to be in the "wandering" stage that precedes pupation. Caterpillars can be returned to the environment in which they were found.

You can order larvae such as the Greater Wax Moth, *Galleria mellonella*, from biological supply houses. Before ordering any insects to be shipped interstate, check for permits and information from: Veterinary Services, Animal Plant Health Inspection Service, USDA, Federal Building, Hyattsville, Maryland 20782.

Do not release wax moths into the environment—they are destructive to bees. Sacrifice the larvae humanely after the investigation by freezing for 24 hours.

Mealworms, *Tenebrio molitor*, can be obtained from bait shops or biological supply houses. Both mealworms and wax moth larvae can be used to feed fish, amphibians, or reptiles.

Insect cage: Poke holes in the lid of a one-gallon jar or make a cylinder of screening material capped on both ends with metal pie plates. For mealworms, a plastic sweater box will do.

Appropriate food for selected insect larvae: Information on food for those larvae ordered from suppliers should come with the shipment. Collected species can be assumed to feed on whatever they were found on. For more information on housing, maintaining, and feeding insect larvae, refer to F. Barbara Orlans, *Animal Care from Protozoa to Small Mammals,* Addison-Wesley Publishing Company, 1977. Be sure that fresh food is continuously available for the entire time larvae are to be maintained.

Text and Lab Correlation
Chapter 33 on insects. Other behavior labs include 30, 31, 32, 37, 38, and 40.

Prelab Preparation
Students should read Chapter 33 to understand the stages of insect development. Clarify that insects have different behaviors at separate stages of their life cycle because the insects' needs are different. Ask the students to give examples of how specific behaviors enable animals to survive.

Students can prepare the paper circles for the orientation steps before the laboratory period so that they will have more time to perform the trials.

Procedure
For Part I, remind students how to use a compass to orient the circles. Caterpillars will often curl up when touched. It may take them a few minutes to uncurl and move. Suggest that the students neither touch nor blow on the caterpillars. Use this time to discuss reflex behavior with the students.

To save time, different teams can perform the trials under different conditions, then share the information at the end of the investigation. Students can also create shade with pieces of cardboard and bright light with artificial illumination.

Do not allow the students to release the caterpillars. Make sure, if they have been collected locally, that they are released in the same location. Otherwise, sacrifice rather than introduce non-native species into the environment.

Postlab Analysis
Encourage students to compare the results of their studies. Ask them if they observed unexpected behavioral responses. If they did, ask them to infer the causes of these responses.

Caterpillars, especially those of the more attractive species of moths and butterflies, can be fascinating to follow through the stages of development. They can be raised in insect jars or cages, according to preferences of the species. Students should observe the behavior of each stage. You can release moths or butterflies into the environment only if the caterpillars were collected locally.

Further Investigation

1. Insect larval T-mazes can be made from cardboard and tape.

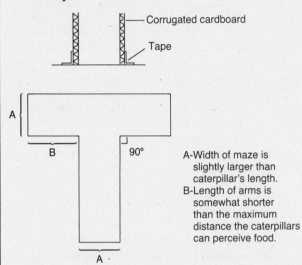

A-Width of maze is slightly larger than caterpillar's length.
B-Length of arms is somewhat shorter than the maximum distance the caterpillars can perceive food.

Remind students to give food rewards on the same side of the T. Other rewards can be used but must be based on the behavior of the larvae of the species used. Some species may respond to light or dark, heat or cold. The object is to see if the trained larva will repeat the behavior—choice of a particular side of the T—even when the reward is not given. Make sure students don't withhold food for long: the larvae are at a stage when they require large amounts of food.

Answer Key

1. This is an example of instinctive behavior.
2. Possible answer: If the larvae feed on tree foliage but pupates in the soil, it will change its responses to light and gravity when it becomes ready to pupate.
3.–4. See students' data charts.
5. Answers will vary.
6. Answers will vary. Some foliage-feeding caterpillars will move toward light and orient themselves so the light source is of equal intensity on both sides of the body.
7. See students' data charts.
8. Answers will vary, but most insect larvae move faster when warm. Other factors can include the stage of growth of the larva. It will move rapidly and restlessly during the wandering stage that precedes pupation and become more sluggish when ready to pupate.
9.–10. See students' data chart.
11.–12. Foliage-feeding insect larvae should respond negatively to gravity. (Response may be positive in evening conditions.)

13.–14. Answers will vary. Mealworms should respond positively to lack of light.
15. The response to light and gravity reflects the larva's needs for food.
16. Larvae need to grow and exhibit behaviors that favor finding an appropriate food source.

Investigation 34

Notes for the Teacher

Time
40 minutes.

Materials
Sea stars: Live sea stars and prepared sections can be obtained from biological supply houses. Place closed plastic shipping bags into marine aquarium for about one hour to allow animals to adjust to the temperature. Then open the bags and let the animals crawl out. Give the sea stars time—1 or 2 days, at least—to adjust to their new habitat before students begin handling them. Sea stars will live in a marine tank kept at 16–18°C for several months.

Sea stars will eat seafood and most meats. If possible, obtain live bivalves like soft-shelled clams or mussels. As well as providing live food for the sea stars, these filter-feeding organisms will aid in maintaining cleanliness of the aquarium. Attempted feeding once per week is adequate. They may or may not be willing to eat at that time. Remove any meat that is not ingested from the tank. If the water becomes contaminated by microbial growth, you will need to completely clean the tank and use fresh sea water.

In general, monitor the tank for evaporation and add only fresh water as it evaporates. Adding more salt water will increase the salinity.
Bivalve extract: Remove the meat from a clam, oyster, mussel, or other bivalve and homogenize (in a blender, if possible) or chop, mince, and grind. Note: Bivalve must be alive or fresh.

Text and Lab Correlation
Chapter 34 on echinoderms. Lab 35 on comparing invertebrates.

Prelab Preparation
Remind your students of the fundamental principles of a hydrostatic skeleton. Explain what happens when force is applied to a liquid. Describe the function of valves in such a system in maintaining unidirectional flow. Although muscles can only contract, the muscles of the tube feet and ampulla do stretch. This is because the force applied to water by muscle contraction in one part of the system will be transmitted to expand other areas. In this way, muscles can be stretched without sets of opposing muscles being involved.

Encourage your students to observe the sea star behavior even on days you are not performing this investigation. If you feed the sea stars live bivalves, encourage students to note when they eat and how many mussels or clams are eaten in a specific time period. Students may note that the sea stars and mollusks cluster. You may use this as a basis for discussion of the advantages (protection) and the disadvantages (competition) of clustering.

Procedure
Encourage your students to use the cross sections to enhance their understanding of the water-vascular system. Encourage them to visualize it in 3 dimensions. If you have time, you might wish to vary the stimuli for the arms and include responses to light and tactile sources.

The pedicellariae may not respond to the touch stimuli. However, you can discuss with students what might happen if they did respond and what might cause them not to respond, for example, they may not recognize the probe as "food" or "hazard." Make sure students do not allow the sea stars to dry out, and have students replace them in the aquarium as soon as possible.

Postlab Analysis
Stress the function of the water-vascular system in a variety of aspects of echinoderm life and its relationship to echinoderm natural history.

Further Investigations
1. You can suggest that students design a system (perhaps, using rubber bulbs, tubing, and clamps) to demonstrate how hydrostatic pressure works in the sea star water-vascular system.
2. One idea might be to keep a sea star in the dark, then flash a light on an eye spot and see if there is a change or if the sea star reacts.

Answer Key
1. See students' work, showing the flow of water, on an illustration of the water-vascular system.

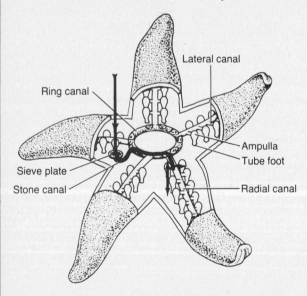

2. The water flows into the tube foot, which extends.
3. The water flows into the ampulla, which expands.
4. See next column.
5. The tube feet may appear to "step" individually. However, they are coordinated, and all the tube feet in an arm move in the same direction (although not necessarily simultaneously).

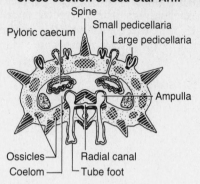

Cross-section of Sea Star Arm

6. Should respond positively to bivalve extract and attempt to move the sea star toward the source.
7. Should respond negatively to bleach and try to move the body away from it or push the bleach water away from the body.
8. One or 2 arms twist until the podia contact a substrate. Then the rest of the body is pulled over.
9. Pedicellaria response will vary. Outside touches tend to result in opening of the jaws. Jaws close up on inner stimulation.
10. There are usually 2–3 plates per pedicellaria.
11. Tube feet grasp bivalve shells and pull the halves apart. No one foot is very strong, but hundreds of tube feet can weaken the bivalve enough to allow the sea star to insert its stomach.
12. Spines, pedicellariae, sieve plate, and anus. Other organisms that settled on the sea star would affect respiration, the water-vascular system, and digestion.

Investigation 35

Notes for the Teacher

Time
40 minutes each for Parts I and II.

Materials
You may want to order materials for this lab, especially live organisms, at the same time you order materials for Labs 29–34. If properly treated, preserved specimens can be reexamined year after year. If other species of animals are substituted for those in the materials list, be sure to alert students. Stress that such a substitution will not affect their observations.

Snails: If snails are not available, use limpets or slugs. Point out that slugs do not have shells. Discuss what effect the lack of a shell may have on the organism's morphology and behavior.

Metridium: If you do not have an operating marine aquarium for maintaining sea anemones, preserved specimens can be substituted in Step A.

Text and Lab Correlation
Chapter 35, as well as Chapters 29–34. Materials from Investigations 29–34 may be used in this lab.

Prelab Preparation

Encourage students to review what they have learned about invertebrates and to imagine themselves in possession of certain biological features. Introduce the notion of basic body plans.

Procedure

The observations may be done quickly, but it is better not to rush your students. Linking the observations to evolutionary trends may take more time and thought.

The success of this lab depends on your guiding students to think about 3 major themes of form and function—locomotion, symmetry, and cephalization—and the limits that evolutionary antecedents might place on those themes developed.

An alternative way to set up this lab is to provide stations around the room. Each station should either have one of the organisms or 2 organisms if their comparison is essential.

Postlab Analysis

Students may need help with the 0, +, ++, +++ rating system for the data chart. Zero means absent or underdeveloped, and +++ means highly developed; + and ++ mean intermediate development.

Locomotion efficiency favors streamlining, which promotes a shift to bilateral symmetry—a long, slim design. A motile animal is better off having its sensory elements concentrated in the anterior region so that it can sense the environment it is approaching rather than the area it is leaving. Hence, developments in locomotor efficiency, bilateral symmetry, and cephalization are developments that parallel each other as evolutionary trends in the invertebrates.

Further Investigations

1. Invertebrates chosen to study will vary. In general, burrowing invertebrates lack appendages and have an enhanced streamlined shape.
2. You might want to have students build models of the invertebrates they design.

Answer Key

1. Students' drawings will probably depict a streamlined shape.
2. Anterior, to sense the oncoming environment.
3. Appendages would be balanced, to allow for forward movement.
4. No, the *Hydra* moves about rarely.
5. Neither invertebrate seems to have a head.
6. Tentacles are arranged radially.
7. Cylindrical shapes, with radial symmetry.
8. The gastrovascular cavity is compartmentalized, so digestion is more efficient.
9. Yes.
10. Yes, there are eyespots.
11. Yes, the *Dugesia* is flat and long.
12. No. Oxygen enters the body through the surface and diffuses to the cells that pack the interior, so the body shape must be flat.
13. Vinegar eels are more active.
14. In *Ascaris*, the muscle cells are aligned against the body wall and parallel to the main body axis. In *Dugesia,* there are muscle fibers in three directions—longitudinal, transverse, and dorsoventral. The muscles are more developed in Nematoda. In nematodes, coordinated contraction along opposite sides of the body enables the worm to bend and straighten, but it cannot twist or elongate.
15. Muscles along one side contract; pressure is transmitted through a fluid-filled pseudocoel, stretching opposing muscles.
16. Yes, the nematode is round. Pseudocoel fluid might take on some circulatory functions.
17. The clam is streamlined because it burrows; the squid is streamlined because it swims. The squid is most highly cephalized because it is an active predator. Shells are protective and are good burrowing tools, but they hinder swimming.
18. The sea worm probably crawls and has the more conspicuous head. Burrowing reduces the need for some sensory organs and increases the need for streamlining; parapodia interfere with burrowing.
19. Both are segmented.
20. Some segments expand to anchor the worm and allow the rest of the body to be pulled along. This movement is peristaltic and very different from the eel's, whose form did not change during activity.
21. Both muscles and the body cavity are segmented into many chambers.
22. Segments.
23. The insect body covering is hard, so respiration across this surface would be impossible.
24. Answers will vary. There is greater segmentation in the insect abdomen. Insect appendages are jointed and hard, and range from antennae to jumping legs; those of annelids are fairly uniform.
25. See students' data chart.

Comparison of Invertebrate Body Plans

	Cephalization	Locomotion	Body Cavity	Segmentation	Appendages	Bilateral Symmetry
Hydra/Metridium	0	0	0	0	0	0
Dugesia	+	+	0	0	0	+
Vinegar Eels/Ascaris	+	++	+	0	0	+
Clam	0	+	++	0	0	+
Squid	+++	+++	++	0	0	+
Snail	++	++	++	0	0	+
Nereis	++	++	++	+	+	+
Lumbricus	+	++	++	+	0	+
Grasshopper	+++	+++	++	+	++	+

26. Answers will vary.
27. Answers will vary. Animals without respiratory systems tend to use body surfaces for gas exchange with the environment. This necessitates moist, permeable surfaces (increasing water loss), increased surface area, and a generally small body size.

Investigation 36

Notes for the Teacher

Time
60 minutes. This investigation will exceed one 40-minute class because of the time it takes to settle the fish and adjust the temperatures. You can do the prelab preparation and demonstrate the technique the day before this lab. You can stop between steps D and E if the students use the same fish, or you can stop before Part II (Step G).

Materials
Goldfish: If you do not live in an area with convenient pet stores, arrange with a supplier a month in advance for an adequate shipment. Bring in the goldfish a week before the lab in order to give them time to recover from the trauma of their change of environment.
Aquarium: Have a large tank of dechlorinated water ready; spring water or filtered pond water is preferred, but tap water can be allowed to stand for two days to remove the chlorine. Water plants can help season the aquarium water. Keep the tank well aerated, and do not feed the fish more than they can eat in a few minutes at a single feeding. If small white patches of fungus appear on the fish after a few days, immediately purchase treatment for "ick" from a pet store. If necessary, use an aquarium heater to maintain constant temperature.

Text and Lab Correlation
Chapter 36 on chordates and fishes and Chapter 43 on the respiratory systems. Lab 43.2 on pulse and respiration rates.

Prelab Preparation
Demonstrate to students how to pick up the fish with the net so that they do not injure it.

See the table (next column) of oxygen content and goldfish ventilation at various temperatures of water.

Procedure
Suggest that 1 team member be responsible for monitoring temperature and another for changing the temperature of the water jacket. The other 2 team members should observe the respiration rates and pool their data; if their values are too far apart, they should repeat that observation point.

Stress that proper handling of the fish is expected from students. Tell the students not to touch the scales, since the slimy film that covers the fish protects it from bacterial and parasite infection. Water temperature for fish should never exceed 30°C. Goldfish tend to adapt well to cooler temperatures.

Students may have difficulty decreasing the water temperature to 5°C. They may have to add some ice directly to the test beaker. If the beaker tends to fog, allow students to count the respiration rate for 15 seconds and extrapolate to one minute. Raise the water temperature slowly. If at any time a fish shows signs of stress or has spasms, do not continue to experiment on that fish. If you allow more time for the procedure, the students will observe that in order to let the fish settle down they must be very quiet and not jar the table. They will have difficulty counting the respiration rates if they rush, and the fish become agitated.

Keep a bucket near the aquarium so excess water can be siphoned off as the students pour their fish back, following the investigation. They should gently pour fish and water together, holding the beaker close to the surface of the water in the aquarium.

Postlab Analysis
Make sure that students understand that, in warmer water, a fish has to respire at a greater rate to get adequate oxygen (since less oxygen is dissolved). The fact that all metabolic processes occur at a faster rate at higher temperatures in cold-blooded animals will also increase the fish's need for oxygen.

The following table shows published values for oxygen content and goldfish ventilation at different temperatures. These data on oxygen content can be graphed as a further investigation.

O_2 Consumption				
Tempera-ture (°C)	O_2 Con-tent of of Water (mL/L)	O_2 Consumption (mL/kg/hr)		Ventila-tion Rate (L/kg/hr)
		resting	active	
5	9.0	8	30	1.3
15	7.0	50	110	9.0
25	5.8	140	255	32.0
35	5.0	225	285	60.0

Source: F.E.J. Fry and J.S. Hart, *Biological Bulletin 94:* 66, 1948.

Further Investigations
1. Most species of fish will die if the temperature of the water reaches 24–37°C. The goldfish is an exception. It has a very high tolerance for warm water (up to 42°C). Thermal death for trout and Pacific salmon occur at 25°C.

Answer Key
1. 1.11 liters.
2. See students' data charts.
3. Small fluctuations are averaged out by the long observation period.
4. You want to determine the effect of temperature changes, not of excitement at being transferred.
5.–7. See students' data chart.
8. Warm water will have already lost much of the dissolved gases before chilling.
9. Answers will vary, but respiration rate should show a significant increase.
10. The increased breathing rate moves more water across the gills to meet the demand for more oxygen during activity.
11. Yes, deeper respiration or greater opening.
12. The increased suction should cause faster flow or increased volume of water across the gills.
13. Deeper inhalation for the athlete permits a greater volume of air to be exchanged at each inspiration.
14. See students' graphs on separate paper.

15. A linear increase of respiration rate within the temperature range of the experiment.
16. The individual goldfish may vary in size, age, and basic metabolic rate. Other variables are the care in handling the goldfish, time to reach temperature equilibrium, and time of day.
17. Answers will vary, depending on student graphs and on whether students already know about thermal lethal points. Students may assume a linear increase continues.
18. Cannot increase temperature indefinitely. Warm water is deoxygenated. Enzyme rates for metabolic reactions will be speeded up to a dangerous level in fish's tissues (since fish are cold-blooded).

Investigation 37

Notes for the Teacher

Time
40 minutes each for Parts I and II.

Materials
Live frogs: You may order live frogs from supply houses or purchase them from local pet shops. *Do not* use any poisonous varieties of toads. If you use animals from other places, do not release them into the local environment when the investigation is completed. If you use indigenous species, it is all right to release the animals into an appropriate habitat. If you must purchase them, place your order early in case supplies run out. The supply of live frogs is greatly reduced during the winter months, especially February and March, when they are emerging from hibernation. Try to order from a nearby source to avoid losses in shipping. Information on maintaining a terrarium is available from many sources. You may obtain information on conditions necessary for the maintenance of particular amphibian species you use in *Animal Care from Protozoa to Small Mammals* by F. Barbara Orlans, Addison-Wesley, 1977.
Insects: Most pet stores stock live crickets and/or mealworms for animal food.
Preserved frogs: These are available in several sizes from biological supply houses.

Text and Lab Correlation
Chapter 37, on amphibians. Comparisons of reptiles and amphibians are made in Lab 38.

Prelab Preparation
Approximately 5 million frogs are destroyed each year to provide specimens for student dissection. An effective way to minimize the number of preserved frogs used is to purchase one large preserved bullfrog and perform one demonstration dissection, allowing students to explore the internal organs carefully.

Part II, the optional dissection of a preserved specimen, is provided because many high school biology teachers feel that it is important for students to acquire hands-on experience with animal dissection as an integral and traditional part of the study of biology. Alternatives to dissection include filmstrips, plastic replicas, diagrams, slides, videotapes, computer software, clay-model building, and making plastic overlays. Replicas and models are expensive but are a once-only purchase for a school. Try to preview any models, films, or programs before purchasing to be sure they fit your standards and needs. More information on these materials is available in the front matter.

Further Investigations
1. More information on the available alternatives to dissection may be found in the front matter.
2. Neoteny is an interesting and unusual condition in vertebrates. Students should focus on those organ systems that mature and those that do not.
3. Students could raise their own amphibians and learn about their habits for this investigation. You can collect frog eggs at the edges of ponds and lakes in the early spring. Collect them about one month before this lab and keep the eggs in a clean, well-aerated aquarium. When the tadpoles start changing, provide floating materials for them to sit on at the edge of the water.

Answer Key
1. Answers will vary.
2. You can learn about behavior in the live frog vs. anatomy in the preserved specimen.
3. Possible answer: The well-developed hind leg muscles allow the frog to leap a distance many times its own body length.
4. A human has a muscular diaphragm that efficiently pulls air into the lungs through negative pressure. A frog draws air into the mouth and forces it into the lungs with positive pressure.
5. The frog's long, sticky tongue curls around the insect, pulling it into its mouth, where it is swallowed whole. The frog "blinks" to help push the food into the gullet, using pressure from the eyeballs.
6. The nictitating membrane functions to protect the eye underwater and to keep its surface swept clean.
7. Answers will vary.
8. The frog's webbed feet and strong leg muscles help propel the frog. Amphibians with tails wiggle their bodies to propel them without using their legs.
9. The light abdomen makes the frog less visible to predatory fish below them in water, and the dark dorsal skin makes the frog less visible to land predators like birds and snakes; it provides camouflage.
10. Answers depend on hypotheses for Question 6.
11. The front legs are much shorter, with less muscle development and no webbing. The front legs are only for balance.
12. The eyes, tympani, and external nares are located on top of the head. When floating or swimming at the surface, most of the frog is underwater, but these structures remain above the water.
13. See students' frogs.
14. The teeth are not used for chewing since they are only on the upper jaw. The frog swallows its prey whole, and the teeth hold the prey in position.
15. They lead to the tympanic membrane; they equalize pressure within the ear.
16. The internal nares lead to the external nares. (They

allow normal respiration at the water's surface.)

17. The glottis carries air to the lungs, and the muscular gullet leads to the throat and stomach.
18., 21. See labeled illustration.

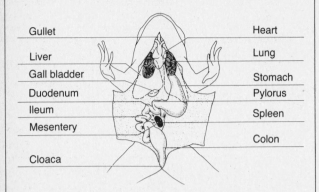

Gullet	Heart
Liver	Lung
Gall bladder	Stomach
Duodenum	Pylorus
Ileum	Spleen
Mesentery	Colon
Cloaca	

19. The blood vessels of the mesentery supply oxygen to the spleen and intestines, as well as absorb nutrients produced by digestion.
20. The spleen is part of the immune system.
22. The membrane on the floor of the mouth.
23. The tadpole's heart is like that of a fish. The adult frog's three-chambered heart is required because of the extra circulatory pressure required and generated.
24. The skin also helps regulate salt/water balance.
25. The lungs, the muscular legs, and highly developed vision appear.
26. During winter, they burrow into the mud or live under the water; some frogs hibernate. Under drought conditions, their physiological reactions slow down.
27. Hypotheses will vary. With one exception (the marine toad), amphibians do not live in salt water. The skin is osmoregulatory but cannot cope with the concentrated salt in sea water.
28. The human lungs are larger and the diaphragm pumps air through more efficiently. However, the frog can supplement its smaller lungs with a respiratory skin. The human heart is more efficient than the amphibian heart in separating the oxygenated blood from deoxygenated blood.
29. The folded cerebrum is the largest structure in the human brain. The optic tectum and olfactory bulb are well developed in the frog's brain. The frog relies heavily on sight and smell. It has little cognitive ability due to the small, smooth cerebrum.

Investigation 38

Notes for the Teacher

Time
Two 40 minute periods

Materials
Reptiles: Try to obtain at least 3 different species. Select from: lizards (fence lizards, American chameleons, or skinks); turtles (box, Blandings, or painted); and snakes (young water, garter or DeKay's brown snake).

Caution: Do not use any poisonous or endangered species. Some states restrict use of turtles; check your state's regulations through their fish and wildlife agency.

Reptiles may be obtained from zoo loans, reptile societies, and biological supply houses. If you obtain snakes from pet shops or the wild, make sure a reputable herpetologist identifies the species. Young water or garter snakes are easy to handle and become accustomed to human handling. Note: Wild reptiles (newly captured) will probably be more responsive than tame ones (those that you may already have had in your lab for some time).

You can design and build a terrarium that is adapted to each kind of reptile's needs. For example, all reptiles need light and shade to maintain body temperature. Aquatic reptiles need more water than land reptiles. Feed the animals with the correct live or fresh food, not pet store dried food. For more information on care, feeding, and housing of reptiles, refer to F. Barbara Orlans, *Animal Care From Protozoa to Small Mammals,* Addison-Wesley, 1977.

Text and Lab Correlation
Chapter 38 on reptiles. Class discussion of this investigation may be correlated with Chapters 36, 37, 39, and 40 on characteristics of fish, amphibians, birds, and mammals.

Prelab Preparation
Make sure students understand and follow precautions about washing their hands after handling the animals. It is important that they learn to respect the reptiles as living things and don't hurt or tease the animals. Encourage students to handle or touch the animals but do not force them if they are afraid. Instead, let them observe the other students handling the animals.

Procedure
Most of this investigation relies on observation. Encourage students to keep notes on their observations, even to make drawings, and to look up information about reptiles that they cannot find out by observation.

If students work in pairs or teams, they will not need to handle the animals as much, and can design the experiment in Part II as a team. Urge your students to be creative in their designs. Try to get several approaches. Remind your students to have patience when working with live animals and reassure them that they have not done anything wrong if the animals do not respond as hoped.

Postlab Analysis
Ask students whether they all obtained similar results. If results were different within the same species of animals, ask them what factors may have led to the differences.

Further Investigations
1. The feeding experiment will necessarily take place over a period of time, especially for snakes, which eat only every few days. You might suggest that your students try this on turtles.
3. This experiment will work better with lizards,

which have better vision. Animals that have been in the lab for a long time might not respond as well as ones that have been recently caught. Although a shadow that resembles a natural predator would seem more realistic, similar results will probably be obtained with the shadow from a cardboard disk.

Answer Key

1. They may be found in forests, parks, marshes, fields, near rivers, or any undeveloped area.
2. Examples: skin, means of locomotion, ways of eating.
3. Either answer may be correct, depending on the student's reasoning.
4. The snake has specially adapted scutes that provide traction. Lizards and turtles have legs. Lizards are generally fast-moving and some have sticky foot-pads that allow them to climb easily. The turtle's short legs and shell cause it to move slowly but its shell helps protect it from predators.
5. Snakes and turtles move like amphibians; lizards can move like amphibians or birds, depending on the species.
6. In general, except for obvious specializations, the dry, tough skin is very similar among all species.
7. Scales are keratinized skin tissue, more like the skin on bird's legs and feet.
8. The lizard has external ears; snakes and turtles have internal, primitive ears. All the reptiles have eyes. Most reptiles depend on their sense of smell. Reptiles smell with chemoreceptors, located on their tongues.
9. Like amphibians and birds, they can see, but not as well. Snakes and turtles cannot hear as well as birds. The chemoreceptors on reptiles' tongues are more like those of amphibians.
10. They have protective coloration. Lizards and snakes have speed to escape predators. Turtles have shells. Some species, such as snapping turtles and poisonous snakes, can defend themselves against predators. The tough, dry skin, which prevents moisture loss due to the heat and sun.
11. Protective coloration is also found in amphibians and some birds. Different species of amphibians and birds use speed to escape danger. Some amphibians produce poisons. Several species of birds will attack if provoked.
12. No, turtles don't have teeth; they have beaks more like birds.
13. Snakes' jaws open wide, so they can swallow food whole. Their teeth are used to grasp food and guide it into their esophagi. Turtles' beaks allow them to grab at food and, when necessary, break off manageable pieces, or crush it.
14. Turtles' mouths are more like birds, while lizards and snakes have mouths more like amphibians.
15. On rough surfaces, the animals should move normally. On smooth surfaces, they will have difficulty. However, some animals may not move at all.
16. Possible design: Take three identical boxes and place a small amount of food in one of them. If the

animal consistently finds the box with the food, the conclusion may be drawn that the reptiles can smell.
17. Do not let it see where stimulus is located.
18. To test sound response, make a loud, sharp noise. If the animal is startled, it must have heard the noise. Vibration sense can be tested by gently shaking the animal's container and observing whether the animal responds.
19. The body will still move around that space.
20. Snakes and lizards may go around or over the obstacles. Turtles may be blocked by the obstacles.
21. Examples: scaly skin, fast-moving or protected by shell, legs on some species, egg-laying and air-breathing only.

Investigation 39

Notes for the Teacher

Materials
Specimens should include: hawk, owl or similar bird of prey, loon or grebe, quail or partridge, wood thrush or robin, pelican, hummingbird, heron or egret, duck, kingfisher, and dove or pigeon. Stuffed and mounted specimens are especially useful for direct, hands-on observation of a bird's beak and legs. Specimens may often be borrowed from museums or environmental education centers with educational loan programs. Wall charts, wildlife posters, and other illustration sources are possible substitutes.

You will need to provide bird guides so that students can validate their predictions about food types and habitats. Some suggestions are: *Complete Field Guide to North American Wildlife*, Eastern or Western editions, Harper and Row, N.Y.; Roger Tory Peterson, *A Field Guide to the Birds* (Eastern or Western), Houghton Mifflin, Boston; Richard H. Pough, *Audubon Bird Guide* (Eastern or Western), Doubleday, Garden City, N.Y.

Text and Lab Correlation
Chapter 39, Lab 41 on comparing vertebrate skeletons, and Lab 51 on owl pellets.

Prelab Preparation
All students should find the commonly known characteristics (e.g., feathers, forearm modifications as wings), and the more persistent students will find others by consulting a variety of sources. You may find it convenient to scour the school library ahead of time and bring reference materials to your classroom.

Postlab Analysis
The beak and legs of each bird represent adaptive traits that developed over a long period due to selective pressure by environmental factors (food supply, terrain, availability of water, nesting sites, and the like). In all populations of a species, those individuals best adapted to the environment tend to be best nourished, healthiest, and most likely to produce offspring that will inherit and carry on similarly favorable traits.

Further Investigations

1. A coping saw will easily cut through chicken bones that have been boiled for about an hour and then dried overnight.
2. Use of dissecting microscopes will allow students to better observe feather structure. Placing collected feathers in a freezer overnight and storing with moth crystals will take care of any parasites such as mites.
3. Local taxidermists and bird-watching clubs are other sources of this information.

Answer Key

1. See students' data charts.

Characteristics of the Class Aves

Characteristic	Adaptive Function
Feathers (modified layer of epidermis; homologous with reptilian scales).	Quill feathers - flight down feather - nest lining and egg incubation pin feathers - insulation of body, filling out the plumage
Forelimbs (arms) modified as wings; 3 fingers at ends reduced and partly fused.	Enables flight (vestigial or reduced in flightless birds, e.g., ostrich, rhea, emu and kiwi bird).
Head periscopic, prehensile and equipped with a forceps-like beak and partly fused.	Enables finding, gathering manipulating and consuming food (forelimbs cannot be used in feeding).
Well developed eyes located on either side of the beak.	Places the eyes near the food-reaching beak to assure accuracy.
Compact, hollow, lightweight bones.	Decreases body weight for more efficient flight ability.
Relatively large heart for body size.	Provides adequate circulation for the tremendous exertion of flight and migration.
Sternum (breastbone) enlarged and boat-shaped.	Provides more surface area for attachment of breast and flight muscles.
Large intenstine (particularly rectum) reduced in length.	Minimizes body weight by eliminating wastes frequently.
Teeth absent in all modern birds (were present in fossil	Reduces body weight (teeth are dense structures); food grinding occurs in stomach (gizzard), using small, ingested pebbles.

2.–6. See students' data charts.(Answers on next page.)
7. It is adapted to sip nectar while hovering. Other adaptations are fast-moving wings, short, lightweight legs, and small body size.
8. Sphinx moths and several types of butterflies. Hummingbirds and insects have long, extended mouthparts, similar body shape, and rapid wing-beat flying.
9. Suggestions might include pointed, strong beak to tear flesh of the prey; scissor-type cutting action to sever flesh; and long, pointed spearing beak. Suggestions might include sturdy legs with claws that

could grab and carry live food; and legs with hook-like or spear-like claws. Strong wings for fast flying and hunting; special vision or hearing to locate prey; and large size.
10. Answers will vary.
11. The loon swims on the water's surface for food and then dives into the water. The heron wades on long legs in shallows to spot fish. The kingfisher perches on low shrubs or trees at the water's edge on the lookout for fish that it can swoop down on, spear, and eat at leisure back at the perch.
12. Each bird has a distinctive way of fishing and catches its prey (which may be different species of fish) in different parts and depths of the body of water; their niches do not overlap and they can co-exist.
13. Hawks and owls hunt in the same habitat for the same prey, but hawks hunt during the day and owls hunt during the night; therefore, each occupies a separate niche.
14.–15. Answers will vary.
16. Habitat destruction and/or pollution may be responsible. Habitats are changing faster than the populations of birds can adapt to the changes or adapt to different, suitable habitats.

Investigation 40

Notes for the Teacher

Time
Steps A and B of Part I may be done as homework. Step C will take 20 minutes in class. Allow 25 minutes for Part II plus 25 minutes for students to discuss and compare results.

Materials
Mammals: Use small rodents that can be handled easily, such as mice, rats, hamsters, gerbils, or guinea pigs. Have 2 groups of the same species of rodent that have not lived together. The rodents can be obtained from laboratory animal suppliers or from pet stores. Try to select healthy animals. Maintain them with clean bedding, adequate food, and water.

Text and Lab Correlation
Chapter 40 on mammals. Other animal behavior labs include Investigations 30, 33, and 38.

Prelab Preparation
Make sure that students understand that they are only to observe, not to participate, in any of the interactions.

Procedure
Do not let students handle gerbils by their tails. If rodents are treated gently and calmly, they are less likely to become frightened. Students should work in teams with the animals so that they do not have to handle the animals so much. Also, students who are not afraid should handle the animals first. If an animal becomes excited or nervous, you should remove it from the experiments because it might bite. For Question 12, note that although studies are ambiguous about significant differences in the departure signals between the sexes,

27. Limb A should be adapted for flight, while Limb B should be adapted for swimming.
28. Homologous skeletal structures were observed in all the vertebrate classes.
29. The classification of humans and cats into the Class Mammalia reflects their similarity of structure.
30. Animals A and B are in Class Mammalia. Animals within the same class can vary greatly in diet, habitat, and method of locomotion. However, their skeletal structures should show similarities in numbers and kinds of bones.

Investigation 42

Notes for the Teacher

Time
40–50 minutes

Materials
Tubing: Use regular laboratory rubber tubing such as for a Bunsen burner.
Weights: Two student textbooks may be used.

Text and Lab Correlation
Chapter 42 on the muscular system and Chapter 43 on the circulatory and respiratory systems. Lab 43.2 on the effect of exercise on pulse and breathing rates.

Procedure
Demonstrate the hand flexing so all students will use the same technique. Students will do these activities in pairs. Encourage good-humored cooperation. Note: Make sure students do not tie rubber tubing so tightly that they completely cut off blood circulation.

Copy the model graph (below) for Step E on the chalkboard. Work with the class to fill in each student's data, and analyze the results.

Bar Graph

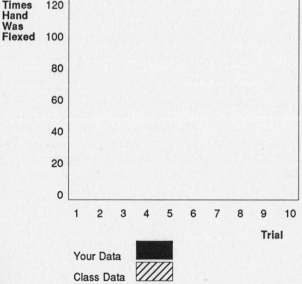

Further Investigations
2. American Heart Association
7320 Greenville Ave.
Dallas, Texas 75231

National Heart, Lung, and Blood Institute
9000 Rockville Pike
Building 31, Room 4A21
Bethesda, Maryland 20892

Answer Key
1. The muscle contracts and bulges.
2. Example: Muscles will tire quickly when students have not eaten for a while.
3. See students' data charts.
4. Answers will vary.
5. See students' data charts.
6. Answers will vary.
7. One explanation might be that the dominant (writing) arm is stronger, therefore tiring less easily.
8. The cold temperature may slow the circulation of blood to the muscles in the hand.
9. Many students will have more difficulty writing after exposure to the ice.
10. Possible answers: Muscle fatigue occurred after extending the weight with the non-writing arm.
11. The restricted blood flow reduces the oxygen reaching the muscles.
12. The better the cardiovascular system functions, the more oxygen is available to the muscles, preventing or delaying fatigue.
13. Answers will vary.

Investigation 43.1

Notes for the Teacher

Time
40 minutes. Allow time for the class to pool data.

Materials
Solutions:
   Adrenaline (by weight), 0.001%, 0.0001%, 0.00001%. (Order epinephrine, Carolina #85–9810)
   Coffee and/or cola drink (by volume), 10%, 20%, 30%.
   Ethyl alcohol (by volume), 2%, 5%, 10%.
   Aspirin or other headache remedies (by weight), 0.5%, 1.0%, 2.0%.
Other chemicals: acetylcholine, antihistamines, nicotine.
Daphnia: Daphnia may be collected from ponds, ordered from biological supply houses, or obtained from aquarium supply stores. *Daphnia magna* is the largest species and most easily observed. Daphnia may be maintained in pond water for 1 or 2 weeks if kept out of direct sunlight. For longer culturing, they may be fed with a suspension of yeast in water (enough to make the culture milky). You may want daphnia for the hydra feeding lab (Investigation 29). Rotifers are an excellent alternative to daphnia.
Microscopes: Binocular dissecting microscopes often work best.

Text and Lab Correlation
Chapter 43, Section 43.3. Class discussion of this lab may be connected with Chapter 47 on the endocrine system and Chapter 48 on drugs. This lab can be a lead-in to Investigation 43.2, which relates pulse and breathing rate to levels of student exercise.

Prelab Preparation
Let students first work in groups of 4 to design techniques. Ways to help include allowing students to contribute ideas out loud before the procedures are begun or giving teams a time limit of 5–10 minutes before you give them constructive suggestions. Discuss range of human heartbeat—60–80 beats per minute.

Procedure
One method that allows students to record the very fast heartbeat of the daphnia (as high as 350 beats per minute) is to tap a pencil point on a piece of paper, keeping the rhythm of the beats. Pencil dots can be counted when finished. One student in each team should be the clockwatcher and should signal another student, the observer, when to start counting heartbeats and when to stop. To facilitate testing with varying concentrations of the solutions, you might assign different concentrations to each group so that here are no repetitions. Treated daphnia should not be used for further experiments on the same day. Copy the table below on your blackboard for students to enter their data: the chemicals they used and at what concentrations and the average heartbeat rates for their daphnias

Model Class Data Chart for Daphnia

| Normal | Experimental | | |
Avg. No. of Beats/Min.	Chemical	Concentration	Avg. No. of Beats/Min.
1			
2			
3			
4			

Postlab Analysis
Allow enough time so that your whole class can pool data, interpret results, and share in error analysis. You should expect ambiguity and contradictions to students' predictions. An important part of this lab is the error analysis, getting students to think about what may have gone wrong and about the complexities involved.

Answer Key
1. Find pulse at wrist or neck. (See Inv. 43.2 for procedure.) Count the number of beats in 15 seconds and multiply by 4.
2. 3 or more times.
3. 10.–12., 15. Answers will vary.
4. See students' data charts.
5. Answers will vary. Adrenaline, caffeine, and nicotine are stimulants; alcohol is a depressant; antihistamines have a depressant effect; and aspirin, an analgesic, is neither stimulatory nor inhibitory.

6. See students' data tables.
7. The water flea is its own control, observed before adding the solution.
8. See students' data tables.
9. The water flea might have been harmed or killed.
13. No. Genetic variability, age, health.
14. Possible answers: difficulty in counting very rapidly; added more than one drop of solution; water on slide dried up; daphnia died.

Investigation 43.2

Notes for the Teacher
Time
40 minutes. Allow enough time for students to fill in and analyze the class chart.

Materials
If you wish to include blood pressure readings, you will need a sphygmomanometer. This may be another option under resting, mild exercise, and strenuous exercise conditions.

Text and Lab Correlation
Chapter 43 on circulation and respiration. Lab 43.1 on the effects of different chemicals on *Daphnia* hearts.

Prelab Preparation
Prior to this lab, demonstrate how to take a pulse rate and how to determine respiration rate. Have students practice on partners at the start of this lab while they relax before taking resting rates. They might also practice on friends or family members the evening before class.

If any students have physical problems that preclude their participation in strenuous exercise, such as asthma or emphysema, suggest that they avoid those trials and share data for information they might need. Caution students to be seated when hyperventilating and to stop any trial if they feel dizzy or faint. They then should sit with their heads between their knees and immediately call you. Ask each student to be aware if a partner appears to be having a problem during a trial.

Procedure
Any time students are measuring pulse and breathing rates, they may count for 15 seconds and then multiply by 4 to get the rate per minute. If the numbers seem to be over too broad a range, have the students repeat the trial 3 times and then average the numbers. In Steps E and G, have the students take pulse and respiration rate measurements every 30–60 seconds until they obtain values the same or close to the resting rates (Steps A and B). Copy the class Data Chart (see next page) on the board.

Postlab Analysis
It is important that students compare and share the results of the tests. Explain that other factors may be involved in a student's data and that there should be no individual judgments about a student's performance. Ask the class to think of some situations that might affect performance in these trials on a particular day; for example, if they are especially tired, have a cold, or suffer from asthma.

Class Data Chart

Student Names

A	RP	___	___	___	___	___
B	RB	___	___	___	___	___
C	HB	___	___	___	___	___
	HB	___	___	___	___	___
D	BR	___	___	___	___	___
	PR	___	___	___	___	___
E	RT	___	___	___	___	___
	HB	___	___	___	___	___
F	BR	___	___	___	___	___
	PR	___	___	___	___	___
G	RT	___	___	___	___	___
	HB	___	___	___	___	___
	PR	___	___	___	___	___
H	BR	___	___	___	___	___
	HB	___	___	___	___	___
I	BR	___	___	___	___	___

RP = Resting pulse (in beats per minute)
RB = Resting breath rate (in breaths per minute)
HB = Holding breath time (in seconds)
BR = Breathing rate (in breaths per minute)
PR = Pulse rate (in beats per minute)
RT = Pulse return time (in seconds)

Answer Key

1. A pulse is a throbbing in the arteries due to the contractions of the heart.
2. About the same in both wrists. Either a difference in arterial resistance between the 2 arms or a mistake in counting or timing.
3. –5. See students' data tables. Pulses: between 60 and 100 pulses per minute. Breathing rates: about 12 breaths per minute. Breath holding: between 45 and 90 seconds.
6. There should be an increase in the rate.
7. There should be an increase; the pulse increases as the heart beats faster to speed up the exchange of O_2 and CO_2 in the lungs and muscle cells.
8. A shorter time.
9. –10., 13., 15. See students' data tables.
11. It should take longer after strenuous exercise.
12. A faster rate after 2 minutes of running.
14. It increases the breathing rate.
16. Increase carbon dioxide and lower oxygen.
17. Breathing rates should increase but the breathing should be shallower.
18.–19. See students' data tables.
20. Breathing is shallower.
21. See class data chart on board.
22. Answers will vary.
23. Regular exercise will cause the pulse rate to return to normal sooner; the heart is more efficient.
24. Exercise increases the carbon dioxide concentration and lowers the relative oxygen concentration, which triggers the breathing center of the medulla.
25. It builds up heart muscle and increases the volume of blood pumped per ventricular contraction.
26. The heart does not have to beat as often to push the same amount of blood through the body because it is more efficient, therefore the pulse rate decreases.
27. Hyperventilation increases the oxygen concentration.
28. Breathing and pulse rates, as well as blood pressure, increase, due to a decreased ability to bring oxygen to the blood and remove excess carbon dioxide from it.
29. Answers will vary.

Investigation 44

Notes for the Teacher

Time
Part I, Steps A and B, 40 minutes. The interviews (Step C) are to be completed as a homework assignment. Part II, 40 minutes for data analysis. It is important to allow sufficient time to discuss the results during class.

Text and Lab Correlation
Chapter 44 on infectious diseases. Class discussion may be correlated with Chapter 19 on viruses and Chapter 20 on bacteria.

Prelab Preparation
This group activity fosters cooperative effort and encourages lively discussion. Students are required to question the veracity of commonly held beliefs.

Students may write the statements during class or as a homework assignment. It is important that students research the statements. Emphasize the importance of writing well-researched statements to ensure that the survey is valid.

It is not essential to restrict the entire class to interviewing two groups. You may want to divide the class into teams and have each team interview two groups. Explain that only students who are interviewing the same category of subjects can pool their results.

Procedure
Before students do the interview, emphasize the importance of not revealing the answer in their reading of the statement. Decide whether or not they may use the telephone.

Postlab Analysis
Data analysis is an important aspect of this investigation and may lead into a general discussion of polls and surveys. Point out the importance of knowing the background of the people who respond.

Answer Key
1. Answers will vary.
2.–4. See students' survey sheets on separate paper.
5.–6. See students' data charts.
7. 1 × 5 × number of students in class. See data charts.
8. 12 × 5 × number of students in class. See data charts.
9.–10. See students' data charts.
11. See students' bar graphs.
12.–15. Answers will vary.

Investigation 45.1

Notes for the Teacher

Time
40–80 minutes, not including homework time for prelab preparation and discussion time for postlab analysis.

Materials
The problems and possible solutions to unhealthy diet habits are clearly and briefly described in the report *Dietary Goals for the United States*, U.S. Government Printing Office, Washington, DC 20402, Stock Number 070-04376-8. "Nutrition and Your Health: Dietary Guidelines for Americans" is a 20-page pamphlet available free from the Office of Governmental and Public Affairs, U.S. Dept. of Agriculture, Washington, DC 20250.

Text and Lab Correlation
Chapter 45 on the digestive system and Chapter 4 on carbohydrates, proteins, and lipids. Lab 3.2 on calorimetry.

Prelab Preparation
The students will be required to supply their own raw data based on a day's consumption of food.

Be sensitive to the pressures that might cause some students to misrepresent themselves in their data. An overweight person may wish to evade any public discussion of a painful problem. Perhaps more serious, a student with eating disorders such as bulimia or anorexia may be very skillful at hiding his or (more likely) her behavior. By talking in a general way about the importance of a balanced diet, about the need for sufficient calories for active and growing teenagers, and by describing the range of eating disorders, you may provide the support needed to amend a problem. Make available a list of local resources that would help both over- and under-eaters.

Procedure
Students are asked to organize their items under several categories and they can pool group data in doing this. It is not necessary that all items be listed, only that commonly eaten foods be discussed and rank ordered. Encourage discussion in the groups; by arguing over the choices, the students' food habits and values can be brought out and challenged.

Postlab Analysis
This is an opportunity to convince the students that science is an outlook that can be applied to everyday life. Evaluating a diet is a process based on accurate data which must come from careful observation; it is essential to distinguish between what you think is happening and information that shows otherwise.

Once a problem in eating habits has been identified, a plan to make a change needs to be formulated. Encourage class discussion on the difficulties of breaking habits and ways to achieve new goals. Try to modify any overzealous reactions. Changes in diet should always be gradual. A balanced diet will be based on variety, moderation, and realistic goals.

Further Investigations
1.–2. Check to make sure the procedure includes some way to evaluate the foods from school lunches that are thrown away by students.
3. Frances Moore Lappé, *Diet for a Small Planet*, Ballentine Books, Inc., 1985, is a classic in presenting complementary proteins.
4. A good source for information on nonnutritive additives is Michael Jacobsen, *The Complete Eater's Digest and Nutrition Scoreboard*, Anchor Press/Doubleday, 1985. Books (including the Jacobsen book), pamphlets, and software are available from Center for Science in the Public Interest, 1501 Sixteenth Street, NW, Washington, DC 20036.

Answer key
1. See homework assignment.
2. Answers should include all food items in 5 categories with the numbers of servings for each item.
3.–4. Answers will vary.
5. Percentages will be calculated from the answers to Question 3.
6. See students' bar graphs.
7. See students' data charts.
8. Answers will vary.
9. Possible answers: tooth decay, high blood pressure, and overweight.
10. Possible answers: add high-fiber whole grain foods and subtract salty snacks.
11.–12. Answers will vary.

Investigation 45.2

Notes for the Teacher

Time
40 minutes. Steps A and B may be done with Prelab Preparation the day before. You will need time a day ahead of this lab to prepare dialysis tubing and to make up solutions.

Materials
0.3% Starch suspension: Add 2.5 g table salt (NaCl) to one liter of water. Bring to a boil. Add 3 g arrowroot or potato starch to a small amount of cold water, mixing into a paste. Dissolve in the boiling NaCl solution, stirring constantly. Cool.

80% Glucose solution: Add glucose to 100 mL water until solution is saturated. Take 80 mL of saturated glucose solution and dilute with water to a final volume of 100 mL.

Iodine solution: Purchase I_2KI (Lugol's) solution or prepare you own: Dissolve 1 g iodine (I_2) and 2 g potassium iodide (KI) in 20 mL water. Bring to 1,000 mL with 25% KCl.

1% Diastase solution: Order Diastase Malt (Carolina Biological Supply Company #85-7538). Dissolve 5 g diastase powder in 500 mL water. Make fresh and store in the refrigerator.

Dialysis tubing: Use dialysis tubing (25 mm or one inch flat width) precut to 15 cm. Cut the tubing at a slight angle so that it will open more easily. Soak tubing in distilled water overnight to soften, or in distilled water in the refrigerator for a longer time.

Rubber bands: They make better seals than string for the ends of the dialysis tubing. Have the students cut them and wrap them around the ends of the tubing a few times and then tie securely.

Clinistix (sugar test) strips: These are available at your local pharmacy for use by diabetics. Testape is an alternative product, available from pharmacies or biological suppliers.

Text and Lab Correlation
Chapter 45. Students should review Chapter 4 and Lab 4 for information on enzymes, as well as Chapter 6 on diffusion and semipermeable membranes and Lab 6 on dialysis bag models.

Prelab Preparation
A demonstration of the handling of dialysis tubing the day before the investigation will allow better use of lab time. You might want to compare results of Clinistix or Testape strips and Benedict's reagent. Benedict's reagent is not sensitive enough to detect sugar when it is diluted in the water of the beakers. Check that Clinistix are fresh before using them in this investigation.

Procedure
Without contamination, no starch should diffuse out of the dialysis tubing. You should get a positive glucose test in the beaker containing the glucose dialysis bag (III) and the beaker with the dialysis bag of diastase plus starch (IV). Encourage students to be honest in their predictions (predictions might also be assigned for homework the night before).

Further Investigation
1. The Further Investigation provides a good opportunity to review and apply information and lab techniques from an earlier investigation (Lab 4). You may have the whole class cooperate in designing experiments, with each group varying a different enzyme activity factor. Buffers (pH 3.4, 6.8, and 8.0) for varying the acid-base environment of the enzyme can be obtained commercially. Combinations of ice and warm water can be used to vary the temperature of the dialysis bag/beaker set-ups.

Answer Key
1. The contents of the small intestine of the digestive system.
2. The cells with semipermeable membranes lining the intestine.
3. The water in the beaker. It absorbs and transports digested molecules.
4. Based on size of molecule, shape of molecule, or kind of molecule.
5. The molecules in the beaker water will start to diffuse back into the dialysis tubing until equilibrium is reached. (There is no active transport system to keep them out.).
6. See students' data charts. Iodine turns blue-black in starch, but stays reddish in sugar. Answers will vary for color change in Clinistix or Testape.
7. Benedict's test or the Clinistix strip.

8. Answers will vary.
9. I—Glucose can diffuse out because of its size.
   II—Starch cannot diffuse out because of its size.
   III—Glucose does not move starch out of the tubing.
   IV—Diastase breaks starch down to glucose or some other simple soluble and diffusable sugar, which can then diffuse out because of its size.
10.–11. See students' data charts.
12. Starch does not pass through the dialysis tubing.
13. Sugar does diffuse through the dialysis tubing.
14. Answers will vary.
15. Diastase breaks down starch into smaller size sugar. The body must have the starch in this smaller size form, i.e., as a simple sugar, or it cannot be used.

Investigation 46

Notes for the Teacher

Time
Allow 40–60 minutes for Part I and 40 minutes for Part II, including Postlab Analysis.

Materials
If you have difficulty finding paper straws, a facsimile can be made by rolling a sheet of typing paper around a pencil. It is easiest to start rolling at a corner. Fasten with a small piece of tape and remove the pencil.

Text and Lab Correlation
Chapter 46 on sense organs. Lab 30 on flatworms and magnetotaxis.

Procedure
Try all of the demonstrations so that you will be able to give hints to students needing assistance. The afterimage exercise is highly successful with the brightly colored rings of an infant's rocking stack of rings. Write the data as a color wheel (a disk divided into equal portions of red—orange—yellow—green—blue—violet) so that students can see the opposite and complementary relationships.

Color Wheel

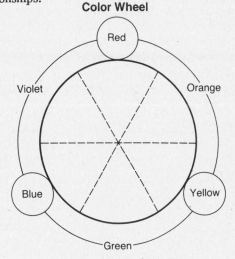

Mix circled primary colors
to create secondary colors inbetween

Opposite colors
produce afterimage

If students have difficulty finding the dominant eye, have them look through the viewfinder of a camera or microscope. The eye used to sight the picture is dominant.

Further Investigations

1. A red shift occurs when an object is receding.
2. Problems usually occur in detecting amplitude. Loss in detecting amplitude can be addressed readily with amplifiers (hearing aids, magnifying lens).

Answer Key

1. Senses might have evolved into vision responsive to wavelengths in the range of x-rays or infrared (for example, pit vipers can "see" mammal heat) or hearing of ultrasonic sounds (porpoises and bats can "see" with soundwaves).
2. Answers will vary—left or right.
3. Answers will vary.
4. See students' responses.
5. The cross reappears.
6. See students' responses.
7. High amplitude will cause constriction of the iris muscles and the pupil will be small. Relative darkness causes the muscles to pull the iris open and the pupil widens. Intense interest in the image also results in widened pupils.
8. As an observed object nears the center of the face, the pupils converge and grow smaller in size.
9. The dominant eye should be more accurate but both eyes may be equal.
10. Possible answer: Glasses having uneven modifications in the lenses can influence the reflex.
11. Red produces a green afterimage, yellow a violet afterimage. Green produces red and blue produces orange.
12. Yes.
13. Frequency (pitch) goes up as the straw is shortened.
14. In the first experiment, the louder (second) fork sounds closer than it is. In the second experiment, it is easier to tell when the 2 forks are apart. The perception in the first experiment is an illusion caused by our misjudging amplitude.
15. Brightness.
16. Nerves that have become accustomed to regular stimulation will fire for some time afterward.
17. Your brain would perceive colors as different pitches of sound and music as patterns of light.

Investigation 47

Notes for the Teacher

Time
40–80 minutes, depending on how many slides you want students to observe and how many diagrams you want them to draw. Allow 40 minutes for 4–5 slides and 80 minutes for 10 slides.

Materials
Suggested slides: **unfertilized egg, zygote, 2-cell stage, 4-cell stage, 8-cell stage, 16-cell stage, 32-cell stage, 46-cell stage, blastula, early gastrula, middle gastrula, late gastrula, young sea star larva.**

Text and Lab Correlation
Chapter 47, and Lab 34 on live sea stars.

Prelab Preparation
Remind students that the sequence of early embryonic development is only the very beginning of development. In humans, the process from zygote to blastocyst takes about 7 days.

Procedure
Be sure each student observes the following as a minimum: zygote, 2–8 cell stages, blastula, and late gastrula.

Further Investigations
1. Prepared slides of chick development may be purchased from biological supply houses.

Answer Key

1. The first cleavage is vertical and divides the egg into approximately 2 equal-sized cells.
2. Examples: mammal cleavage is slow, echinoderm cleavage is rapid. Mammal forms blastocyst, echinoderm forms blastula.
3. See student diagrams.
4. The zygote and the blastula are the same size. When the blastula develops into the gastrula, the embryo is bigger than the zygote.
5. When the gastrula forms, the cells look different.
6. The cells, up until the gastrula stage, are roughly spheres. From the first cleavage until the gastrula, the cells get progressively smaller since they do not grow.
7. The nuclei are the same size as the stage progress but with less and less cytoplasm around them.
8. The same number of chromosomes in each cell. In adult organisms, however, all cells would have the same number of chromosomes except for sperm and eggs which are formed by meiosis.
9. A blastocoel allows invagination to occur during gastrulation, when organs begin forming.
10. Check drawings for labels.
11. The sea star embryo and larva exhibit bilateral symmetry. The adult exhibits pentaradial symmetry. No, both human embryo and human adult exhibit bilateral symmetry.
12. Specialization of cells and organization of cells into tissues, organs, and systems must occur.
13. As a model of cell differentiation similar but not identical to early human development.

Investigation 48

Notes for the Teacher

Time
40 minutes for Part I and 40 minutes for Part II

Materials
Use meter sticks that are smooth and free of splinters.

Text and Lab Correlation
Chapter 48 on tobacco, alcohol, and other drugs and Chapter 46 on nervous system and sense organs. Lab 43.1 on the effects of chemicals on Daphnia heart rate.

Prelab Preparation

Review the procedure for using the meter stick and allow the students a few minutes to practice. Discuss reaction time and how it plays an important role in their lives.

Procedure

The table converting distance of fall to reaction time is based on the formula

$$distance = 1/2 \, g(t^2)$$

where $g = 980$ cm/sec/sec, the acceleration due to the force of gravity.

Reaction time can also be determined by using a computer to generate the stimulus and record the time of the response. One such program is available through HRM Software, 175 Tompkins Ave., Pleasantville, N.Y., 10570.

For Step G, copy the class data table onto the blackboard.

Number of Students Whose Average Reaction Time Falls in Each Time Range

Reaction Time (sec)	Stimulus		
	Visual	Auditory	Auditory with Distractions
.02 – .10			
.12 – .20			
.22 – .30			
.32 – .40			
.42 – .50			

You may fill in the figures as students raise their hands to indicate the time range that their reactions indicated, or perhaps a student might do it. For Step H, make an enlarged copy of the model bar graph and key (in lab pages) on the chalkboard. Divide each column into 3 sections: the bar on the left will be the visual stimulus; the second will be the auditory, and the third, the auditory with distractions.

The students should see that there is variability in the data but that most individuals fall within a narrow range of reaction times.

Postlab Analysis

Use this opportunity to stress the possible biological consequences of drug use and abuse.

Further Investigations

In a further investigation into the effect of coffee or cola drinking, fatigue, diet, or regular exercise on reaction times, students relate well to these variables and are likely to be highly motivated. Be sure that their experimental design does not include anything that might be injurious to their health, such as smoking a pack of cigarettes.

Answer Key

1. Answers will vary.
2. Receptors, such as the eyes, receive stimuli, which are transferred along sensory neurons to the spinal cord. The spinal cord transmits information to the brain for processing. The brain transmits messages down the spinal cord to motor neurons, which cause muscles to move.
3. Interfere with the normal functioning of the nervous system, increasing or decreasing reaction time.
4. A person working alone will anticipate the drop of the meter stick.
5.–7. See students' data tables.
8. Each conversion from distance to time results in a small amount of error due to rounding. Using the average distance of fall avoids compounding this error.
9.- 10. See students' data tables.
11. Reaction time may increase as a result of the distraction.
12.- 13. See students' data tables.
14. See students' graphs.
15. The graph reveals the variability of the data as well as any trends in the data.
16. Possible answer: Reaction time increased over time. Fatigue may have been a cause.
17. Reaction to a visual stimulus is generally somewhat faster than to an auditory stimulus. This may indicate that more brain area is devoted to this processing.
18.–19. Answers will vary.
20. Psychoactive drugs interfere with the normal function of the nervous system. They may speed or slow the transmission of nerve impulses across the synapse or interfere at the molecular level in the function of the nerve cell. Disorientation and reduced ability to concentrate are common symptoms associated with the use of some drugs.
21. Answers will vary.
22. Possible answers: Driving, athletics, eating (digestion).

Investigation 49

Notes for the Teacher

Time

20 minutes for Part I. (You may want to set up the two hatching trays with the help of just a few students.) 40 minutes for Part II (Steps C–F) 2–3 days after Part I. For Part III (Steps G–H), 15 minutes every 3 days for 2 weeks; 40 minutes on last day.

Materials

Brine shrimp (*Artemia*) eggs: You can obtain viable dried eggs in pet and aquarium stores or from biological supply companies. Spread a small sample out on a paper towel to check for clumping. If they have become wet and caked, they will not hatch well. Dry them out for about an hour on a flat, absorbent surface before setting them in solution to hatch.
Hatching trays: You can make these from shallow enamel or glass oblong dishes. Fill the trays half full

with the salt solution. Use a fish-feeding ring or the cut-off rim of a styrofoam coffee cup to contain the eggs that will remain on the surface. Place the eggs in the ring and cover the tray so that one fourth of the surface is exposed to light. Direct a light toward the open end. The larvae will swim toward the light. This way, the students won't mix up the eggs and the tiny larvae. An oven baster can be substituted for a pipette or dropper. For more information on hatching brine shrimp, refer to F. Barbara Orlans, *Animal Care from Protozoa to Small Mammals,* Addison-Wesley, 1977.

Saline solutions: You can use rock salt, but don't use iodized salt. Distilled water is preferrable. Tap water may be used but remove chlorine by letting a wide-mouthed container of water sit uncapped for 24 hours. For 5% saline, dissolve 50 g sodium chloride in 1 liter water. For 10% saline, dissolve 100 g in 1 liter. For 20% saline, dissolve 200 g in 1 liter.

Text and Lab Correlation
Section 49.3 on aquatic biomes, Section 6.1 on osmosis, and Chapters 15 and 16 on evolution. The organisms from this laboratory can be used to feed and maintain hydra for Lab 29. Be sure to rinse the salt water off the brine shrimp before introducing them into a freshwater environment.

Procedure
Students can help set up the hatching trays for Steps A and B before the class period. To decrease hatching time, try to maintain temperatures of 24–25°C (75–80°F).

Students should work in pairs for Part II. Each team can prepare 1 set of 8 culture dishes. Encourage students to divide the labor. You can suggest variables for dishes 7 and 8 to the students if they can't think of any. Each team can select different variables and share results. Suggest using some of the variables listed in Further Investigations. Another variable might be water turbulence, achieved with a fan positioned to blow over the water. The water pH can also be adjusted. The amount of solution can be reduced, making the water very shallow.

Supervise the continued care of the brine shrimp for best results. Overfeeding will cause mortality. Keep the water levels constant by adding distilled water *only.* Do *not* continue to add saline solutions.

At the end of the 2 weeks, students should be allowed time to plot growth curves. They can prepare a class chart on the chalkboard to compare the survival of brine shrimp under different conditions.

Brine shrimp larvae are very small. The difficulty of counting them accurately may affect the students' results. Encourage the students to pay attention to qualitative attributes such as size and activity levels of larvae and unhatched eggs or dead larvae.

Further Investigations
1. The organism takes advantage of a relatively short reproductive period to provide as many adults as possible, then produces eggs that will survive adverse environmental conditions to maintain the species. Students can use plastic ice cube trays to separate egg-bearing females.

Answer Key
1. Possible answer: Adaptations of osmoregulatory system to tolerate high salinity.
2. The habitat is wet but typically dries up for periods of time.
3.–4. Answers will vary.
5. Possible answer: 5% salinity.
6. Brine shrimp can withstand a saturated solution of salt, but student results may vary.
7. Brine shrimp can survive in water with widely ranging salinity, but short-lived experiments may give varying results.
8. Possible answers for best condition: light, warmth, non-moving water, and about 5% salt water.
9. The brine shrimp will grow in less favorable conditions than those necessary for hatching. The eggs remain dormant until specific conditions stimulate hatching, giving the larvae a better chance of surviving to reproduce.
10. See students' data charts.
11. See students' graphs on separate paper.
12. Trends might include: increased mortality, decreased activity, slower growth, increased food consumption.
13. 5% salinity.
14. The brine shrimp should grow better in light and warmth, as long as the warmth is not excessive.
15. Answers will vary. Salinity, still water, warmth, light, oxygenation, and shallow water such as you would find in salty, briny pools or salt lakes.
16. The eggs survive long periods of drying, a characteristic which would ensure the survival of a species living in an environment characterized by infrequent and unpredictable rainfall.

Investigation 50

Notes for the Teacher

Time
2 lab periods, 1 week apart, for setup, and then a few minutes observation time daily over 3–4 weeks for Step D. Allow a week to elapse between Part I and Part II. This will enable the algae to establish in the jar and give the seeds time to sprout.

Materials
Plants and Animals: These are up to you and the students and may be obtained locally (earthworms, crickets, seeds) from pet stores (Anacharis, guppies, platies, mealworms, pond snails), and biological supply houses (*Daphnia, Chlamydomonas,* duckweed, isopods).
Water: If you use tap water, leave it in an open container for at least 24 hours to allow chlorine gas to escape.
Glass jars: Gallon jars are practical. Obtain them from the school cafeteria or from local delis and sandwich shops. Other containers students could use include 2 L plastic soda bottles (cut the narrow tops off and use plastic wrap as covers); inverted bell jars; deep, clear-plastic tubs used by delis; and old, leaky aquariums.

Text and Lab Correlation
Chapter 50 on structure of ecosystems, Lab 22 on algal blooms, Lab 24.2 on oxygen production in plants, and Lab 49 on abiotic factors influencing brine shrimp.

Prelab Preparation
Encourage the students to think about the ecological roles that organisms have—as primary producers, herbivores, carnivores. Decide at this point how many jars will be devoted to the exercise. Will the entire class address the same hypothesis? If so, perhaps a few jars in which one component has been varied will prove useful in testing the hypothesis. Do you wish to study variation within similar populations? In that case, observing several jars that have identical initial conditions could be instructive.

Procedure
Recording the number of organisms (Question 7) may be tricky in some cases, and estimates may be required. Discuss the sources of error in estimates *versus* direct counts.

Postlab Analysis
Part of an extra lab period may be required to discuss what the graphs reveal.

Further Investigations
1. The ecosystems may be kept for the rest of the school year and then given to students to take home over the summer.

Answer Key
1. Some organisms (primary producers) use the energy of the sun to produce their own food. Other organisms (herbivores or primary consumers) eat plants; still others (carnivores or secondary consumers) eat other animals.
2. Answers will vary.
3.–4. Answers will vary depending upon answers to Question 2.
5. The week gives seeds time to sprout and enables the algae to be established.
6. More than 1 jar set up in an identical manner would approximate baseline variation. Varying 1 component among several jars would assess the influence of that component or interaction.
7. Answers depend upon the ecosystem that students are working with.
8. Possible answers: organisms died, were eaten, or had young.
9. See graph on separate graph paper.
10. In a predator-prey relationship, the number of predators usually increases as the number of prey decreases; the trend reverses as the number of prey bottoms out.
11.–16. Answers will vary.

Investigation 51

Notes for the Teacher

Time
40 minutes. To do a good job, this investigation may require two 40-minute periods, one to dissect the pellets and another to identify and analyze the findings.

Materials
Owl pellets: These may be found in roosting and feeding areas, old farm buildings, woodlands, and parks. However, it is easier to buy pellets, already dried and fumigated, from biological supply companies. Usually the pellets are accompanied by information and identification guides. Soaking the pellets in water for about 2 hours before the lab begins softens the mucilage and makes the dissection easier. If you collect your own pellets, dry them and fumigate them in polyethylene bags with naphthalene to destroy insect eggs.

Text and Lab Correlation
Chapter 51. Lab 52 on sampling a population and Lab 41 on identifying different kinds of bones.

Prelab Preparation
Reassure students that, while it may sound unpleasant to dissect something owls cough up, it is really nothing more than examining animals' bones and tissues "cleaned" by stomach enzymes. Review the use of field guides with students. The layout and use of such guides may need some explanation.

Procedure
Warn students to handle the pellets with care. They should try to preserve the teeth and skulls, especially.

Encourage students to predict what kinds of animals they will find. They should not worry if they cannot find every small bone, as long as they can account for most of the major bones. Any extra major bones should be matched and considered parts of one or more animals. If students have more than 4 skulls to chart in Step B, they may add extra columns to their charts.

Encourage students to check small materials under the dissecting microscope to see if they are invertebrate parts. Owls have been known to eat amphibians, reptiles, and fish.

See the model in the pupil lab for the class chart to be placed on the chalkboard for Question 13.

Postlab Analysis
Ornithologists are not sure how long it takes owls to produce pellets. However, for the sake of demonstrating the effects of owls as predators, you could "guesstimate" that owls produce one pellet a day. If you then multiply each population in one pellet by 365, you can calculate the annual decimation of this population by just this owl. It would be inaccurate to make too many generalizations.

Further Investigations
1. Students who are fascinated by this investigation might enjoy assembling a set of bones from one of the pellet animals into a complete skeleton. This

3. The mice will eat the grain, and the lead will concentrate in their tissues. More lead will concentrate as they eat more contaminated grain. The owl will eat many mice, thereby concentrating the lead even more in its body.
4. There are usually many more prey animals, and they usually reproduce at faster rates than the predators in balanced ecosystems. Many predatory animals prey on the young and feeble and the old, slower animals. This leaves the most healthy, adult members of a population to reproduce and saves food supplies for these most reproductive animals.

Answer Key

1. Use reference books to look up common foods of owls and also check on behaviors and habitats of small mammals, and relative sizes.
2. Possible answer: The owls then don't expend energy pushing large quantities of indigestible materials through their digestive systems. This might allow the owl to devote more energy to catching prey.
3. See students' data tables.
4. Possible answers: Bones, hair, and feathers.
5. Possible answer: The chitin of the exoskeletons of invertebrates is very different form the bones and hair or feathers of vertebrates.
6. See students' data tables. Answers will vary.
7.–8. Answers will vary.
9. Answers will vary. Skull shapes vary widely. Owls' diets consist of a variety of species.
10. Answers will vary.
11. See students' lists.
12. Yes.
13.–14. See class data chart.
15.–16. Answer will vary. Possible answers: mice and moles. Both are rodents and are active at night.
17. Predation.
18. The most common prey probably had the largest population of all the prey animals.

Investigation 52

Notes for the Teacher

Time
Part I, 40 minutes to half a day to perform the field study. Step E should be completed in the lab the same day as Step D. Step F can be assigned for homework. Part II, 40 minutes to analyze results and soil samples. Conduct this lab when the ground is not frozen and is soft enough to dig out soil samples. You may want to arrange a field trip to perform Part I.

Text and Lab Correlation
Chapter 52 on populations. Investigation 51 on owl pellets and predator-prey relationships will give students an introduction to another type of population study.

Materials
Stakes: Wooden stakes can be obtained at a hardward store or lumberyard. Be sure that the sticks are sturdy enough to be pounded in the ground.
Identification guides: The Peterson Field Guide Series, Audubon Field Guides, and Golden Guides are excellent field guides for identification of trees, shrubs, birds, insects, and mammals. The mammal guides contain pictures of animals tracks.
Screens: Fine mesh screen sieves come in different sizes from biological supply houses. You may also find appropriate-sized screening from a hardware store. The size of the mesh that you will need depends on the type of soil you will sample and should be small enough to retain animals more than 1 mm in length.

Prelab Preparation
With input from your students, select an ecosystem that will give students a variety of plants and animals to study. If you are in a city, study a city park. An urban ecosystem study may be expanded to include a discussion of the different types of plant and animals life that are found in a city.

As you discuss the field study setup, you may wish to incorporate some geometry to ensure that students stake out a perfect square. Suggest that the easiest way to stake out a perfect square is to use a 3,4,5 triangle to form the angles of the square. The angle will be right when the 5-m string connects the 3-m point to the 4-m point.

How to Make a Perfect Right Angle

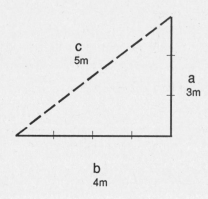

You should also take the time to familiarize students with field guides. Suggest that students look through the table of contents and browse through each guide they will use to see how they are organized.

Tell students to be sure to read the procedure carefully before they go to the site. This will help prevent confusion and wasted time. Caution them to use care if the habitat contains potentially harmful animals and plants. Before going into the field, discuss some of the more dangerous organisms they might find and how they may avoid them. Show them pictures of poison ivy and other plants to avoid. If possible, tell students to wear long pants.

Procedure
To make a profile in Part I, the students do not need to identify every plant and tree in the selected area. They

should try to identify the dominant species. If they cannot determine the identity while they are in the field, you may pick a small branch, if they are abundant, to be identified in class. Remind students not to pick leaves and plants. If they do, they are altering and destroying part of the ecosystem. In Step B, encourage students to select quadrats that will represent different portions of the ecosystem. The total numbers then will be more representative of the entire ecosystem.

It is inaccurate to collect insects or count birds from such a small study. It is therefore more reasonable to perform a qualitative study of these mobile organisms. For this purpose, carry field guides and remind students before beginning the field study just to observe these animals.

In Step D, soil samples are preserved not only for storing specimens until they may be analyzed, but also to protect students from any harmful organisms that might be hidden in the soil. Caution students not to inhale the formaldehyde directly or get it on their skin. They should wash off any spills immediately. They will be rinsing the samples and so will not be inhaling formaldehyde while working on the samples.

In Part II, students should be able to identify classes of organisms and common insect species by using a guide. They should be able to differentiate nematodes, oligochaetes, insect larvae, and molluscs.

Further Investigations
3. State or local fish and wildlife agencies abide by state regulations, setting limits on fishing and hunting to protect the environment. This is a good way for students to comprehend the practical value of studying populations.

Answer Key
1. Biases might include availability of location, ease of accessibility, and perceived presence or absence of organisms.
2. *Dominant organisms*, those present in greatest numbers; *herbaceous*, having little or no woody tissue and persisting usually for a single growing season; *deciduous*, having leaves that are shed seasonally.
3. Answers will vary.
4. See students' charts on separate paper.
5. Answers will vary and depend upon geographical area.
6. Based on students' charts indicating counts and dominant species.
7. Answers will vary and depend upon geographical area.
8. See students' charts.
9. Answers will vary and depend upon season and geographical area.
10.–12. Answers will vary, depending upon season and geographical area. Example: Many moth and butterfly larvae will be present in a heavily wooded area.
13.–16. Answers will vary. Possible answers: many oligochaetes in rich soil, and many insect larvae.
17.–18. See students' charts.
19. Specific answers will vary. Plants are producers.

20. Answers will vary. If prey are limited, intraspecific competition among the predators will intensify. The result is that some individuals will starve or be forced to move; others will stay and thrive.

Investigation 53

Notes for the Teacher

Time
40-minute class period to complete Part I and 20 minutes on each day when detailed observations are made. Allow 2 days to elapse between Parts I and II. Lab will be completed 7 to 10 days after setup.

Materials
Seeds: Suggested seeds are beans, peas, and corn, since they germinate within 2 to 3 days and are generally more resistant to mold. Sort through the seeds and remove any that are broken.

Solutions: You can use household bleach to prepare the mold inhibitor. For a 1% solution, combine one part bleach to four parts water. Each class of 25 students will need about 600 mL.

Prepare stock solutions of 10% sulfuric acid (10 mL concentrated H_2SO_4/90 mL distilled water) and 1% sulfuric acid (1 mL concentrated H_2SO_4/99 mL distilled water). Use these stock solutions in the preparation of water solutions with pH values of 1, 3, 5, and 7.

To prepare solutions with pH 1 and 3, add 10% sulfuric acid drop by drop to 500 mL distilled water, until you reach the desired pH. Use pH paper or a pH meter to determine the pH. Use 1% sulfuric acid to make solutions with pH 5 and above. Note: Distilled water usually has a pH lower than 7. You will have to boil the distilled water to reach a pH of 7 or use the pH of the distilled water as the upper pH level. You can have students check the pH of their solutions with pH paper before they use them.

You can prepare nitric acid the same way to use as an additional variable. If you wish to test your local rainwater, collect at least 500 mL and check the pH.

Text and Lab Correlation
Chapter 53 on protecting life and Chapter 27 on seed germination. Review Lab 3.1 on acids and bases.

Prelab Preparation
You may want to review pH with your students. Discuss differences in germination among the different types of seeds (monocot and dicot) the students will use.

Procedure
For Question 9, put a large data chart (modeled on the Team Data Table) on the chalkboard so that students may pool their data and get class results.

Have the students make their first observations after 2 days. Have the final observations made 7 to 10 days after the start of the investigation.

Prepare a model graph for Question 18 on the chalkboard if your students have difficulty following the directions. Generally, plant growth is better in a pH range of 4.5 to 6.5.

Postlab Analysis

Ask how the local rainwater (if used) compared to the other pH solutions. Discuss what impact increased acidity might have on local crops.

Addresses of organizations studying the effects of acid rain:

> National Wildlife Federation
> 1412 16th Street N.W.
> Washington, DC 20036

> The Acid Rain Foundation, Inc.
> 1630 Blackhawk Hills
> St. Paul, MN 55122

Further Investigation

1. The germinated seeds could be transplanted and growth measurements continued. Other seeds could be grown and the results compared to data from this investigation.

Answer Key

1. Conditions include proper levels of moisture, oxygen, temperature, and pH. Moisture could be controlled by measuring exactly how much solution is used. Oxygen could be controlled by leaving the same quantity of air in the bag whenever it is opened for observation. Temperature could be controlled by keeping all bags in one place at the same controlled temperature. The pH could be controlled by frequent checks with pH paper.
2. Collect rainwater or snow and test with pH paper or similar indicator.
3. Possible answer: Very acidic solutions are expected to be injurious to seed germination.
4. See students' data tables. Answers will depend upon the types of seeds used.
5. Possible answer: 0% of seeds at lowest pH and 100% at highest pH will germinate.
6. Answers will vary. Generally, the seeds will have softened and expanded in size. Some discoloration may be observed.
7.–8. See students' data tables. Answers will vary. A few fast-growing seeds may have germinated after 2 to 3 days.
9. See class chart.
10. More seeds will have germinated in water solutions of the 4.5 to 6.5 pH range.
11. Predictions will vary. Possible answer: More seeds in all solutions will germinate within the next 24 hours.
12. See students' data tables.
13. Answers will vary, but seeds will have gotten larger, softer, and more will have germinated.
14. Possible answers: split seed coats, emerging roots and shoots, branching roots.
15.–16. See students' data tables and class chart.
17. See students' graphs.
18.–19. Answers will vary.
20. Successful germination and continued growth should be better in the pH range of 4.5 to 6.5. The highly acidic (pH 1) should be the least beneficial.

Modern Biology
Laboratories

Consumable Edition

HOLT, RINEHART AND WINSTON

AUSTIN NEW YORK SAN DIEGO CHICAGO TORONTO MONTREAL

Developmental assistance by Ligature, Inc.

Printed in the United States of America   89012 022 543

ISBN 0-03-013923-6

Table of Contents

Introduction to Inquiry and Process

This is an exciting time to study biology. Scientists are learning more about people and the world around them than they ever have before. As the media report on the search for a cure for AIDS, on cancer research, and on the effects of acid rain on the environment, we all share in following the efforts of many scientists. As a biology student, these reports will have special significance to you. In your class work, you will learn many of the concepts that will help you to understand scientific research. In the laboratory, you will also learn about and, most importantly, practice many of the basic processes that scientists use to help them find answers.

Fundamental to any scientist's work is the ability to identify and ask important questions. Once scientists have clearly identified what needs to be found out, they use specific methods to help them investigate the problem. As you perform the investigations in *Modern Biology Laboratories,* you will also use these methods. You will discover some scientific facts in much the same way that scientists learned them when they were still unsolved problems. These are the kinds of questions you will explore in this course:

- What makes a pond die?
- How are genes turned on and off?
- How have reptiles adapted to life on land?
- How do you make sauerkraut?

Listed below are the scientific process skills you will use during each lab. In addition to explaining the skill, we have also provided strategies, or hints, about how you might use them. It is important to read these skills before you begin laboratory work and to review them as you progress through the course. This way, you can keep track of your progress. Chapter 2 of your textbook also discusses process skills and how scientists use them.

Observing is the use of one or more of the five senses to perceive objects or the course of events as they occur. Data are collected as specific information is gathered and recorded based on observations.

- To make reliable observations:
  Think about possible sources of error before making observations.
  Make several observations.
  Have different classmates make the observation.
  Know precisely what you are looking for.
  Use descriptive language.
  Avoid distractions.
  Check any equipment before and after making observation.
  Record observations as soon as they are made; use diagrams.
  Do not recopy your record of observations.

- To evaluate firsthand observations:
  Gather information about the observer, the equipment, and
      the conditions.

- To use more of your senses than just sight:
  Look over the observations you have made. Take away those that
  use sight. How many are left? Think about each of your other senses in
  turn and try to make as many observations as possible. (Always use
  caution in smelling and never taste anything in the lab.)

Measuring is the process of gathering quantitative data by making precise observations involving numbers. Measurements usually have units such as millimeters, grams, milliliters, seconds, pH, or calories.

• To make accurate measurements:
    Check that you are using the unit of measure specified in the procedure.
    Record the units of your measurement as you record the numbers.

Organizing Data is the process of placing observations and measurements in logical order. This includes sequencing, grouping, and classifying information by setting up tables and charts, plotting graphs, and labelling diagrams. More information can be seen at one time when data are shown in a visual format such as a time line, chart, or graph.

• Use visual ways to organize information:
    graphs (bar and line)
    flow charts
    concept maps
    charts

• To set up a chart:
    Arrange the data for the chart with the control data on the left and the experimental data on the right. This enables you to make an easier comparison between the two sets of data.

Classifying is the process of grouping or ordering according to the similarities and differences among the items. This grouping can either use established taxonomic classes or develop new organizational patterns. Classifying can be used to help find information (as in a library search) and to help understand an individual once it is known to belong to a group.

• To classify
    skim items and note similarities and differences
    list possible categories and choose best ones
    group items in categories
    revise categories

Hypothesizing is the process of making statements that may answer questions about observations. A hypothesis is a possible explanation of why something happens. These explanations can be tested. Evidence can either support or refute a hypothesis but can never prove it true beyond all doubt.

• To make hypotheses:
    List relevant information
    List possible hypotheses or explanations
    Judge each hypothesis according to the information
    Choose the best hypothesis

• When testing a hypothesis:
    Check to be sure you have the means to do an experiment.

Predicting is the process of stating what the most likely outcome of an experiment will be. Predictions are based on hypotheses and specific data. Predictions often take the form of an *If/then* statement. (*If* the hypothesis ___ is true, *then* expect to observe ___.)

The process of predicting helps scientists understand relationships between variables. By making a prediction and then testing it through an experiment, more evidence is gathered, and the hypothesis is supported or refuted.

- To make a prediction:
    List possible predictions.
    List reasons for and against the most likely predictions.
    Rank the importance of each reason by evaluating the underlying
        hypothesis and available information.
    Select the prediction most likely to occur.

Experimenting is the process of designing data-gathering procedures as well as gathering data for the purpose of testing a hypothesis. A controlled experiment is one in which the scientist regulates the variables affecting the phenomenon. An independent variable is a factor that is manipulated by the experimenter. A dependent variable is a factor that responds to changes in the independent variable.

- To design an experiment:
    Look for all factors that might affect the experiment.
    Think of ways to control variables so that you will look at only one
        factor at a time.
    Include a control as a basis for comparison.
    Make observations as small an intrusion as possible.
    Predict what results are expected.
    Think back over other experiments you have performed previously. Try
        to identify the parts of the experimental approach that would apply.
        Once you have outlined the approach, list the resources—supplies and
        equipment—that you will need to carry out your experiment.

- To execute an experiment:
    If one method doesn't work, don't abandon it; try a variation to see if it
        will be more successful.

Analyzing Data is the process of determining how reliable the data are and whether they support or refute a hypothesis or prediction.

Scientists analyze data in many ways, including interpreting graphs, using statistics, comparing data to those gained in other studies, determining the relationships between variables, and finding sources of errors. Analyzing data involves asking questions about the data, which may lead to the construction of new inferences and new hypotheses. It involves interpreting and evaluating, and comparing actual data to predicted results.

- To analyze data
    Compare the results of measurements and calculations.
    Make comparisons by noting similarities and differences between the
    experimental data and the control data.
    How did the results change as one factor was changed?
    Once a pattern is established, check to see if that pattern fits
    all observations.

- To make a generalization
    Choose a large sample that represents all members of the class.
    Organize data based on accurate observations.
    Formulate a general statement based on evidence from the sample.

- To determine sources of error:
    See *Observing* for related strategies.

Inferring is the process of drawing conclusions about objects or events based on related facts or premises but not on direct observation. Inferences are based on data collected, reasoning, or past experience. Inferences are much like hypotheses, but unlike hypotheses, they are not testable because they are too small, too far away, or from an earlier time.

• To recognize an inference:
   Say "It makes sense to think that..."
   Ask whether it can be tested.

• To make inferences:
   List possible explanations for observations about objects
   or events.
   Think of what evidence would support or not support each inference.
   Gather evidence and choose the most likely inference.

•To make better inferences:
   Base them more on observations, less on assumptions.

Modeling is the process of making visual or verbal constructs that help explain data. To create models, data are simplified, patterns are pulled out, and a framework is built to make difficult concepts more easily understood. Examples of modeling in the labs include dialysis bags as models for osmosis (Investigation 6) and for digestion (Investigation 45.2); Velcro models of nucleic acids (Investigation 8); groups of different beans to model gene pools (Investigation 10).

A model is a representation or a simulation of a process that occurs in nature. Some models, such as a model airplane, are physical. If the model is a faithful representation, it may be able to fly much like the real thing. Other models are symbolic. A computer-generated graph showing the growth of a population of a species of animals on an island consists of lines and numbers based on assumptions concerning the island environment. Although no actual animals are part of the model, it may give a very accurate prediction of the growth pattern expected for this population. Still other models may be a combination of symbolic and physical. Plastic pieces can be used to symbolize a glucose molecule. This model shows the actual number, type, and arrangement of the atoms that make up the real molecule.

• To develop models
   Think of a number of different things the subject is like. What does
      it remind you of?
   Think of common experiences and concrete examples that would be
         familiar to most people.
   Choose and analyze the most promising ideas. Select the best model.

Communicating is the process of sharing information with others in writtten or spoken form.
   Scientists often give speeches at national or international science meetings and discuss the results of their latest research. Scientists also publish the results of their research in scientific journals. Communicating scientific data allows many scientists to work together to solve problems.

• To communicate effectively
   Read your report aloud to hear how it will sound to the audience.
   Ask a friend to read your report or describe your diagram aloud while
   you listen. Is the language specific and detailed? Are ideas clearly
    explained and organized? Think of other ways they could be stated.

• To evaluate communication
   Keep in mind that often only one side of an issue is presented by a scientist or by a journalist. Recognize that a writer or speaker has his or her own opinion which can lead them to leave out some information or to placing too much emphasis on other information. When you read or listen, be critical. Are you being presented with a fair, balanced picture?

Working as a Biologist

As a biology student, you will spend much of your time in the laboratory. The topics of your labs will vary. Sometimes the lab might deal with the behavior of an organism. Sometimes it might deal with the function of an organ. Whatever the lab topic, your purpose will always be to find the answer to a question using some of the methods employed by scientists.

Whenever you confront a question or a problem, the best way to solve it is to break it into parts and proceed step by step.

1. State what you are trying to find out.
2. Think about what you already know about it.
3. Think about what you need to do to solve it.
4. Solve the problem.
5. Check to see that your solution answers what you were trying to find out.

The labs in *Modern Biology Laboratories* have been designed to follow this problem-solving method by following a standard format. Although the topics and activities of the labs will change, this format will stay the same throughout the lab manual.

Investigation Question
Each lab begins with a question which clearly states what you are trying to find out. Your efforts throughout the lab should focus on answering that question in your own words, through your own experience in the laboratory.

Materials
The materials you will need to get from your teacher (or sometimes bring in from home) are clearly listed on the first page of each lab.

Learning Objectives
The learning objectives highlight the most important information you will expect to learn in the investigation as you answer the question.

Process Skills and Strategies
Throughout the year, you will develop skills used by biologists. The process skills that you will use are listed for each lab. Explanations of these process skills can be found on page *v* and in Chapter 2 of *Modern Biology*. Think about improving these skills as you perform each lab. Strategies are hints and guidelines to help you use the process skills as you work through the lab. The strategies are located in the margin next to the point where the process skills are used.

Introduction
The introduction should help you think about what you already know about the question. Your knowledge can have come from a textbook or from everyday experience. Sometimes, the introduction defines unfamiliar words or phrases which you will need to know during the investigation.

Prelab Preparation
Before every lab, you will read through the Prelab Preparation and answer the questions. For many of the labs, the Prelab Preparation will tell you to read through the entire lab before you get into the laboratory. This way you have a better understanding of the Procedure, and can be more aware of any possible pitfalls or safety hazards you may encounter while doing it.

For some of the labs, the Prelab Preparation will tell you to review chapters in your textbook so that you will feel comfortable with the biological terminology used in the investigation. You may also use your textbook knowledge as a basis for making hypotheses and predictions.

The prelab time is a good opportunity to ask your teacher to explain anything you might not understand about the Procedure. Be as prepared as possible for the lab. This way you won't waste time and will carry out the lab effectively and safely.

Procedure

Throughout any procedure, you should be alert to potential dangers to yourself or others, including animals. Pay close attention to the safety symbols in each lab and perform those steps cautiously and responsibly. If you are unsure of a step in the procedure, reread it and ask your teacher for guidance.

As you perform the lab, you will be asked a number of questions which you should answer as you get to them. Your answers, or data, include all the observations you make during the lab. Although the results of your work may not always be what you expect, remember that because living things vary, results will vary. Seemingly "wrong" answers can be valuable if they are analyzed carefully.

In many of the labs, you will work in teams. Be sure to vary the tasks within the group. Also, you should feel free to discuss your observations with team members, but it is important that all team members contribute to the group effort. If work as a team or group is to be truly cooperative, all members of the team should be thinking about all answers; don't just copy your lab partner's answers.

It is important to record your data carefully and accurately. In dealing with structures (the parts of living things), you may record your observations by labeling a drawing that has been provided or by drawing your observation and labelling it. A label should always be printed and have a straight line running to the structure being named. These lines should be clear and should not cross each other. Your drawings should show the size, shape, and location of structures found in an organism.

Data charts are probably the most common means of recording data. You may use data charts provided in this lab manual, but you should also be able to construct your own. If you are creating your own chart, choose a title for your data chart and then make a list of the types of data to be collected. This list will become the headings of your columns and rows. When filling out a chart, always be careful to place the information in the correct column or row. You can always ask new questions about the data, but you cannot get new data without repeating the experiment. Do not clean up your experiment until you are sure you have recorded all the data you need.

Data are often presented in graphs. For example, how much people in different groups know or don't know about infectious diseases can be illustrated with a bar graph. (See Investigation 44). Line graphs can be used to analyze data for trends, as, for example, in the goldfish lab (Investigation 36), where the fish breathing rate will be plotted as a function of water temperature.

Postlab Analysis

The Postlab Analysis is the time for you to determine the answer to the Investigation Question. After you have completed the lab, analyze your findings carefully. Review the Procedure and evaluate your performance of it. This is a good time to discuss common problems that were encountered and to review and revise the hypotheses or predictions you may have been asked to make before the lab.

Further Investigations

The Further Investigations section offers alternative or more open-ended investigations for you to pursue on your own. Many are extensions or branching activities that come from the main body of the lab.

Humane Treatment of Animals

Some of the most worthwhile and memorable experiences you will have in the biology laboratory will involve the study of live animals. Live animals give you a unique opportunity to observe in a living organism what you have studied in your textbook. As you observe these animals, you will be able to distinguish their "personalities," their relationships with other animals and with their environment, and finally, the impact that humans have on them.

Modern Biology Laboratories will give you the opportunity to observe the behavior of a number of live animals, both invertebrate and vertebrate. You will have the opportunity to discover the answers to questions such as these:

- Are flatworms sensitive to magnetic fields?
- Can frogs hear?
- Do earthworms react to colored light?
- What happens to a fish's breathing rate in cold water
- Do reptiles prefer certain types of surfaces as they crawl?
- Does caffeine stimulate a water flea's heart rate?

Animal Care

While it is important to have live animals in the classroom, it is also an added responsibility for both you and your teacher. As a biology student, you must assume a commitment for the responsible care of the animals. It is important that they not be mistreated. You should learn how to handle the animals properly, for the protection of both the animal and yourself.

Before you bring animals into the classroom, be sure that a proper environment can be created and maintained for each species. Work with your teacher on providing the proper temperature, housing, and feed.

Wild animals are not appropriate for use in the classroom. Their long term needs usually cannot be met in a classroom environment. Discuss with your teacher which small, native animals can be collected and whether natural habitats can be simulated in the classroom (for example, frogs, toads, salamanders, insects, spiders, and earthworms.) Handle wild animals infrequently, and return animals to their natural habitat after observation. Animals obtained from pet stores or biological supply companies are usually more appropriate for classroom care than are wild animals.

All animals should be treated humanely. Do not disturb the animals unnecessarily by making loud noises or tapping on the cages or glass enclosures.

Feeding

Keep a record of who is feeding the animals and how much they are being fed. For example, do not leave fish food by the fish tank so that anyone is able to feed them. Most fish will eat until the food is gone. After a period of time, this overfeeding will kill them. Also, be sure that the animals always have access to water and that the type and amount of food they receive is consistent with their nutritional needs.

Weekends and School Vacations

Live animals must be cared for over the weekends and school vacations. If a proper level of care for animals cannot be maintained during periods when school is not in session, the animals should go home with you, your teacher or student volunteers.

Humane experiments

Any experiment to be done with liveanimals should have clearly defined objectives and should be directly supervised by your teacher. No experiment should cause pain or suffering. Follow lab procedures carefully, paying close attention to Animal Safety and Animal Care symbols and cautions.

Dissections

Many teachers find dissections of preserved animals a valuable part of a biology course. Other teachers question the need to kill animals for high school labs. There are pros and cons to both approaches; we encourage you to discuss the controversy with your teacher. We also recommend that you research the issue by reading current periodicals in your library. Contact individuals and groups with varying points of view on the subject of animal dissection. Explore when and why animal research is necessary in modern science. Also, explore if animal research can be avoided and what alternatives exist. You may want to prepare a written report for your teacher and/or an oral report for your classmates.

Safety

The investigations in this lab manual are designed to give you "hands-on" experience with the various specimens and chemicals you are studying. In order to make the laboratory a place of learning and discovery, it must be a *safe* place in which to work. Safety in the lab is as much a part of the scientific process as is organization of observations and logical thinking. Safety is your responsibility and that of every student in the lab. Unsafe practices endanger not only you but your classmates as well. If you have any questions about safety or about laboratory procedures, be sure to ask your teacher.

Where potentially hazardous situations may exist, *Modern Biology Laboratories* has included various safety symbols that will alert you to be especially careful. The word CAUTION, followed by an explanation, appears in the lab procedures at such points where you should pay special attention to this information. Begin thinking about safety now by carefully reading the following safety guidelines and by learning what each safety symbol means.

Electrical Safety

Never handle electrical equipment with wet hands. Work areas, including floors and tables, should be dry.
- Never overload an electric circuit.
- Make sure all electrical equipment is properly grounded.
- Keep electrical cords away from areas where someone may trip on the cords, or where the cords can tip over laboratory equipment.

Chemical Safety

- Never taste any substance in the laboratory. Do not eat or drink from laboratory glassware.
- Do not eat or drink in the laboratory.
- Properly label all bottles and test tubes containing chemicals.
- Never transfer substances with a mouth pipette; use a suction bulb.
- Never return unused chemicals to the original container.

Clothing Protection

- Wear laboratory aprons in the classroom.
- Confine loose clothing.

Caustic Substances

- Alert your teacher to any chemical spills.
- Do not let acids and bases touch your skin or clothing. If a substance gets on your skin, rinse immediately with cool water and alert your teacher.
- Wear your laboratory apron to protect your clothing.
- Never add water to acids; always add acids to water.
- When shaking or heating a test tube containing chemicals, always point the test tube away from yourself and other students.

Explosion Danger

- Use safety shields or screens if there is a potential danger of an explosion or implosion of apparatus.
- Never use an open flame when working with flammable liquids such as ether or alcohol.
- Follow a water bath procedure to heat solids. Never risk an explosion by heating materials directly.

Eye Safety

- Wear approved safety goggles in the laboratory.
- Make sure an emergency eye wash station is available in the laboratory.
- Never look directly at the sun, even for short periods of time. Laboratory goggles will not protect your eyes from the sun's rays.

Fire Safety

- Make sure that fire extinguishers and fire blankets are available in the laboratory.
- Tie back long hair and confine loose clothing.
- Wear safety goggles when working with flames.
- Never reach across an open flame.

Heating Safety

- Use proper procedures when lighting Bunsen burners.
- Turn off hot plates, Bunsen burners, and other open flames when not in use.
- Heat flasks or beakers on a ringstand with a wire gauze between the glass and the flame.
- Store hot liquids only in heat-resistant glassware. Heat materials only in heat-resistant glassware.
- Turn off gas valves when not in use.

Water Safety

- When working near water, always work with a partner or adult.
- Always wear a life jacket.
- Do not work near water during stormy weather.

 Gas Precaution

- Do not inhale fumes directly. When instructed to smell a substance, wave fumes toward your nose and inhale gently.
- Use flammable liquids only in small amounts and in a well-ventilated room or under a fume hood.
- Always use a fume hood when working with toxic or flammable fumes.
- Do not breathe pure gases such as hydrogen, argon, helium, nitrogen, or high concentrations of carbon dioxide.

 Glassware Safety

- Check the condition of glassware before and after using it. Inform your teacher about any broken, chipped, or cracked glassware; it should not be used.
- Air-dry glassware; do not dry by toweling. Do not use glassware that is not completely dry.
- Do not pick up broken glass with your bare hands.
- Never force glass tubing into rubber stoppers.
- Never place glassware near edges of your work surface.

 Hand Safety

- Always wear gloves when cutting, fire polishing, or bending glass tubing.
- Use tongs when heating test tubes. Never hold test tubes in your hand while heating them.
- Always allow heated materials, including glassware, to cool before handling them.

 Hygenic Care

- Always wash your hands after the laboratory.
- Keep your hands away form your face and mouth.
- Use correct sterile technique when transferring bacteria or other microorganisms from one culture to another or to a microscope slide.
- Do not open a petri dish to observe or to count bacterial colonies inside.

 Proper Waste Disposal

- Clean up the laboratory after you are finished; dispose of paper toweling, etc.
- Follow your teacher's directions regarding proper procedures for waste disposal, especially for chemical disposal and microbial disposal.
- Place broken glass in a specially designated container.

 Plant Safety

- When collecting plants, be aware of poisonous plants that grow in the area. Wear gloves as a precaution.
- Never put your hands near your face and mouth when handling unknown or potentially poisonous plants; wear gloves.
- Handle pollen-producing fungi carefully so that pollen and spores are not widely dispersed.

 Animal Safety

- Wear leather or thick gloves when handling laboratory animals, especially rodents.
- When working in the field, be aware of poisonous or dangerous animals in the area. Do not touch or approach wild animals.

 Animal Care

- Handle all animals gently, with consideration for their comfort. Do not poke or otherwise injure their skin, scales, or body parts.
- Assure proper and regular cleaning of cages and tanks, including disposal of wastes.
- Assure adequate food, water, ventilation, and exercise for laboratory animals.
- Assure that social and behavioral needs of the animals are met. Do not keep incompatible animals together.

Reviewing Laboratory Safety

Room Map

1. Draw a map of your biology laboratory. Identify and label the location of the fire extinguisher, eye wash station, fire safety blanket, safety shower, and first aid kit. Also, indicate and label all exits, containers for disposal of broken glass and solid wastes, chemical storage area, and the closest fire alarm box. Locate the electrical outlets, gas outlets, and water faucets nearest your work area.
2. Your teacher will demonstrate how to use the most important pieces of safety equipment in your laboratory. In your own words, briefly describe correct procedure for using each of the following:
   a. fire extinguisher
   b. eye bath
   c. fire safety blanket
   d. safety shower
3. What would be the safest and fastest path from your work area to each of the pieces of safety equipment? Indicate your work area with an X on your room map and draw lines showing the paths.

General Safety

Each of the phrases below describes possible student behavior in the laboratory. On a separate piece of paper, list whether each phrase describes *safe* or *unsafe* behavior. Remember that safety equipment, such as goggles and laboratory aprons, will not protect you unless you are using them properly and when necessary.

1. Eating a candy bar while your lab partner assembles the materials for the day's investigation.
2. Holding a test tube with your thumb and forefinger while heating over a Bunsen burner.
3. Aiming your microscope directly at the sun to get more light.
4. Using a suction bulb on the top of a pipette when transferring substances rather than pipetting by mouth.
5. Making each member of your lab team aware when you are lighting a burner.
6. Using chipped or cracked glassware if it looks like just minor damage.
7. Disposing of a broken beaker in the wastepaper basket.
8. Cleaning all glassware and equipment thoroughly before putting them away.
9. Removing your safety goggles for a moment, during the lab, to relieve discomfort.
10. Bringing three test tubes, each with a different solution provided by your teacher, back to your work area without labeling them first.

Emergency Procedures

What if there were a laboratory accident? How would you deal with the emergency? For each situation described below, write what you would do immediately to deal with the emergency Then, write an explanation of how the situation could have been avoided if safety guidelines had been closely followed. Discuss your answers with your teacher and your classmates. Remember, all accidents should be reported immediately to your teacher.

1. You are cutting a flower in half with a scalpel and you cut your thumb.
2. You are using a mercury-filled thermometer as a stirring rod and it breaks.

3. You spatter hydrochloric acid on your shirt and arm.
4. Your lab partner reaches past a Bunsen burner and the sleeve of her sweatshirt catches on fire.
5. You are heating leaves in a beaker of alcohol over the Bunsen burner and the alcohol ignites.
6. Your lab partner tells you he feels faint or dizzy while he is doing a pulse and breathing rate lab with you.
7. You are in the middle of a timed procedure when you hear the fire alarm sounding.

1 *What Is Life?*

Learning Objectives
- To become familiar with the characteristics of living organisms.
- To learn how to examine unknown specimens.

Process Objectives
- To observe a series of unknown specimens.
- To classify specimens as living, dead, or inanimate.

Materials
For Class
- 10 Unlabeled specimens at different stations
- Dissecting microscopes
- Compound microscopes

How can you determine whether something is alive, dead, or inanimate?

Introduction

Whenever we speak of life, we must think in terms of cells. Even though we cannot see cells without a microscope, they are the basic unit of life and they exhibit all of the characteristics of living organisms. They can exist individually, as do bacteria, or they may work together, taking on specialized tasks to create a more complex organism. However, all living organisms share certain characteristics, which are discussed below.

Cells are made of **protoplasm**. Protoplasm is a specially organized solution of salt, nutrients, and complex molecules in water. The composition of protoplasm may vary among different types of cells, but its purpose is the same—it provides a medium in which complex reactions can occur.

For complex reactions to occur, cells need a source of **energy**. Energy may come to the cells in different forms, but in all cases it is ultimately derived from the energy of the sun.

Living organisms also **respond** to specific **stimuli**. For example, a plant responds to a light stimulus by growing toward it.

Besides responding to stimuli, populations of living things **evolve** over time so that they **adapt** to their surroundings. Adaptations are changes that result from selective pressures placed on the organisms.

Living things are also able to **develop** and **grow**. Individuals go through a series of specific stages until they arrive at the one stage common to all organisms—death, or the end of life. Species do not become extinct with the death of its individual members because individuals can **reproduce**.

With all these characteristics, it may seem simple to decide whether or not something is alive. However, this determination is not always straightforward. Viruses have some, but not all, of the qualities of living organisms, and scientists have long argued whether or not they should be considered as living organisms. (You will learn more about viruses in Chapter 19.)

In this exercise you will determine whether specimens are inanimate, alive, or dead.

Prelab Preparation

Think about the characteristics that are common to all living things.

1. List ten processes that occur while an organism is alive but cease when it is dead.
2. How do plants differ from animals in their life activities?
3. How does an organism's growth compare to the growth of an icicle?

Look at the table in which you will record your data to see what categories you will use to classify each specimen.

Procedure

A. Visit each of the stations in turn. First, look at the specimen without touching it, and examine its size, color, smell (if possible), and shape. Then follow your teacher's instructions at each station for examining the specimen further. As you look at each specimen, ask yourself whether it has some or all of the characteristics of a living organism.

B. Work with your team to compile a list of what you know about each specimen.

   4. Record in your data table all the characteristics your team observed that would classify each specimen as living or as nonliving.

   5. What was the most surprising observation that you yourself missed?

   6. What functions or processes were you not able to observe for some specimens? Can you think of a way to obtain this information?

C. With the information you have gathered, go back to the specimens for a second look.

   7. Did you see anything that you did not notice during your first inspection?

Postlab Analysis

   8. Which specimens are alive or contain living organisms?
   9. Which are dead?
   10. What specimens were once part of, or made by, a living organism?
   11. What characteristics do the nonliving materials have in common?
   12. In which specimens did you observe all the characteristics of living organisms?
   13. What extra experiments would you perform to resolve any questions about whether a specimen is living?

Further Investigations

1. Carry out some of the experiments you mentioned in Question 13.
2. The first samples of the moon's surface to be brought back by Apollo astronauts were carefully examined for signs of life. What sort of evidence do you think scientists looked for? Were there any signs of life? Check your answers in a library resource.
3. Why is it so hard to define viruses as living or nonliving? (Read Chapter 19 in your textbook and check your library for more references.)

Strategy for Observing
Try to use several of your senses to examine the specimens. Touch, smell, and hearing can sometimes be more valuable than sight. (Do not taste the specimens; it is *never* advisable to taste an unknown.)

Strategy for Classifying
Keep the characteristics of living organisms discussed in the Introduction in mind as you observe each specimen. Note whether the specimens meet all, some, or none of the criteria. Give particular attention to any characteristics that seem to go directly against the criteria.

Investigation

1 | *What is Life?*

1. _____
_____
_____
_____
_____
_____

2. _____
_____

3. _____
_____

4. Enter your answers on the data chart.

Specimen	Characteristics		Description
	Living	Nonliving	
1			
2			
3			
4			
5			
6			
7			
8			
9			
10			

Team Data Table

Life Categories
1. Alive
2. Alive, but dormant
3. Dead (once alive)
4. Product of a living organism
5. Never alive

5. _____

6. _____

_____

_____

7. _____

8. _____

9. _____

10. _____

11. _____

12. _____

_____

13. _____

2 | *Introduction to the Microscope*

Learning Objectives
- To understand how to use a compound light microscope.
- To learn how to measure the size of a specimen you see under the microscope.
- To know the relative sizes of the low and high power fields.

Process Objectives
- To observe specimens under the microscope.
- To gather data on the sizes of specimens you observe under the microscope.
- To communicate your observations through diagrams.

Materials
Parts I and II
For Group of 2
- Compound microscope
- Microscope slide
- Coverslip
- Water
- Eyedropper or dropping bottle
- Lens paper or Kimwipes
- 70% Ethyl Alcohol
- Dissecting needle or pin
- Forceps

Part I
- Newspaper
- Colored sewing thread

Part II
- Transparent plastic ruler with millimeter scale
- Potato
- Iodine
- Single-edged razor blade or scalpel
- Cork

How do you use a microscope?

Introduction

In almost every type of biological research, the microscope plays a fundamental role. Scientists in each field rely on it to study the fine structures of cells and tissues—things too small to see with the unaided eye. Were it not for the microscope, our understanding of life would be far different from what it is today.

In this lab, you will learn how to use a compound light microscope to observe structures too small to see with the unaided eye. In future labs, you will observe microscopic organisms using the techniques you learned in this lab.

Prelab Preparation

It is essential that you learn how to use and care for your microscope properly. Study the diagram of the microscope and learn the names of all its parts.

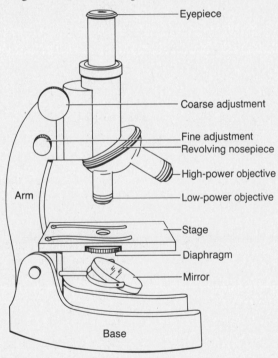

Standard Compound Microscope

Your microscope is expensive and fragile. It is important for you to use it correctly to avoid damaging it and to avoid breaking slides or destroying specimens. When you are using your microscope, it should rest securely on your table or bench, away from the edge. When you carry your microscope, always use two hands. Hold its base with one hand and its arm with your other hand as shown.

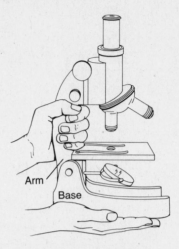

How to Hold a Microscope

Always use an appropriate light source. If your microscope has a lamp, plug it in and turn the lamp on. If your microscope has a mirror, adjust it to get a good amount of light through the eyepiece. CAUTION: Never use direct sunlight as your light source. Make sure the diaphragm is sufficiently open so enough light can get through. (This will be especially important if you look through the eyepiece and see nothing).

Always keep both eyes open as you look into the eyepiece. This is important because it reduces eyestrain. If you find this difficult, cover your other eye with your hand. This may feel awkward at first but it will become easier with practice.

Keep the lenses on your microscope clean. Never touch them with your fingers. If the eyepiece or objective lenses get dirty, clean them with a piece of lens paper moistened with alcohol (or xylene). Wipe the lens in a light circular motion and change the lens paper as it picks up the dirt. Make certain that you leave no streaks on the lens. NOTE: Cleaning the lens with anything other than lens paper, or wiping too hard, will scratch the lens.

The purpose of the microscope is to magnify your specimen. Microscopes use two lenses—the eyepiece and an objective—to magnify the image. The magnification is the number of times the size of an object appears increased. If the magnification of an object is $10\times$, it will appear 10 times larger than it really is.

The magnification of your microscope is equal to the product of the separate magnifications of the eyepiece and the objective. (The magnification of each lens is written on the lens case.) If the eyepiece is $10\times$ and the low power objective is $10\times$, then the magnification under low power is $100\times$. In equation form, this is written:

$$\text{(Eyepiece magnification)} \times \text{(Objective magnification)} =$$
$$\text{Total microscope magnification}$$

1. If the magnification of the eyepiece is 10x and the magnification of the high power objective is 40x, what is the total magnification under high power?
2. How many times larger than life will a specimen appear under this magnification?

If you have a scanning power ($4\times$ objective), note that it gives a very low magnification. This is useful for locating a specimen on the slide, but in many cases it is not appropriate for observation.

Procedure

Part I: Using your microscope

A. Place your microscope on your bench or table, away from the edge. Make certain that you are familiar with all of its parts and that it is functioning properly.

   3. What are the magnifications of your microscope under low and high power?

The first thing you will observe under the microscope is the letter R. It is very easy to focus on and it helps demonstrate what a microscope does to the image of an object.

B. Prepare a wet mount of an R by first cutting a capital R out of a newspaper. (Do not use one from a headline.) As shown in the illustration, place one drop of water on the slide. Add the R to the drop. Place one edge of a coverslip on the slide, in the water, next to the R. Use a dissecting needle or pin to gently lower the coverslip onto the R. To get rid of bubbles, raise and lower the coverslip until any trapped air is released. Do not press directly down on the coverslip. If there is not enough water under the coverslip, add water at the edge.

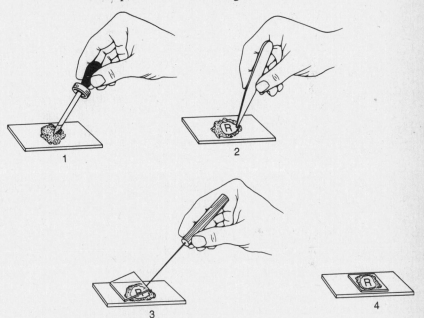

C. Adjust the diaphragm and the lamp or mirror to obtain the appropriate light for viewing. (Remember, do not use direct sunlight.)

D. Become familiar with the high and low power magnifications of your microscope. Put your wet mount on the microscope's stage and clip it in place. Turn the nosepiece so that the low power objective is in place. Focus under low power, using the coarse focus first and then the fine focus. You should be able to clearly see the letter R on the slide.

E. To use the high power objective, first focus on the letter R under low power. Without changing the focus, turn the high power objective into place. If your microscope is properly adjusted, the image should be almost in focus. Use only fine focus on high power. Move the slide around and focus to see small irregularities in the R. NOTE: Slides are easily broken by using the coarse focus with high power.

   4. Which magnification, low or high, is more appropriate for looking at the R?

   5. How does the R appear to be oriented compared to the way it looked before you put it under the microscope?

6. Move your slide to the left and then right. Which way does the image of the R appear to move?

7. Move your slide toward you and then away from you. Which way does the image of the R appear to move?

8. What can you conclude about the way the image moves as compared to the way you are actually moving the slide?

F. Center the R in the middle of the low power field. Now switch to high power and focus. (Remember to use only the fine focus under high power.)

9. Do you see more or less of the R on high power than on low power?

G. To learn how to see the depth of a specimen, obtain a piece of colored thread about 2 cm long. Place the thread in a drop of water on a clean coverslip. Hold the thread in place with forceps and fray one end of it with a dissecting needle or pin. Without adding a cover slip, clip the slide in place on the stage. Focus up and down using both low and high powers.

10. Why do only parts of the thread appear sharp and clear at any given time?

Adjusting the diaphragm opening changes the contrast and aids in making more accurate observations.

H. With the slide from step G in place, put your high power objective in place. Observe the thread carefully as you slowly open and close the diaphragm.

11. What differences did you observe under the various diaphragm settings?

I. When you are finished, put the low power objective in place. If necessary, return your microscope to its proper storage place. Clean your slide and cover slip with water and paper towels.

Part II: Microscopic measurement

You will now measure the diameters of the low and high power fields. This will enable you to estimate the actual size of the specimens you will observe later.

J. Put a clear plastic ruler on the microscope stage so that you can see the millimeter scale under low power. Place one millimeter marking of your ruler at the far left hand side of the low power field. You should see one other millimeter marking in your field of view. This means that your low power field is between 1 and 2 mm in diameter.

12. To determine the diameter of the low power field of your microscope, approximate, to the nearest 0.1 mm, what fraction of the second millimeter is in your field of view.

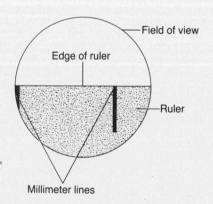

Strategies for Measuring

If your high power magnification is 400× and your low power magnification is 100×, you would divide 400× by 100× to obtain a result of 4. This would mean that the diameter of the low power field is 4 times larger than the diameter of the high power field.

If your low power field is 2 mm wide and the low power field is 4 times larger than the high power field, then you would divide 2 mm by 4 to obtain a diameter of 0.5 mm for your high power field.

If the diameter of the high power field is 0.5 mm, it would be 500 microns wide. The average size of animal cells is 10 to 20 microns.

If there are too many grains to count conveniently, make an estimate by counting the number of grains in one quarter of the field and multiplying by four.

13. Divide the high power magnification of your microscope by the low power magnification to determine how many times larger your low power field is than your high power field.

14. Now determine the diameter of your high power field by dividing the diameter of your microscope's low power field by the number you obtained above.

K. Many of the specimens you will observe under the microscope will be smaller than 1 mm in size. Because of this, microscopic measurements are often expressed in **microns** (μm or micrometers). One millimeter equals 1,000 microns.

15. What are the diameters of your low and high power fields in microns?

For the remainder of this lab, express all of your measurements in microns. For example, if you measure a cell to be 0.2 mm wide, then report it as 200 microns. Since nearly all cells are less than 1 mm in diameter, you can see that it is better to think of them in terms of microns than millimeters.

Part III: Microscopic observations

In addition to whole cells, you will often wish to observe the parts of a cell. A good example of a part of a plant cell is a grain of starch. Starch grains appear white in the normal plant cell and are not easy to see. To make them more visible, you will stain them with iodine, which turns them black. Stains are commonly used in microscopy to make certain types of cells or cell parts easier to observe.

L. Prepare a wet mount of starch grains by first cutting a piece of a potato. Then, use a single-edged razor blade or scalpel to gently scrape the surface of the potato (not the skin). You should see a whitish liquid substance on the edge of the blade. CAUTION: Be careful to not cut yourself with the razor blade or scalpel.

M. Place a drop of this fluid onto a clean slide and spread the drop to make a circle about the size of a dime.

N. Add one drop of iodine and place a coverslip on top of the iodine. Examine your slide under low power. You will see numerous black circles and ovals. These are starch grains from the potato.

O. Move the slide around to find a field that has about 100 starch grains.

16. Count and record the number of starch grains on one field of view under low power.

17. Without moving the slide, switch to high power. Count and record the number of grains in one field of vision.

18. How does the number of grains you have counted under high and low power relate to the relative sizes of the high and low power fields? How do you explain this relationship?

To observe the size and shape of plant cells you will examine a piece of cork. Cork is a mass of dead plant cells in which only the cell wall remains. These cell walls will appear as compartments under the microscope.

P. Slice a small piece of cork with a single-edged razor blade or scalpel. Make your slice as thin as you possibly can. Place your slice on a slide but do not add water or cover it with a coverslip. CAUTION: Use extreme care as you work with a single-edged razor blade or scalpel.

Q. Observe the thinnest part of your cork under low power. (If your cork slice is extremely thin, you might be able to make useful observations under high power as well.)

Strategy for Observing

When you observe the cork cells, draw what you see, not a prepared illustration or a copy of someone else's drawing. When you make observations it is essential that you record only what you actually see. Do not attempt to record anything that you cannot observe on the basis that "it's supposed to be there."

Strategy for Communicating

Exchange drawings with your lab partner. Check whether everything that you saw under the microscope is represented in your partner's drawing. Also, look to see whether your partner has included anything that you did not observe. Discuss your analysis with your partner.

19. Draw a diagram of what you see. Label the cell wall and cell cavity. Make your drawing as accurate as you can and indicate the magnification in the lower right hand corner of your drawing.
20. Measure the average size of a single cork cell. First determine how many cork cells are needed to reach from one side of the visual field to the other. Divide this number into the diameter of your low power field. This number is the average size of a single cork cell.

Postlab Analysis

21. What range of specimen size is appropriate for observing under your microscope?
22. Why is it necessary to make your drawings as accurate as you can?
23. What is the purpose of indicating the magnification on your drawing?
24. How are observations made with a microscope usually communicated?

Further Investigations

1. Read about the history of the microscope, modern uses of the microscope, or different kinds of microscopes.
2. Look at prepared slides or make your own slides of things you would like to see in more detail. For example, insects, parts of plants, pond water, or samples of almost anything are very interesting when seen in fine detail.

Investigation

2 | *Introduction to the Microscope*

1. _____
2. _____
3. _____
4. _____
5. _____
6. _____
7. _____
8. _____
9. _____
10. _____
11. _____
_____
12. _____
13. _____
14. _____
15. _____
16. _____
17. _____
18. _____
19.

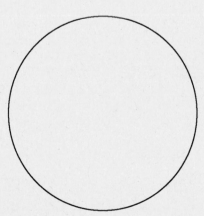

20. _____

21. _____

22. _____

_____

23. _____

_____

24. _____

_____

3.1 | *Acids and Bases*

Learning Objectives

- To relate the pH scale to how acidic or basic a solution is.
- To explain how a buffer affects the pH of a solution.

Process Objectives

- To classify biological solutions as acids or bases.
- To measure the pH of a solution.

Materials

For Group of 2
Parts I and II
- Lab apron
- Safety goggles

Part I
- 10 Common substances (such as distilled water, vinegar, baking soda, shampoo, orange juice, tomato juice, grapefruit juice, soda water, tap water, cola, fabric softener, household ammonia, bleach, liquid detergent, salt water, milk)
- Full-range pH paper
- 10 test tubes
- Distilled water in wash bottle
- Glass rod
- Grease pencil or marker
- Test-tube rack

Part II
- Bromothymol blue indicator
- Egg white solution
- 125-mL Erlenmeyer flask
- 0.1 N Hydrochloric acid
- Dilute basic solution
- Medicine dropper
- Full-range pH paper
- Glass rod

How does pH affect biological solutions?

Introduction

Organisms are very sensitive to how acidic or basic (alkaline) their environment is. For example, some bacteria and fungi can grow only in acidic solutions while some marine organisms can only live in a slightly basic environment. The reason for this sensitivity is that the enzymes used to control metabolic functions can operate only within a narrow range of pH.

How acidic or basic a solution is depends on the number of **hydronium ions** (H_3O^+) that it contains. The hydronium ion forms naturally when a few molecules of water ionize:

$$2\,H_2O \rightleftharpoons H_3O^+ + OH^-$$

Because pure water has the same number of hydronium ions and **hydroxide ions** (OH^-), it is considered neutral. If a solution has a greater concentration of H_3O^+ than OH^- it is considered acidic. If a solution has a lesser concentration of H_3O^+ than OH^- it is considered basic.

A special number scale called the pH scale uses numbers to indicate the relative concentration of H_3O^+. Human blood must be at a constant pH of 7.4. If the blood pH drops to 7.0 or rises above 7.8, death results.

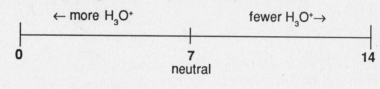

Prelab Preparation

Review Section 3.2 on acids and bases in your textbook.

1. Why are some substances acids? What are some of their characteristics?
2. Why are some substances bases? What are some of their characteristics?

In this lab you will measure acidity using pH paper. Do not handle the paper strips too much or the chemicals on your fingers will cause the pH paper to react incorrectly. Always wash glassware with distilled water (which is neutral). This will help to give valid and consistent results.

3. What does it mean to say that distilled water is neutral? What pH should it have?

The pH paper is made by combining paper and chemical indicators such as litmus and phenol red. The color of an indicator changes as the pH changes.

4. How is an indicator different from a dye?

In Part II you will be working with **buffers**. A buffer is a mixture of chemicals that keeps the pH of a solution relatively constant. A buffer system acts by taking up or releasing H_3O^+ or OH^- in a solution so that sudden changes in pH are prevented. Human blood has a buffer system to keep its pH within the narrow range.

5. Where else in the human body might a buffer system be useful or important?

As you work in your groups be sure to label all experimental solutions to avoid confusion. Record your observations and measurements immediately as some of the color changes may fade in a short time. Become familiar with the pH chart so you can easily determine the pH number.

6. For the following solutions, how would you describe the H_3O^+ concentration? Label them as acidic, basic, or neutral.
Solution A—pH 10
Solution B—pH 7
Solution C—pH 3

Procedure

Part I: Acid-Base Classification

CAUTION: Strong acids and bases are harmful to skin or clothing. Avoid letting them contact your clothing and skin. Be careful about protecting eyes.

A. Label 10 test tubes in a test-tube rack so there is one test tube for each of the test substances. Add a few drops of each substance to the corresponding test tube.

B. Use a glass rod to touch one drop of a test substance to a strip of pH paper. Hold the strip of pH paper up to the standard pH color chart to determine the pH. If the result is between two numbers, estimate to the nearest tenth.

7. Record your results in your data chart for Part I.

C. Rinse the glass rod with distilled water and proceed to test the other substances.

8. Record these results in the data chart.

D. After all substances have been tested with pH paper, test each substance with red and blue litmus paper. The pH paper indicates the range of the pH with a corresponding range of colors. The litmus paper has only two colors: red and blue. Red litmus turns blue in the presence of bases; blue litmus turns red in the presence of acids.

9. Record the litmus results in the data chart.
10. Why would some substances not change the color of the litmus paper?
11. Use the diagram of the pH scale in the Introduction to rate each substance in your data chart as having a high, low, or medium concentration of H_3O^+ and to classify each substance as an acid or a base.

Part II: Buffers

E. Fill a 125-mL Erlenmeyer flask with 25 mL of the dilute basic solution (water adjusted to a pH equivalent to that of egg white). Add 5 drops of bromothymol blue indicator and swirl the flask until the indicator is completely dissolved. Use pH paper as in Step B above to determine the initial pH.

12. Record this in your data chart for Part II.

F. Slowly add 0.1 N HCl drop-by-drop to the solution in the flask, keeping count of the drops. NOTE: Be extremely careful that you add one drop at a time and that each drop enters the solution and does not adhere to the inner surface of the flask. Swirl after each drop to completely mix. Stop

Strategy for Measuring
In order to maintain consistency, have the same person read each of the strips of the pH paper over the standard spectrum.

Strategy for Classifying
To determine whether the test substances are acids or bases, compare pH and litmus color results. Look for a pattern. Then make a generalization about your criteria for classifying these substances.

when there is a significant color change that does not disappear after 30 seconds. Use pH paper to determine the new pH. Record how many drops of HCl were required to make this change. CAUTION: Do not allow the HCl to contact your skin or clothing.

13. Enter observations and measurements on your data chart.

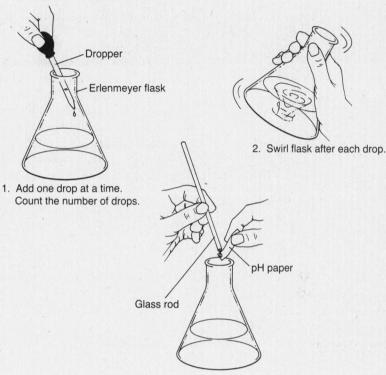

Dropper

Erlenmeyer flask

1. Add one drop at a time. Count the number of drops.

2. Swirl flask after each drop.

pH paper

Glass rod

3. Determine pH after color of indicator changes permanently.

G. Wash out the flask with distilled water. This time fill it with 25 mL of egg white solution. Add 5 drops of bromothymol blue indicator to the egg white and swirl the flask until it is completely dissolved. Use pH paper to determine the initial pH. Repeat step F.

14. Enter all new observations and measurements on your data chart. Record the initial and final pH of egg white as well as the number of drops of HCl required to make a significant color change.

15. Which solution, the weakly basic one or the egg white, required the most drops of HCl to change its color and pH? Which solution required the least?

Postlab Analysis

16. From the data you collected in Part I, what generalizations can you make about the pH of a solution and the effect of a solution on litmus paper?

17. Using data from Part I, make a list of those substances you tested that are acidic, in the order from strongly acidic to weakly acidic. Make a similar list for the basic substances, in the order from strongly basic to weakly basic. What do acids have in common? What do bases have in common?

18. In Part II, look for a delay in pH change when adding HCl. Which substance would you consider a buffer?

19. What biological purpose can buffering serve?

Further Investigations

1. The liquid that results after you boil red cabbage can be used as an indicator. Set up a range of colors in a series of test tubes using the common substances listed under Materials and mixing each with several drops of red cabbage liquid. Determine how the colors of cabbage water correspond to the colors of pH paper.

2. Try using milk, buffered aspirin, nonbuffered aspirin, or antacids in place of the egg white in Step G. Which substance does the best job of absorbing extra hydronium ions?

3. Retest your 10 substances from Part I with a pH meter. Compare your results to the results with pH paper.

4. Why must some backyard gardeners add lime to "sweeten" their soil while others must add sulfur? County extension agencies often do soil pH testing and can help you with your research.

5. In many locations, the water from the tap is *not* neutral. Find out what dissolved substances there are in tap water that affect its pH.

Investigation

3.1 | *Acids and Bases*

1. _____

2. _____

3. _____

4. _____
   _____

5. _____

6. _____
   _____
   _____

7.–9. Enter your answers on the data chart.

Part I Data Chart

Substance							
pH							
Red Litmus							
Blue Litmus							
H_3O^+ Concentration							
Acid or Base							

10. _____
    _____

11. Enter your answers on the data chart above.

12.–14. Enter your answers on the data chart.

Part II Data Chart

	Initial pH	Number of Drops of HCl	Final pH
weakly basic solution (plus indicator)			
egg white solution (plus indicator)			

15. _____

16. _____

_____

17. _____

_____

_____

_____

_____

_____

18. _____

19. _____

_____

3.2 | *Calorimetry*

Learning Objectives
- To find out how much energy is contained in different kinds of food.
- To use mathematical formulas to make your raw data meaningful.

Process Objectives
- To measure, indirectly, the amount of energy released during combustion of different foods.
- To predict which foods will produce the most energy.
- To organize and analyze data so as to compare your results with your predictions.

Materials

For Class
- Metric balance accurate to 0.1 gram

For Group of 2–3
- 250-mL Beaker of cool water
- 10-mL Graduated cylinder or pipette
- Ring clamp and ring stand
- Test-tube tongs
- Large cork with pin to support food sample
- Square of aluminum foil
- 2 15-mL Test tubes
- Matches
- Thermometer with 0.2°C gradations
- 6 Test foods, such as puffed rice cereal, potato chips, popcorn, cookies, chocolate, raisins, shredded coconut, shredded wheat, dry pasta, or marshmallows.

Strategy for Predicting
Base your predictions on your knowledge of which foods are more fattening.

How can we measure the energy contained in foods?

Introduction

Plants have evolved processes that convert light energy into the chemical bonds of complex molecules. The chemical bonds in carbohydrates, fats, and proteins store energy until needed by the plant. The plant can then release the energy by breaking the appropriate chemical bonds.

Every animal maintains its life processes by consuming complex molecules that store energy. The processed plants and animals we eat as foods contain varying amounts of proteins, carbohydrates, and fats. Because each of these types of foods contains varying amounts of energy, these foods will release varying amounts of energy when they are used by cells. Within our bodies, the energy is released slowly by a series of chemical reactions.

Prelab Preparation

By burning pieces of food, the chemical energy stored in molecular bonds is released as heat and light. The heat can be measured in units called **calories.** A calorie is the amount of heat (energy) required to increase the temperature of one gram of water by 1°C. This process is the basis of the technique of **calorimetry.**

The more calories a food contains, the more heat it gives off when burned. Foods high in calories will release large amounts of energy. One gram of a protein will release far fewer calories than one gram of fat. You will study foods with different proportions of protein, fats, and carbohydrates to see how much energy (calories) they release.

Read the procedure and answer the following questions.

1. The burning of the food sample releases energy in the form of light and heat. Explain the technique for measuring heat in calories?
2. Does the water need to be at the same temperature at the start of each new test?

Procedure

CAUTION: Since there will be an open flame in the classroom, take appropriate precautions.

A. Assemble 6 food samples. Each should weigh 1–5 grams.

3. Predict which foods, when burned, will raise the temperature of water the most and the least.

B. Study the figure that shows the apparatus used in this exercise. Assemble the ring stand and clamp so that a test tube placed in the holder will be one cm above the food sample.

C. Place 5 mL of water in the test tube and put the test tube in the holder.

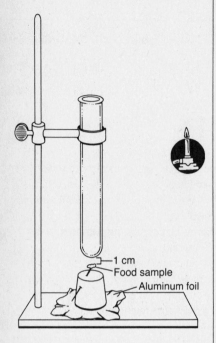

—1 cm
Food sample
—Aluminum foil

Measure the temperature of the water and hold the thermometer in place until the final reading in Step F.

4. Record this temperature on your data table.

D. Weigh a sample of a test food.

5. Record the weight on your data table.

E. Use the pin to affix the sample to the cork. Place the cork on the table away from the test tube. Then strike a match and set the food sample on fire. Immediately move the sample under the test tube.

F. After the food sample is completely burned, measure the temperature of the water again.

6. Record the reading on your data table.

G. Weigh the amount of sample remaining.

7. Record the reading as "ash weight" on your data table.

CAUTION: The test tube may be very hot after the experiment. Use test tube tongs to handle it.

H. Subtract "ash weight" from initial "food weight" for "weight change."

8. Record your calculation for weight change on your data table.

I. Discard the water and allow the tube to cool. Use a fresh, clean tube for your next test and repeat the procedure for 5 or more food samples.

J. Calculate the temperature difference for each sample by subtracting the initial water temperature from the final water temperature.

9. Record your calculations on your data table.

K. To estimate the calories in each food sample, use this formula:

Total kilocalories per sample = weight of water (in grams) × difference in water temperature × 1kg/1000g × specific heat of water

The specific heat of water is 1 kilocalorie/kg°C and the density of water is 1g/mL. You will see that all units of measurement except kilocalorie cancel each other out of the equation.

10. Calculate the kilocalories produced by each sample. Record your calculations on the data table in the column entitled "Total Kilocalories."

L. To figure out the number of kilocalories per gram in each type of food, use the following formula:

$$\text{kilocalories/gram} = \frac{\text{Total Kilocalories Measured in Sample}}{\text{Change in Mass of the Sample}}$$

For example, if you measured 0.52 kilocalories in your test sample, and the weight loss was 0.1 gram, then the sample has 5.2 kilocalories per gram.

11. Calculate kilocalories per gram for each sample and record on the data table.

Postlab Analysis

12. What are the problems with this technical design, which might be sources of error?

13. Compare the caloric values obtained by your group with those of the rest of the class.

14. Which food(s) gave off the most calories/gm? The least?

15. How do these results compare with your predictions?

Further Investigation

1. Can you build a more efficient calorimeter that produces results closer to standard calorie tables? Think of ways to use insulation to increase the efficiency of the calorimeter. Check your design with your teacher before you try it out.

Strategy for Analyzing
If your data do not support the predictions you made in Step A, consider whether your assumptions should be revised or whether there are sources of error you have not taken into account.

3.2 | *Calorimetry*

1. _____
_____
_____
_____

2. _____

3. _____
_____
_____

4.–11. Enter your answers on the data chart.

Data Chart for Food Calorimetry Measurements

Specimen	Weight	Water Volume	Temp. at Start	Temp. At End	Temp. Difference	Ash Weight	Weight Change	Total Kcal	Kcal per Gram
1									
2									
3									
4									
5									
6									

12. _____
_____
_____
_____
_____
_____

13. _____
_____
_____

14. _____

_____

_____

_____

15. _____

_____

_____

Investigation

4 | *Catalase*

Learning Objectives

- To demonstrate the activity of an enzyme in living tissues.
- To learn how changes in temperature and pH affect the activity of catalase.

Process Objectives

- To experimentally test for the presence of catalase in living tissues.
- To analyze factors affecting enzyme activity.

Materials

Parts I and II
For Group of 2-4

- 40 mL 3% Hydrogen peroxide solution
- 10-mL Graduated cylinder
- 10 Test tubes (small)
- Test tube rack
- Stirring rod
- Scissors, straight-edged razor blade, or scalpel
- Forceps
- Thermometer

Part I
For Group of 2-4

- 3 Pea-sized pieces of fresh liver (chicken or beef)
- Slice each: fresh potato, fresh chicken meat, fresh apple or carrot

Part II
For Class

- Boiling water bath (100°C)
- Warm water bath (approximately 37°C)
- Ice bath (0°C)
- Room temperature water bath (approximately 22°C)
- 4 Thermometers

For Group of 2-4

- Test-tube holder
- 1N HCl (in dropper bottle)
- 1N NaOH (in dropper bottle)
- pH paper
  [Continued]

How do enzymes work in living tissues?

Introduction

What would happen to your cells if they made a poisonous chemical? You might think that they would die. In fact, your cells are always making poisonous chemicals. They do not die because your cells use **enzymes** to break down these poisonous chemicals into harmless substances. Enzymes are proteins that speed up the rate of reactions that would otherwise happen more slowly. The enzyme is not altered by the reaction. You have hundreds of different enzymes in each of your cells. Each of these enzymes is responsible for one particular reaction that occurs in the cell.

In this lab, you will study an enzyme that is found in the cells of many living tissues. The name of the enzyme is **catalase** (KAT-uh-LAYSS); it speeds up a reaction which breaks down hydrogen peroxide, a toxic chemical, into 2 harmless substances—water and oxygen. The reaction is as follows:

$$2 \ H_2O_2 \longrightarrow 2 \ H_2O + O_2$$

This reaction is important to cells because **hydrogen peroxide** (H_2O_2) is produced as a byproduct of many normal cellular reactions. If the cells did not break down the hydrogen peroxide, they would be poisoned and die.

In this lab, you will study the catalase found in liver cells. You will be using chicken or beef liver that your teacher purchased in the supermarket. It might seem strange to use dead cells to study the function of enzymes. This is possible because when a cell dies, the enzymes remain intact and active for several weeks, as long as the tissue is kept refrigerated.

Prelab Preparation

Before you begin this lab, review what you learned about pH in Lab 3.1. Recall that pH is the measure of the acidity or alkalinity of a solution. An acidic solution has many hydrogen ions (H^+) and a pH below 7. An alkaline, or basic, solution has very few hydrogen ions and a pH above 7. A neutral solution has a pH of 7.

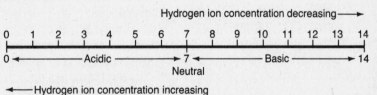

Review what you learned in Chapter 4 about enzymes. Recall that the **substrate** is the molecule that the enzyme acts on, and the **products** are the molecules produced by the reaction. Review why enzymes are reusable.

> **1.** In the reaction that you will be studying in this lab, what is the enzyme? What is the substrate? What are the products?

Under certain conditions enzymes are denatured. An enzyme is denatured when the protein molecule loses its proper shape and cannot function. Some

- 8 Pea-sized pieces of fresh liver (chicken or beef)
- 5 Drops 5% hydroxylamine solution
- 2 pieces of graph paper

Strategy for Experimenting
Be sure to measure all solutions accurately. Avoid contaminating your tools or samples with materials from any other steps in the experiment.

Strategy for Analyzing
Be sure to record your observations before continuing with the next step in the experiment. Compare your observations with the rest of your group and discuss together what the observations have shown you.

things that can denature an enzyme are high temperatures, extremes of pH, heavy metals, and alcohol.

Procedure

Part I
Normal Catalase Activity
NOTE: Be sure to clean your stirring rod (and your test tubes, if necessary) between steps.
A. Place 2 mL of the 3% hydrogen peroxide solution into a clean test tube.
   2. Is the hydrogen peroxide bubbling?
B. Using forceps and scissors, cut a small piece of liver and add it to the test tube. Push it into the hydrogen peroxide with a stirring rod.
   3. Observe the bubbles; what gas is being released?
   4. Throughout this investigation you will estimate the rate of the reaction (how rapidly the solution bubbles) on a scale of 0–5 (0=no reaction, 1=slow,...,5=very fast). Assume that the reaction in step B proceeded at a rate of "4" and record the speed on Chart II of the data sheet as the rate at room temperature.
C. Recall that a reaction that absorbs heat is **endothermic;** a reaction that gives off heat is **exothermic.** Now, feel the temperature of the test tube with your hand.
   5. Has it gotten warmer or colder? Is the reaction endothermic or exothermic?

Is Catalase Reusable?
D. Place 2 mL of 3% hydrogen peroxide solution into a clean test tube and add a small piece of liver.
   6. What is happening in your test tube?
E. Pour off the liquid into a second test tube.
   7. What is this liquid composed of? What do you think would happen if you added more liver to this liquid? Why?
F. Add another 2 mL of hydrogen peroxide to the liver remaining in the first test tube.
   8. Can you observe any reaction? What do you think would happen if you poured off this liquid and added more hydrogen peroxide to the remaining liver?
   9. Are enzymes reusable?

Occurrence of Catalase
Catalase is present in many kinds of living tissues. You will now test for the presence of catalase in tissues other than liver.
G. Place 2 mL of hydrogen peroxide in each of 4 clean test tubes. To the first tube, add a small piece of liver. To the second tube, add a small piece of potato. To the third tube, add a piece of chicken. To the last tube, add a piece of apple or carrot.
   10. As you add each test substance, record the reaction rate (0–5) for each tube in Chart I of your data table.
   11. Which tissues contained catalase?

Part II
Effect of Temperature on Catalase Activity
H. Put a piece of liver into the bottom of a clean test tube and cover it with a small amount of distilled water. Place this test tube in a boiling water bath for 5 minutes.
   12. What will boiling do to an enzyme?

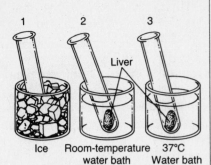

1   2   3

Liver

Ice   Room-temperature 37°C
       water bath   Water bath

I.   Remove the test tube from the hot water bath, allow it to air cool, then pour out the water. Add 2 mL of hydrogen peroxide. CAUTION: Use a test-tube holder when handling the hot test tubes.

    13. What is happening in the test tube? Explain your results and record the reaction rate (0–5) on Chart II of the data sheet.

J.   Put equal quantities of liver into 3 clean test tubes and one mL H_2O_2 into 3 other test tubes. Put one test tube of liver and one of H_2O_2 into each of the following water baths:
     Ice bath (0°C)
     Room temperature water bath (about 22°C)
     Warm water bath (37°C)

K.   After 3 minutes, pour each tube of H_2O_2 into the corresponding tube of liver and observe the reaction.

    14. Record the reaction rates (0–5) in Chart II of your data sheet.

    15. Make a graph of the estimated reaction rate as a function of temperature. You should have 4 points: 0°C (ice water); room temperature (in degrees Celsius); 37°C (human body temperature) and 100°C (boiling).

    16. What is the "optimum" temperature for catalase? (This is the temperature at which the reaction proceeds fastest.)

    17. Why did the reaction proceed slowly at 0°C?

    18. Why did the reaction not proceed at all at 100°C?

Effect of pH on Catalase Activity

L.   Add 2 mL hydrogen peroxide to each of 3 clean test tubes. Treat each tube as follows:
    Tube 1—add a drop of 1N HCl (acid) at a time until pH 3.
    Tube 2—add a drop of 1N NaOH (base) at a time until pH 10.
    Tube 3—adjust the pH to 7 by adding single drops of either
              1N HCl or 1N NaOH as needed.

    CAUTION: Do not let acids or bases contact your skin or clothing. Swirl each test tube after adding each drop and measure the pH of each solution with pH paper. To do this, remove a drop or two of solution from a test tube, using a clean glass stirring rod. (Rinse your stirring rod in tap water and wipe it dry with a clean paper towel before you dip it into each test tube.) Place the drop on pH paper.

    19. Record your results in Chart III of your data table.

M. Next, add a small piece of liver to each test tube. Estimate the reaction rates (0–5).

    20. Record the rates in chart III of your data table.

    21. Make a graph of the estimated reaction rate as a function of pH on a separate piece of graph paper. You should have 3 points on your graph.

    22. Does there appear to be a pH "optimum"? At what pH?

    23. What is the effect of low or high pH on enzyme activity?

Inhibitors of Catalase

Substrates bind to enzymes at a particular region on the enzyme called the **active site.** The reaction then occurs on the surface of the enzyme. After the reaction is complete, the products leave the enzyme, thereby allowing the enzyme to react with another molecule of substrate. Imagine what would happen if there was another molecule in the cell that could fit onto the active site of the enzyme, but could not react. Because there is limited number of enzyme molecules, this molecule would compete with the normal substrate for the active sites. There would then be less enzyme available for the substrate, causing the reaction to proceed more slowly. Such molecules do exist. They are called **competitive inhibitors. Hydroxylamine** (hie-DROK-suh-luh-

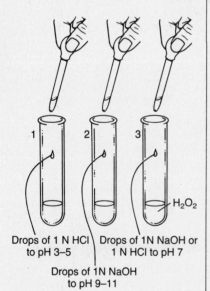

1   2   3

             H_2O_2

Drops of 1 N HCl Drops of 1N NaOH or
to pH 3–5   1 N HCl to pH 7

   Drops of 1N NaOH
     to pH 9–11

MEEN) is a competitive inhibitor of catalase. Hydroxylamine attaches to the catalase molecule and interferes with the formation of the normal enzyme-substrate complex.

N. To test the effect of hydroxylamine on liver tissue, place 2 mL of hydrogen peroxide into a clean test tube. Add 5 drops of 5% hydroxylamine solution. Add a small piece of liver to this mixture and observe the reaction.

24. Record the rate of the reaction (0–5) in Chart IV of your data table.
25. Explain what happened in the test tube.

Postlab Analysis

26. Make an inference about what happens to the heat when the reaction described in Step C occurs in living cells?
27. Name several conditions that slowed down or stopped the action of the enzyme catalase.
28. Compare optimal temperature and optimal pH for catalase to the physiological conditions of a cell.
29. Did you find catalase to be present in many different kinds of tissue? Explain why.

Further Investigations

1. Draw a model of enzyme action using the example of liver catalase and hydrogen peroxide.
2. Discuss how you think a competitive inhibitor might work. Draw a model for enzyme action in the presence of a competitive inhibitor.

Strategy for Analyzing
Did any of the reactions you observed surprise you? Do the reactions form any patterns?

Investigation

4 | *Catalase*

1. _____

2. _____

3. _____

4. Enter your data on Chart II.

Chart I	
Sample	Rate of Enzyme Activity (0–5)
liver	
potato	
chicken	
apple or carrot	

Chart II	
Temperature	Rate of Enzyme Activity (0–5)
0°	
Room Temperature_____ °	
37°	
100°	

Chart III	
pH of Sample	Rate of Enzyme Activity (0–5)

Chart IV	
Sample	Rate of Enzyme Activity (0–5)
Step B—without Hydroxylamine	
Step N—with Hydroxylamine	

5. _____

6. _____

7. _____

_____

8. _____

9. _____

10. Enter your data on Chart I.

11. _____

12. _____

13. _____

14. Enter your data on Chart II.

15. Make a graph on a separate piece of paper.

16. _____

17. _____

_____

18. _____

19. Enter your data on Chart III.

20. Enter your data on Chart III.

21. Make a graph on a separate sheet of paper.

22. _____

23. _____

24. Enter your data on Chart IV.

25. _____

_____

26. _____

27. _____

28. _____

_____

29. _____

_____

Investigation

5 | *Cells*

What do animal and plant cells look like?

Learning Objectives
- To examine characteristics common to both animal and plant cells.
- To identify the differences between animal cells and plant cells.

Process Objectives
- To observe the microscopic characteristics of eukaryotic cells.
- To infer the functional significance of specific cell structures.
- To classify several unknown specimens as plant or animal.

Materials

For Class
- Onion
- Iodine solution
- Sprigs of *Elodea*

For Group of 2
- Forceps
- Microscope slide
- Transparent plastic ruler with millimeter scale
- Straight-edged razor blade or scalpel
- Flat-edged toothpick
- Coverslip
- Compound microscope
- Paper towel
- Prepared slide of mouth epithelial cells
- Prepared slide of human blood
- Prepared slides of unknown specimens

Typical Epidermal Tissue

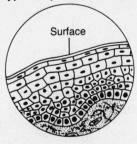

Introduction

Cells are the basic functional units of all living organisms. They may exist singly or in aggregates. When cells join together to take on a specialized function within a larger organism, they form a **tissue.**

There are 2 major divisions into which all cells fall—**prokaryotic** (organized nucleus absent) and **eukaryotic** (organized nucleus present). Bacteria make up the former division while the cells of plants, animals, fungi, protozoa, and algae comprise the latter.

Animal and plant cells share many characteristics, which you will observe in this lab. They also differ in several important ways. In this lab, you will use these similarities and differences to classify unknown specimens.

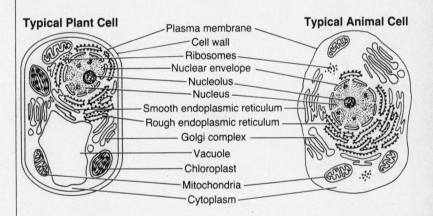

Typical Plant Cell — Plasma membrane — Cell wall — Ribosomes — Nuclear envelope — Nucleolus — Nucleus — Smooth endoplasmic reticulum — Rough endoplasmic reticulum — Golgi complex — Vacuole — Chloroplast — Mitochondria — Cytoplasm — Typical Animal Cell

Prelab Preparation

Review the procedure on how to use a microscope given in Investigation 2 and read Chapter 5 in your textbook.

Both animal and plant cells may occur **unicellularly** or within **multicellular organisms.** Because they often take on specific functions within tissues, animal cells are frequently more specialized than plant cells. **Epithelial** (EP-uh-THEE-lee-ul) cells and blood cells are examples of different tissues.

In this lab, you will look at epithelial cells in both plants and animals. Epithelial cells form the skin of the outer body surfaces and the linings of the inner surfaces. These cells are specialized for transportation of substances and protection. The individual cells of these layers may be shaped like cubes, columns, or be flat—depending on their location and function.

Blood cells, although also common in higher animals, appear very different from other cells due to their specialization for transporting oxygen or fighting infection. Blood is a tissue in which the cells are maintained and transported by a liquid **plasma.** Red blood corpuscles, also known as **erythrocytes** (ih-RITH-ruh-SITES), contain hemoglobin, a protein that transports

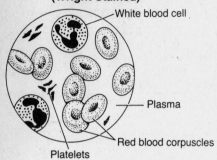

**Typical Blood Cells
(Wright Stained)**

White blood cell

Plasma

Red blood corpuscles

Platelets

oxygen through the body. As an erythrocyte matures, the nucleus and organelles of the cell diminish. Eventually they are forced out of the cell. White blood cells, also known as **leukocytes** (LOO-kuh-SITES), are a diverse collection of cells. **Platelets** are small cell fragments that are important in blood clotting.

1. What is the basic structural feature that distinguishes plant and animal cells from bacteria?
2. How is the degree of specialization of cells in an organism related to the diversity of cell types in that organism?
3. In what way do blood cells resemble unicellular organisms? Why are they then considered a tissue?
4. Can mature erythrocytes be classified as cells? Explain.

Procedure

Part I: Plant Cells

Onion bulbs are organized tissue that, under the appropriate conditions, will give rise to an entire plant. The curved pieces that flake away from a slice of onion are called **scales**. On the underside of each scale is a thin membrane called the **epidermis**.

A. Obtain a piece of onion and remove one of the scales from it. Use forceps to pull away the epidermis from the inner surface. Be careful not to wrinkle the membrane. In a drop of water on a microscope slide, cut a piece of membrane about 0.5 cm square with a straight-edged razor blade or scalpel. CAUTION: Handle the razor blade or scalpel with care. Use a toothpick to straighten out wrinkles. Prepare a wet mount as you did in Investigation 2.

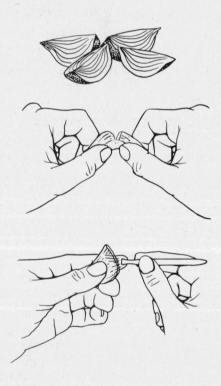

B. Examine the epidermis first with the low power objective of your microscope. Unstained specimens are often seen better with less light. Try reducing the illumination by adjusting the diaphragm. Estimate the size of the cells that you are looking at; refer to Part II of Investigation 2 for instructions on measuring.

5. How many layers thick is the epidermis?
6. What is the general shape and size of a typical cell?

C. To stain your specimen, remove your slide from the microscope stage. Place a drop of iodine on one side of the coverslip. CAUTION: Iodine is toxic. Draw the fluid from underneath with a scrap of paper towel on the opposite side of the coverslip. The stain will be drawn under the coverslip.

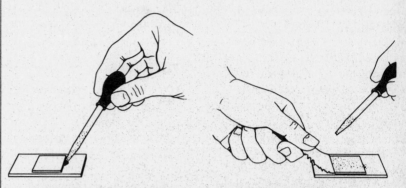

D. Examine your specimen under both low and high power.
7. What does the nucleus look like under low and high power?
8. Within an individual cell, where are the cytoplasm and the nucleus found? What general characteristic of plant cells can be inferred from observations of the cytoplasm and nucleus?
9. Make a diagram of several cells as observed under high power. Label the following structures in one cell: nucleus, cell wall, central vacuole, cytoplasm.

E. Obtain a single leaf of *Elodea* (from the young leaves at the tip) and prepare a wet mount of it.
10. What does *Elodea* look like under low power?

F. Examine the chloroplasts under high power.
11. What does a single chloroplast look like?
12. Are the chloroplasts moving or stationary? Make an inference to explain this.
13. In what ways are the cells of onion epidermis and *Elodea* similar? Different?
14. What observable characteristics can be used as evidence for classifying a specimen as a plant?

Part II: Animal Cells

As you observe the different types of cells, think of how differences in cell structure might be related to specialization of cell function.

G. Obtain a prepared slide of epithelial cells from the human oral cavity. Examine the cells first under low power and then under high power.
15. Inside the mouth, these cells are joined together in a sheet. Why are they scattered here?
16. Draw a few cells and label the cell membrane, nucleus, and cytoplasm.

H. Obtain a prepared slide of blood and examine it under low and high power.

This specimen has been treated with Wright Stain, which causes erythrocytes to appear pink. About 1% of the pink-stained structures are **reticulocytes** (rih-TIK-yuh-loe-SITES), immature erythrocytes that still have their nuclei.

Strategy for Observing
Although animal and plant cells each contain organelles that are particular to that cell type, many of these are too small to see with a light microscope. When determining cell type, only those structures or characteristics that you can actually see should be considered.

The nuclei in these cells will be stained deep blue or purple and the **endoplasmic reticulum** will look like a blue mesh. Leukocytes appear blue and platelets are violet or purple.

17. How many different cell types can you see?
18. In what ways are these cells similar to each other and to the epithelial cells you observed? How are they different?
19. Were you able to observe other organelles in animal cells? Explain.

I. Obtain 3 slides of unknown specimens and examine each under low and high power.

20. Fill in the data table for your unknowns.

Postlab Analysis

21. What did you observe in animal cells but not in plant cells?
22. In what observable ways are animal and plant cells structurally similar?
23. How are they different?
24. How does the structure of the cells that you observed relate to the function of each type of cell?
25. Onions are classified as green plants. Where in the onion plant are the green cells located?

Further Investigations

1. Obtain other specimens from different plants and from different parts of plants, such as the roots or stem. Locate the structures that are common to all plants or to a single plant. Try to relate any differences you observe to special functions that the cells may have.
2. Look up in a histology book the different types of cells that exist in our bodies. If your teacher has slides that contain examples of different tissues, examine them for the features described in the reference book.
3. Look in an immunology book for the different types of white blood cells in our bodies. What is the function of each type? Try to identify each type on your slide. If any slides are available that contain examples of white blood cell abnormalities, examine them as well.
4. Think about cell structures that you were unable to see with light microscopy. Use library resources to locate electron micrographs of these structures.

Strategy for Classifying

As you examine your specimens, look for the characteristics that you observed in animal and plant cells. How are the specimens different? How are they alike?

Strategy for Inferring

To infer the function of a specific structure, think of the cell in its natural location. Based on what you know about the organism, how might that structure help the organism to survive and flourish?

Investigation

5 | *Cells*

1. _____

2. _____

3. _____

   _____

4. _____

   _____

5. _____

6. _____

7. _____

8. _____

9.

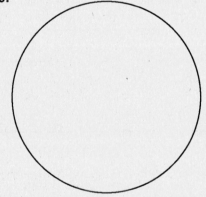

10. _____

   _____

11. _____

12. _____

13. _____

   _____

14. _____

15. _____

   _____

16.

17. _____

18. _____

_____

19. _____

20. Enter your answers on the data table.

Classification of Unknowns		
Unknown Number	Classification (Animal or Plant)	Reasons for Classification
1		
2		
3		

21. _____

22. _____

23. _____

_____

24. _____

_____

_____

25. _____

6 *Osmosis*

How do atoms and molecules move in and out of cells?

Learning Objectives

- To understand the role of osmosis in the movement of molecules.
- To learn how different kinds of solutions move through selectively permeable membranes.

Process Objectives

- To design an experiment to investigate osmosis.
- To predict experimental results from previous knowledge.
- To observe osmosis by using a selectively permeable membrane.

Materials

For Group of 4
Parts I and II
- Metric balance

Part I
- Distilled water
- 50% Glucose solution
- Experimental solutions I and II
- 4 15-cm Pieces of presoaked dialysis tubing (25 mm flat width)
- 8 Rubber bands
- Small funnel
- 4 Test tubes
- 4 500-mL Beakers
- Paper towels
- Labeling tape
- 25-mL Graduated cylinder
- 12 Test tubes

Part II
- Apron
- Safety goggles
- Raw egg
- 200-mL Beaker
- 500-mL Beaker
- 2 400-mL Beakers
- 1 M Hydrochloric acid
- Karo syrup, undiluted
- Tablespoon or tongs
- Paper towels
- Grease pencil or marker

Introduction

In all cells, survival depends on the ability to maintain homeostasis by regulating the movement of molecules across the cell membrane. In osmosis, water molecules move across a selectively permeable membrane toward a solution with a higher concentration of solutes. The molecules continue to move until equilibrium is reached (that is, equal concentrations of water molecules and of solute molecules on either side of the membrane).

The concentration of solutes in water is low or nonexistent. Therefore, water is **hypotonic** to the fluid within the corpuscles. If human red blood corpuscles are placed in water instead of blood plasma, water will enter the corpuscles until they swell, burst, and die. If the corpuscles are placed in a solution with a high concentration of solute compared to the corpuscles, water will pass through the cell membranes into the surrounding **hypertonic** solution, and the corpuscles will shrink, or plasmolyze.

The selectively permeable membranes you will use to study osmosis—dialysis tubing and the membrane within an eggshell—allow water to pass through in response to concentrations of the solutions on either side of the membrane. The water will move toward the more hypertonic, or highly concentrated, solutions.

Prelab Preparation

Diffusion is the process in which molecules spread throughout a space until they are equally distributed. Diffusion and osmosis are two passive processes that cells use to regulate the movement of molecules into and out of the cells. In diffusion, molecules spread throughout a space until they are equally distributed. A closer look at the factors involved in osmosis will give you a better understanding of how cells maintain homeostasis.

Carefully review Chapter 6 in the textbook. Be sure to read completely through the Procedure before you start.

1. What is the difference between diffusion and osmosis?
2. When are molecules in a state of equilibrium across a selectively permeable membrane?

Design your own osmosis experiment for dialysis bag 4 in Part I. You may use solutions your teacher provides or you may use experimental variables you think of, or that are suggested by your teacher.

3. Record your experimental design and ask your teacher to review it before you start.

Procedure

Your team of 4 will be divided as follows: one recorder/coordinator, one "chemist," one dialysis bag assembler, and one weigher. You must cooperate as a team and divide up the work before you begin.

Part I: Dialysis Bags

A. Place 10 mL of each of the following solutions in appropriately labeled test tubes:
   1. distilled water
   2. 50% glucose
   3. distilled water
   4. your experimental solution I

B. Obtain 4 pieces of presoaked dialysis tubing. Close off the end of each piece by wrapping a cut rubber band around a few times and tying securely. NOTE: Tight knots will prevent leaks. Using a small funnel, add to each piece of tubing solution from one test tube. Close the other end of each tube securely with a cut rubber band. NOTE: Make sure there are no air bubbles, but plenty of slack, in the tubing. Always place each dialysis bag on its own labeled paper towel to ensure proper identification. Weigh each bag separately to the nearest 0.1 gram. The same person should do all the weighings.

   4. Record the initial weights of the filled dialysis bags in your Part I data chart.

C. Label 4 empty 500-mL beakers with the solutions in the 4 dialysis bags. Fill the beakers with 350 mL of the following solutions and label with these solutions also:
   1. distilled water
   2. distilled water
   3. 50% glucose solution
   4. your experimental solution II

   5. From what you know about osmosis, predict the outcome for each dialysis bag in Part I and record your predictions in the data chart.

   6. What is the purpose of the dialysis bag containing only water in a beaker of water?

D. Place the dialysis bags in the appropriate 500-mL beakers. Note the time when you begin.

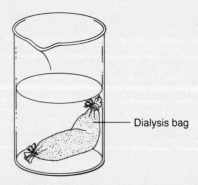

Dialysis bag

E. Weigh the dialysis bags at 10-minute intervals: remove the tubing from the beakers, rinse, and blot off excess water. Carefully weigh each bag. Remove all bags at the same time and return them to their beakers at the same time for another 10 minutes. Remove, weigh, and return once more for 10 minutes. Do not replace the bags after the last weighing.

7. Record the masses at each weighing so you have masses for 10, 20, and 30 minutes in your data chart.
8. Based on the changes in mass that you have recorded on your data chart, in which of your 4 setups did osmosis occur?
9. In each case of osmosis, was the net flow of water into the tubing or out of the tubing?

Part II: Raw Egg

F. Determine the mass of a raw egg to the nearest 0.1 g.

10. Record this as the mass of raw egg and shell in the Part II data chart.

G. Put on your safety goggles and apron. Place a 500-mL beaker on several sheets of paper toweling. Carefully place 300 mL of HCl solution in the beaker. CAUTION: HCl will burn if spilled on your skin or clothing.
H. Using a tablespoon or tongs, carefully place the egg into the HCl solution. Place a 200-mL beaker half full of water over the egg to keep it submerged. Wait 15 minutes.

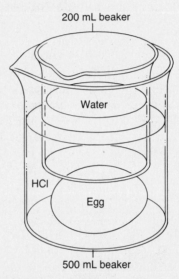

200 mL beaker

Water

HCl

Egg

500 mL beaker

11. Record your observations in your Part II data chart.

Over the sink, remove the egg with a spoon and rinse it well under tap water. Dispose of the HCl solution as directed by your teacher. CAUTION: Be careful not to spatter the acid as you remove the egg and dispose of the acid. Reweigh the egg.

12. Record the mass of the egg in your data chart.

I. Place the egg in a 400-mL beaker. Pour water into the beaker so that the egg is completely covered. Wait 20 minutes.

13. Record observations in your data chart.

Remove the egg from the beaker using a spoon. Reweigh the egg.

14. Record the egg's mass.

J. Now place the egg in another 400-mL beaker and cover it with Karo syrup. Label the beaker with your team's names.

15. Observe the egg in the syrup at the end of the period, then store it overnight in the beaker.

K. The next day, remove the egg from the Karo syrup, rinse, and blot.

16. Enter observations and record the mass of the egg in your data chart.

Strategy for Observing
Recall the definition of *selectively permeable membrane* from Chapter 6 in the textbook. Think about what molecules are passing through the egg membrane as you make your observations.

Postlab Analysis

17. Based on the changes in mass in Part I, how did your predictions compare with your results?
18. In view of your results, how would you modify the experiment you designed in Part I?
19. To what biological structure is the dialysis bag comparable? In what ways are they similar? Different?
20. In the tests where osmosis occurred, look at the differences in mass for 10, 20, and 30 minutes. What conclusion can you draw about the rate of osmosis?
21. In Part II, what effect did the acid have on the egg?
22. What passive process of moving molecules is demonstrated in Part II? In which instance is the egg in a hypertonic solution? In which instance is it in a hypotonic solution?

Further Investigations

1. Using the technique for dialysis tubing from Part I you can determine whether glucose, protein, or starch can move through a selectively permeable membrane. Be sure to test the solutions left inside the dialysis bags as well as the solutions in the beakers. Your teacher can give you the procedures using Benedict's solution to test for glucose, Biuret reagent to test for protein, and iodine solution to test for starch.
2. Place approximately 5 mL of actively growing (24 hours) yeast culture into each of 2 test tubes. Boil one tube for 10 minutes. Set it aside to cool and label it. Add 10 drops of Congo red to the contents of both tubes. Prepare and label slides of the culture from each tube, add coverslips, and examine under a microscope. Draw diagrams and record your observations. From what you have learned about selectively permeable membranes in cells, explain your observations.

Investigation

6 | *Osmosis*

1. _____
_____
_____

2. _____

3. _____
_____
_____

4.–5. Enter your answers in the data chart.

Part I – Dialysis Bags

Contents Dialysis Bag	Contents of 500-mL beaker	Prediction	Beginning Mass	Mass after 10 min.	Mass after 20 min.	Mass after 30 min.
1.						
2.						
3.						
4.						

6. _____
_____

7. Enter your answers in the data chart.

8. _____
_____

9. _____
_____

10.–16. Enter your answers in the data chart.

Part II – Egg

Mass of raw egg and shell	Observations during acid bath	Mass of "acid" egg	Observations during water bath	Mass of egg after water bath	Initial observations of egg in Karo syrup	Final observations of egg after overnight in syrup	Final mass of egg

17. _____

18. _____

19. _____

20. _____
21. _____

22. _____

7.1 | *Chromatography*

Do all leaves contain the same pigments?

Learning Objectives

- To learn that the color of a leaf is caused by several different pigments found in the leaf's cells.
- To learn the scientific technique of paper chromatography and to use it to separate a mixture of leaf pigments.

Process Objectives

- To observe the pigments that give a leaf its color.
- To analyze data obtained from paper chromatography.

Materials
For Group of 2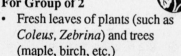

- Fresh leaves of plants (such as *Coleus, Zebrina*) and trees (maple, birch, etc.)
- Fresh or thawed leaves of spinach
- Filter paper or chromatography paper, about 1 cm x 15 cm
- Scissors
- Test tube, 15-cm length or larger
- Paper clip
- Cork (to fit test tube)
- Mortar and pestle
- Sand (optional)
- Safety goggles
- 5 mL Ethyl alcohol
- 10-mL Graduated cylinder
- Glass stirring rod
- 5 mL Chromatography solvent (92 parts petroleum ether to 8 parts acetone)
- Test-tube rack
- Ruler

Introduction

Chlorophyll often hides the other pigments present in leaves. In autumn, chlorophyll breaks down, allowing xanthophyll and carotene, and newly made anthocyanin, to show their colors.

   The mix of pigments in a leaf may be separated into bands of color by the technique of **paper chromatography,** shown in this exercise. Chromatography involves the separation of mixtures into individual components. **Chromatography** means "color writing." With this technique the components of a mixture in a liquid medium are separated. The separation takes place by absorption and capillarity. The paper holds the substances by absorption; capillarity pulls the substances up the paper at different rates. Pigments are separated on the paper and show up as colored streaks. The pattern of separated components on the paper is called a chromatogram.

Prelab Preparation

Gather leaves from several different plants. CAUTION: Avoid poisonous plants. Autumn leaves from deciduous trees are especially interesting. Sort the leaves by kind (maple, etc.) and color. Bring the leaves to the lab.

   1. List the kinds of leaves you have collected and their colors.
   2. List the names and colors of various plant pigments. Refer to text Chapter 7, Section 7.2 if you need to.

Review the diagram of a plant cell in Chapter 7 of your textbook. Find the **grana** in the **chloroplasts** of the cell.

   3. What is the function of grana?
   4. Where are plant pigments located?
   5. How does location of a pigment relate to its function?

Procedure

Leaves brought in by all students should be grouped by kind (maple, etc.) and color. Each team of 2 students will work with a spinach leaf and with one other type. Your teacher will coordinate this. Results will be shared with the rest of the class. CAUTION: Chromatography solvents are flammable and toxic. Have no open flames; maintain good ventilation; avoid inhaling fumes.

A. Cut a strip of filter paper or chromatography paper so that it just fits inside a 15-cm (or larger) test tube. Cut a point at one end. Draw a faint pencil line as shown in the picture. Bend a paper clip and attach it to a cork stopper. Attach the paper strip so that it hangs inside the tube, as shown. The sides of the strip should not touch the glass.

B. Tear a spinach leaf into pieces about the size of a postage stamp. Put them into a mortar along with a pinch or two of sand to help with grinding. Add about 5 mL ethyl alcohol to the leaf pieces. Crush leaves with the pestle, using a circular motion, until the mixture is finely ground.

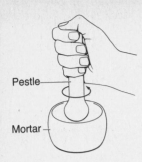

Pestle

Mortar

Strategy for Observing

Some color streaks may be very light and cannot be seen well until the paper dries.

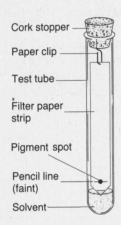

Cork stopper

Paper clip

Test tube

Filter paper strip

Pigment spot

Pencil line (faint)

Solvent

Strategy for Analyzing

Look down the "colors from top" column of the data table. Compare each box. Are the pigments always listed in the same order? Look at the column of R_f values. Are the numbers in each box in increasing or decreasing numerical order?

The liquid in which the leaf pigments are now dissolved is called the **pigment extract.**

C. Use a glass rod to touch a drop of the pigment extract to the center of the pencil line on the paper strip. Let it dry. Repeat as many as 20 times to build up the pigment spot. NOTE: You must let the dot dry after each drop is added. The drying keeps the pigment dot from spreading out too much.

6. Why is it important that the pigment spot be heavy and dark?

D. Pour 5 mL chromatography solvent into the test tube. Fit the paper and cork assembly inside. Adjust it so that the paper point just touches the solvent (but not the sides of the tube). The pigment dot must be above the level of the solvent. Watch the solvent rise up the paper, carrying and separating the pigments as it goes. *At the instant* the solvent reaches the top, remove the paper and let it dry. Observe the bands of pigment. The order, from the top, should be carotenes (orange), xanthophylls (yellow), chlorophyll b (yellow-green), chlorphyll a (blue-green), and anthocyanin (red). Identify and label the pigment bands on the dry strip. Write your names on the strip as well.

7. Record the colors and pigment names in the data chart.

E. Wash the mortar and pestle thoroughly, using a little alcohol to remove any remaining pigment. If you have enough time, repeat Steps A through D using the second leaf type assigned for you.

Each pigment has an R_f value, the speed at which it moves over the paper compared with the speed of the solvent.

$$R_f = \frac{\text{Distance moved by the pigment}}{\text{Distance moved by the solvent}}$$

F. Measure the distance in cm from the starting point to the center of each pigment band. Then measure the entire distance traveled by the solvent. Do this for both chromatograms you made.

8. Calculate the R_f value as a decimal fraction for each pigment, on both chromatograms. Record the R_f values in the data chart.

G. Compare your results for both leaves with those of other students.

9. Are the R_f values for all chromatograms constant for each pigment type, regardless of the species?

Postlab Analysis

10. Based on the class data, which kind of leaf showed the most pigments? For this leaf, which pigments were most visible in the chromatogram? Are these the pigments responsible for the leaf's original color? Which pigments were least visible in the chromatograms? Are these pigments obvious in the leaf's original color?

11. Did the leaves contain any pigments that surprised you?

Further Investigations

1. Paper chromatography may be used to separate other mixtures into their parts. Name some of these mixtures.

2. Plants are an important source of pigments used in making dyes and paints. Use the library to find out some of these plants and the dyes they are used to make.

Investigation

7.1 *Chromatography*

1. _____

_____

2. _____

_____

_____

3. _____

_____

4. _____

_____

5. _____

_____

_____

_____

6. _____

7.–8. Enter your answers on the data chart on the next page.

9. _____

_____

10. _____

_____

_____

_____

11. _____

_____

Chromatography Data Chart

	Leaf (Species)	External Color	Chromatogram Pigments		
			Colors from Top	Pigment Names	R_f Values
Our Work					
	spinach				
	(other)				
Other Students' Work (for comparison)					
	1				
	2				
	3				

7.2

Fermentation

How do you make sauerkraut?

Learning Objectives
- To trace the transformation of cabbage to sauerkraut.
- To study fermentation.

Process Objectives
- To observe the presence of bacteria in the brine as cabbage is converted to sauerkraut.
- To infer the reactants and products of fermentation.
- To organize data into a graph showing changes in pH over time.

Materials

For Group of 2
For Parts I and II
- pH paper
- Microscope
- 2 Microscope slides
- 2 Coverslips
- Methylene blue in dropper bottle
- 2 Toothpicks, flat

Part I
- Plastic container with tightly fitting cover
- Shredded cabbage, enough to fill the container
- 2.5% NaCl (salt) solution, enough to fill the container

Part II
- Container of fermenting cabbage from Part I
- 2.5% NaCl (salt) solution
- 2 Sheets graph paper

Introduction

Have you ever seen what happens when bread rises or when apple cider "goes hard"? These actions are the result of **fermentation,** the process by which an organism breaks down glucose in the absence of oxygen. During this process, 2 molecules of **adenosine triphosphate (ATP)** are made from a glucose molecule (instead of 36 made in the presence of oxygen), and a variety of by-products are produced. Fermentation is important in making many foods, and it produces the alcohol found in beer and wine. In this lab, you will create conditions in which cabbage will become fermented in the presence of certain bacteria and will turn into sauerkraut.

Prelab Preparation

Read Sections 7.1 and 7.3 in your textbook. You should understand how glucose is broken down in the absence of oxygen. When no oxygen is present, **nicotinamide adenine dinucleotide (NAD)** donates the 2 hydrogen (H) atoms it picks up in **glycolysis** to pyruvic acid, the end product of glycolysis. The fate of pyruvic acid depends on the organism. Humans have an enzyme that combines pyruvic acid and 2H to form lactic acid. Some bacteria and yeast form alcohol and carbon dioxide. Others form acetic acid (vinegar) and carbon dioxide. Still others form a variety of chemicals depending on enzyme specificity.

(a) pyruvic acid   lactic acid
$$CH_3COCOOH + 2H \longrightarrow CH_3CHOHCOOH$$
(b) pyruvic acid   ethyl alcohol
$$CH_3COCOOH + 2H \longrightarrow CH_3CH_2OH + CO_2$$
(c) pyruvic acid   acetic acid
$$CH_3COCOOH + 2H \longrightarrow CH_3COOH + CO_2$$

> **1.** When does the human body produce lactic acid?
> **2.** Why is the CO_2 given off during alcoholic fermentation important for making bread?

Procedure

Part I
A. Pack a plastic container with shredded cabbage.
B. Cover with 2.5% NaCl (salt) solution until the container is full.
C. Cover the container tightly. For the next 2 weeks, make sure the container remains full. Replenish with more NaCl solution as needed.

> **3.** For Day 0 in the chart, record the color and texture of the cabbage, the odor, whether any gases are produced (check for bubbles), and the pH of the solution (use pH paper). Repeat these observations and recordings every school day.

Strategy for Observing

Draw your sketches and make your estimates while you are observing the bacteria—do not wait until after you have finished.

Model Graph of Brine pH

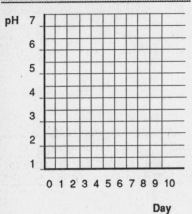

Strategy for Organizing

To organize your data on pH changes into a line graph, review your chart to see how many values for pH you have and what the range is for those pH values. Then plot the points of your graph for each day and draw a line to connect the points.

Strategy for Inferring

To make inferences about gas production and its cause, think about what you have read in your textbook and observed in this lab. List statements that might explain what you have seen in this lab, and choose the most likely one.

On the first day (Day 0) and the last day of the experiment, check for the presence of the bacteria and the condition of the cabbage cells.

Since the day you set up the experiment is designated Day 0, Day 1 occurs after the cabbage has been incubating for one day. These observations should take just a few minutes at the beginning of each class.

D. To check the condition of cabbage cells, make a wet mount of a few cells peeled from the lower epidermis. Stain these cells using methylene blue. Examine under the microscope using low and high power. Look carefully to see if the cell membrane has separated from the cell wall.

   4. Why would 2.5% NaCl solution cause such a separation?

E. To check for the presence of bacteria, make a wet mount by gently scraping the bottom of the cabbage with a toothpick and putting it on a slide with a drop of brine. Stain with methylene blue and examine under high power.

Part II

F. Make daily observations and recordings until no more gas bubbles are produced. This should take about 2 weeks. Remember to replenish with more NaCl solution as needed. On the last day prepare slides for microscopic observation by repeating Steps D and E.

   5. Describe any changes over the course of the experiment in the odor, color, and the texture of the cabbage and in the pH of the brine.
   6. Does the number of bacteria increase between the first and last days of the experiment?
   7. Sketch any bacteria you see. Estimate how many you see.

G. You and your teammate should each make your own graph showing the pH from Day 0 to Day 10 (last day).

   8. Make your graph on separate paper.
   9. Describe any changes in the appearance of the cabbage cells. How might the hypertonic brine and the enzyme activity of the bacteria explain these changes?

Postlab Analysis

The bacteria growing **anaerobically** (without air) in the brine will convert glucose to lactic acid, carbon dioxide, acetic acid, ethyl alcohol, mannitol, dextran, and esters. These combine to produce the characteristic odor and flavor of sauerkraut.

   10. Why do you think you soaked the cabbage in brine (2.5% NaCl) instead of water? What do you think would have happened if you had used water instead of brine?
   11. What do you think caused the gas production?
   12. What happened to the pH over the course of the experiment? What do you think caused this change in pH?
   13. Which equation in the Prelab Preparation section could account for the odors produced during this experiment?
   14. Why did you keep the container tightly covered at all times?

Further Investigations

1. Yogurt is made by different bacteria from those in the sauerkraut. You may make yogurt using commercial yogurt cultures and fresh milk. Look for changes in pH, odor, texture, and the number of bacteria
2. Observe how yeast makes bread rise. Look for changes in odor, texture, and the number of yeast for CO_2 production and for changes in pH.

Investigation

7.2 | *Fermentation*

1. _____

_____

_____

_____

2. _____

3. Enter your answers on the data chart.

Fermentation Data Chart

	Day 0	Day 1	Day 2	Day 3	Day 4	Day 5	Day 6	Day 7	Day 8	Day 9	Day 10 (or last day)
Date											
Color of Cabbage											
Texture of Cabbage											
Odor											
Any Gases Produced?											
pH											
Condition of Cabbage Cells											
Presence of Bacteria in Brine											
Description of Bacteria in Brine											

4. _____

_____

5. _____

_____

_____

_____

_____

7.

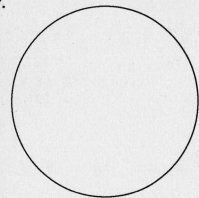

8. Make a graph on a separate sheet of paper.

9. _____

_____

10. _____

_____

_____

11. _____

_____

12. _____

_____

13. _____

14. _____

8 | *Modeling Nucleic Acids*

Learning Objectives
- To recognize the differences between mRNA and DNA.
- To learn how DNA codes for proteins.
- To learn how DNA replicates.

Process Objectives
- To make a model of DNA, mRNA, and amino acid sequences.
- To communicate with classmates in order to form models of mRNA and of DNA.

Materials
For Group of 2–4
- 1 Amino acid pair on two 3" x 5" index cards, assigned by teacher
- 1 Piece of yellow ribbon, 2 cm wide, approximately 24 cm long
- 4 Pieces of blue ribbon, 2 cm wide, approximately 24 cm long
- Velcro strips in 3 colors (e.g. black, beige, and white), a rough strip and a smooth strip for each color.
- Stapler
- Staple remover
- Scissors
- Marking pen

Strategy for Modeling
As you work with your model, think about how it reflects the real structure and how it simplifies the real structure.

How does DNA code for proteins and how does it make copies of itself?

Introduction
The three-dimensional structure of DNA was discovered in 1953 by James Watson and Francis Crick. They accomplished this feat—for which they were awarded the Nobel Prize—by building a model that showed the nature of the DNA double helix. The data on which they based their model was obtained from **x-ray crystallography** studies and research on the chemical properties of DNA. With these data, they assembled the first model of the molecular structure of DNA.

In this lab, you will use strands of ribbon and strips of Velcro to create models of the structures of DNA and RNA. Note that the normal flow of genetic information in cells is:

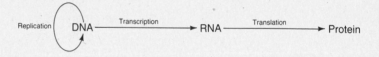

However, in this lab you will reverse this process by beginning with amino acids, then making mRNA (messenger RNA) and DNA. Finally, you will model the replication of DNA.

Prelab Preparation
Read Chapter 8 in your textbook to help prepare for this investigation. Be sure you bring your textbook with you to the laboratory.

The DNA in every living cell is involved in 2 very important processes: **replication** and **gene expression.** Replication occurs during cell division; the double-stranded DNA is unzipped and an exact copy of both strands is made. This results in a duplication of the original DNA and allows for the transfer of genetic information to daughter cells.

When a gene (the message-containing part of DNA) is expressed, one strand of DNA is copied into a molecule of mRNA through a process called **transcription.** This mRNA then migrates from the nucleus into the cytoplasm. In the cytoplasm, the mRNA is used to create a string of amino acids, a peptide, through the process **translation.**

1. Why are 2 different colored ribbons used to represent the DNA and RNA backbones?
2. What are other differences between RNA and DNA?
3. How is RNA produced from DNA?

Procedures

Part I: mRNA Model

You will use different color strands of ribbon to represent the RNA and DNA backbones and Velcro strips to represent the nucleotide bases. The rough and soft sides of the Velcro will be joined together to represent base pairs. The color designations for the bases are shown in the table.

DNA Model Velcro Designations

Base	Color	Rough or Soft Texture
Purines		
Adenine (A)	Black	Rough
Guanine (G)	Black	Soft
Pyrimidines		
Thymine (T)	Beige	Soft
Cytosine (C)	Beige	Rough
Uracil (U) In mRNA only	White	Soft

A. Your class will be divided into groups of 2–4. Your teacher will give each team two 3″ × 5″ index cards taped next to each other. The name of 1 amino acid will be written on each card. Each group will then construct a fragment of mRNA that codes for the assigned amino acids.
Use Table 8–1 on page 121 in your textbook to determine what nucleotide bases would be found in your team's fragment of mRNA. If there is more than one choice, select only one.

4. Record the codons that correspond to your 2 amino acids.

B. Cut a piece of yellow ribbon 24 cm long. Using a marker, draw an upper case "P" within a circle every 4 cm. These represent the phosphate groups that link the nucleotides in the backbone of the RNA.

C. Determine the colors and textures of Velcro you will need to construct the model mRNA that corresponds to your amino acids. Your team should cut six 3-cm x 2-cm rectangles from the appropriate strips of Velcro. Label each piece with the appropriate letter for the nucleotide base. Staple the 6 bases, with its textured surface facing you, to the yellow ribbon. The yellow ribbon represents the mRNA backbone.

5. Record the color and texture of the Velcro pieces you used.

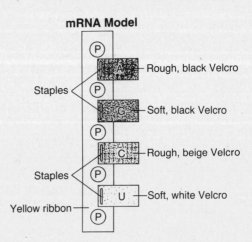

mRNA Model

Staples

Rough, black Velcro

Soft, black Velcro

Rough, beige Velcro

Staples

Soft, white Velcro

Yellow ribbon

Part II: DNA Model

D. Next, you will construct a model of the single strand of DNA from which your mRNA would have been copied by transcription.

6. Based on which bases bind to each other, record the sequence of nucleotide bases for the corresponding fragment of DNA.

E. Use the Velcro designation table to determine the colors and textures of Velcro needed to form the DNA strand.

7. Record the colors and textures of your sequence of Velcro pieces.

Cut 2 pieces of blue ribbon 24 cm long and carefully mark them with phosphate groups as you did in Step B. Cut the Velcro rectangles identified in step E, label them as you did in Step C, and staple them to one of the blue ribbons which represents the DNA backbone. After this model DNA strand has been completed, hook the DNA bases to the corresponding mRNA bases. This joining represents the bonds between DNA and mRNA during the process of transcription.

8. Where would this process occur in the cell?

F. Since DNA is a double-stranded nucleic acid, your team must now make the DNA strand complementary to the one that you have already made in Step E. This strand should be the same as the mRNA strand except that any uracil bases on the mRNA will now be thymine bases on the DNA.

9. Record the nucleotide base sequence for this complementary strand of DNA.

10. Record the colors and textures of Velcro that will form the complementary DNA.

G. After team members have cut the appropriate squares of Velcro and stapled them to the second blue ribbon, hook the strands of DNA together.

11. What type of bonds are represented in this model by the binding of the Velcro pieces to each other?

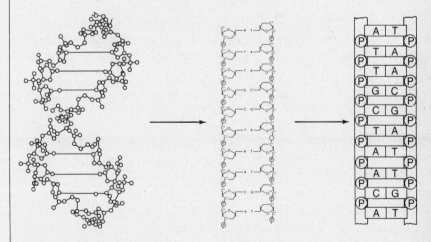

Part III: DNA Replication

H. You should now have a double-stranded stretch of DNA, 6 base pairs long. Since one of the important functions of DNA is its ability to replicate itself, unfasten the base pairs bond by bond. Construct 2 new strands of DNA, each complementary to one of the strands that you already have. Determine the colors and textures of Velcro needed and construct the 2 new strands, using the same methods as in Step G and 2 more pieces of blue ribbon. Hook them to your 2 original DNA strands to form 2 double strands of DNA.

12. How do the 2 pieces of DNA made in Step H compare to the piece of DNA made in Step G?

Part IV: A Model Gene

I. On a long lab bench, staple one of each team's double strands of DNA together to form a long stretch of DNA. Do the same for the mRNA from Step C and also for the pairs of amino acids from Step A. You now have a model for the DNA and mRNA that corresponds to a complete gene and a model of the amino acid sequence that corresponds to an entire protein.

13. How many nucleotide base pairs long is your model gene? How many codons does this correspond to? How many amino acids long is the protein coded for by this model gene?

Strategy for Communicating
Discuss the following question with your classmates: How many ways are there to combine the pieces from all the teams into one long model of a nucleic acid? Decide together what order to use for your class. Be sure that all the pieces of DNA strands read in the some direction: from left to right.

Postlab Analysis

14. If you made a model of mRNA being created from a DNA template, what biological process would you be modeling?
15. What biological process makes a peptide by reading the mRNA molecule?
16. What is the process by which DNA is copied?
17. What would be the sequence of a DNA segment that codes for the amino acids: Ala, Pro, Tyr?

Further Investigations

1. In eukaryotic cells, mRNA does not code for proteins immediately after being transcribed from the DNA. Before it gets to the ribosomes, certain regions are clipped out and the mRNA is thus shortened. Use your model of mRNA to demonstrate this phenomenon.
2. Only 20 different amino acids form the myriad of proteins that are present in an organism. As a model for how this apparently small number of different amino acids can produce this diversity, use 6 different colors of pop-it beads (symbolizing just 6 of the amino acids) to construct as many combinations as possible.
3. Use library materials to learn more about how important modeling was in determining the structure of DNA. One resource is James Watson, *The Double Helix,* Atheneum, 1968.

Investigation

8 | *Modeling Nucleic Acids*

1. _____

_____

2. _____

_____

_____

_____

_____

3. _____

_____

4. _____

5. _____

_____

_____

6. _____

7. _____

_____

_____

8. _____

9. _____

10. _____

_____

_____

11. _____

12. _____

_____

13. _____

_____

_____

14. _____

15. _____

16. _____

17. _____

9.1 *Cell Size*

Learning Objectives
- To measure the rate of diffusion in agar.
- To calculate ratios of surface area to volume.

Process Objectives
- To infer from collected data how the size and shape of a cell affects passage of molecules in and out.
- To model the diffusion of solutions through the cell membrane.

Materials
For Group of 2 or 3
- Surgical gloves
- Agar block containing bromophenol blue
- Plastic spatula or spoon
- Glassplate
- Metric ruler
- Scalpel or single-edged razor blade
- 250-mL Beaker
- 100 mL 0.2 N Hydrochloric acid
- 4–6 Sheets of white paper

Strategy for Modeling
As you work with the model, discuss with your team members the different elements of the model and how they each represent the actual cellular structure and process.

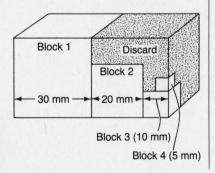

Block 1   Discard
Block 2
← 30 mm → ← 20 mm → ←
Block 3 (10 mm)
Block 4 (5 mm)

Introduction

In this lab, you will construct a simple model that will allow you to examine the relationship between cell structure and cell function. The structure of a cell relates to its function. The difference in structure of a nerve cell and a skin cell, for example, reflects the difference in their functions.

Increasing cell size affects **passive diffusion**—the movement of small molecules across the cell membrane from a region of higher concentration to a region of lower concentration. In this lab, you will see how increasing cell size imposes certain restrictions on passive diffusion. You will do this by modeling the relationship between diffusion and the ratio of a cell's surface area to its volume.

Prelab Preparation

Passive diffusion results from the tendency of molecules in solution to reach a concentration equilibrium across a semi-permeable barrier such as the cell membrane. This is an inexpensive process in terms of metabolic cost, since no energy or special chemicals are needed from the cell. (Review diffusion in Chapter 6 of your textbook.)

You will model passive diffusion using agar, a gelatinous extract of a red alga, that is saturated with an acid-base indicator. Scraps of agar will be supplied by your teacher. You will place the agar in acid and measure the penetration of the acid into the agar. The movement of the acid through the agar models the diffusion of molecules across a cell membrane. Bromophenol blue, the indicator used in this experiment, is bluish purple at a pH greater than 4, and yellow at a pH less than 4. Therefore, you can tell how far the acid has diffused into the agar by watching the color change of the indicator.

This model is useful because it allows you to observe the process of passive diffusion in a simple structure that is easy to manipulate. For example, the size of the agar cubes (5 mm to 30 mm) is much larger than the size of a real cell, thus allowing you to visualize the process.

1. What is the size of an average human cell?
2. In what ways does the model simplify or ignore the features of real cells?

Procedure

For this lab, work in groups of 2 or 3. Wear surgical gloves when working with the agar.

A. Note the diagram showing how to cut the agar blocks. Do not make any cuts until you understand how the large piece will be partitioned. Use a plastic spatula or spoon to place the agar block on a glassplate. Align the ruler alongside the long edge of the agar. Use the scalpel to cut it in half (two 30-mm pieces). One piece will be Block 1; place it in the empty beaker with the plastic spoon. NOTE: Handle the agar only with a plastic spatula or spoon. Take great care not to scrape or gouge the surface. If necessary, wet it with a little water to help it slip along. The other 3 blocks will be cut from the remaining 30-mm piece. To make Block 2, a 20-mm cube, cut 10 mm from 2 adjoining sides, following the diagram. Then turn the block on its side and cut another 10-mm slice. Make Block 3, a 10-mm cube, and Block 4, a 5-mm cube, from the remaining pieces. When you have finished, have your teacher check your preparation. CAUTION: Be careful when handling the scalpel or razor blade.

B. Place all 4 blocks into the beaker. Make sure that none of the edges touch. Add enough dilute hydrochloric acid to cover. CAUTION: Do not let the acid touch your skin or clothing.

   3. Record the time on a separate piece of paper.

C. Gently lift each block to be sure that all surfaces contact the acid; swirl the beaker occasionally to promote contact with fresh acid. At your teacher's instruction, carefully lift out Block 4. (Lay the cut pieces on white paper to compare colors more easily.)

   4. Record the time on your separate piece of paper.

D. Cut Block 4 down the center. Work quickly but carefully. NOTE: Once you have cut into an agar block that has been soaked in acid, be sure to rinse off and dry the scalpel or razor blade before the next cut.

E. Repeat Step C through Step D for each block in order from smallest to largest (Block 3, Block 2, Block 1).

   5. Measure in millimeters the distance the acid has penetrated into each block, and record on the Class Data Chart. Is the distance the same from all edges of the same block?
   6. Record the time elapsed in minutes on the Class Data Chart.
   7. Calculate the rate of diffusion as millimeters per minute and record. Add your data to the Class Data Chart and record the data from the other teams on your chart.

Postlab Analysis

   8. What was the average rate of diffusion for each block?
   9. The volume of a cube or block is defined by $V = l \times w \times h$; the surface area is $A = 6 \times l \times w$. Calculate the surface area-to-volume ratio for each block and record in the chart.
   10. How does this ratio relate to the rate of diffusion?
   11. Using the distance that the acid traveled into the cube, calculate the percentage of the volume of Blocks 1 and 2 that exhibited diffusion. Do this by calculating the volume where the acid did not reach, and subtract that from the total volume.
   12. If a cell has a small surface area-to-volume ratio, how well do you think molecules can pass in and out of it? Relate this to respiration and excretion at the cellular level.

Strategy for Inferring
Using what you have read in your textbook and what you have learned in this lab, think about the types of molecules that are transported across cell membranes.

Further Investigations

1. What shapes would let you pack the most membrane into a given space? Find examples of cells in the human body that have become more efficient by modifying their surface membranes.

2. Suggest some ways in which the restrictions in cell size have affected the organization of cells in tissues, organs, and systems. Think in terms of processes that are necessary for the survival of organisms.

Investigation

9.1 | *Cell Size*

1. _____

2. _____

    _____

3.–4. Record your answers on a separate sheet of paper.

    5.–7. Record your answers in the data chart.

Class Data — Rate of Diffusion

Group	Block 1 Dist. (mm)	Time (min)	Rate (mm/min)	Block 2 Dist. (mm)	Time (min)	Rate (mm/min)	Block 3 Dist. (mm)	Time (min)	Rate (mm/min)	Block 4 Dist. (mm)	Time (min)	Rate (mm/min)
A												
B												
C												
D												
E												
F												
G												
H												

8. _____

    9. Enter your answers in the data chart.

Surface Area/Volume Ratio

		Length	Width	Height	Surface Area A = 6 x l x w	Volume V = l x w x h	Ratio A/V
Block 1	30 mm cube						
Block 2	20 mm cube						
Block 3	10 mm cube						
Block 4	5 mm cube						

10. _____

_____

_____

11. _____

_____

_____

12. _____

_____

_____

9.2 *Mitosis*

Learning Objectives
- To study actively dividing plant cells through the microscope.
- To use colored clay to construct models of chromosomes.

Process Objectives
- To observe different stages of cell division in plant cells.
- To prepare three-dimensional models of nuclear division.

Materials
Part I
For Class
- Fixative
- Pectin solvent
- Carnoy's solution in dropper bottle
- Toluidine blue O stain in dropper bottle
- Distilled water
- Clear nail polish

For Group of 2
- *Tradescantia* root, 1–3 cm long
- Single-edged razor blade
- 2 Microscope slides
- 1 Medicine dropper or pipette
- Pasteur pipette and rubber bulb
- 2 Pairs of safety goggles
- Test-tube holder or clothespin (clamp type)
- Hot plate
- 1 Plastic coverslip
- Forceps
- 2 Paper towels
- Compound microscope

Part II
For Group of 2
- Modeling clay (2 colors)
- Metric ruler
- Small plastic comb
- 2 Paper towels
- 2 Fine spatulas

How does a cell divide?

Introduction

Before a cell divides, the DNA inside the nucleus is duplicated and packaged for division. It is crucial to have a way for the DNA to be copied and packaged without altering the genetic information contained in the DNA. The DNA is copied during **interphase** and divided equally during **mitosis**.

Cell division occurs in all organisms. In unicellular organisms, it is a form of asexual reproduction. In multicellular organisms, cell division occurs most rapidly where growth is taking place: in embryos, in the roots and stems of sprouting plants, and in healing wounds.

Cell division occurred billions of times in your body to transform you from a single cell to the billions you are now. Even now, cell division is going on in your body, replacing cells as they wear out or get injured.

In this lab you will study cell division in root tips—plant parts in which cells are dividing rapidly. The stages of mitosis will be readily visible.

Prelab Preparation

Study Chapter 9 in your text. Review the instructions for use of the microscope in Investigation 2.

1. During which part of the cell cycle is DNA replicated?
2. Name the 4 stages of mitosis. Describe what you expect to observe during each stage.
3. How is **cytokinesis** different from mitosis?
4. Describe the relationship that exists between chromosomes and chromatin?

Bring your textbook with you to class; you will need it to identify cells in the different stages of mitosis.

Procedure

Part I: Cell Division in Tradescantia
A. Remove a fine root from a cutting of *Tradescantia*. Use a razor blade to cut off the tip about 3 mm behind the cap, and place the tip on a microscope slide. CAUTION: Cut away from your fingers.
B. Use a medicine dropper or pipette to cover the root tip with fixative. A fixative is used to kill and preserve a specimen. Let it sit for 5 minutes.
C. Remove the fixative with a pasteur pipette. Use a medicine dropper or pipette to cover the root tip with pectin solvent. Let it sit for 5 minutes. This solution will weaken the plant tissue, allowing the cells to be squashed thin enough to permit viewing with your microscope. CAUTION: Pectin solvent is a corrosive chemical containing hydrochloric acid. Wear safety goggles when handling. If it gets on your skin or in your eyes, flush with water immediately and ask your teacher for help.

Strategy for Observing
Check your slide under low power first. Then compare your slide with that of another lab group. If the slides are very different, look at a third slide.

5. Why do specimens have to be thin to be viewed through the microscope?

D. Remove the pectin solvent with a pasteur pipette. Use a medicine dropper or pipette to add Carnoy's solution. Wait 3 minutes for the tissue to firm up. This will help protect it during squashing. Use a medicine dropper or pipette to remove the Carnoy's solution. Use a clean dropper to add 3 drops of toluidine blue O stain. CAUTION: Be careful not to spill any stain on your hands or clothing.

E. Clip a test-tube holder or a clothespin to the end of the slide and put the slide on the edge of a hot plate set on low heat (without letting the slide drop off). NOTE: Don't allow the stain to boil. As the stain starts to evaporate, remove the slide from the heat for a moment. Repeat this step 3 times or until the slide is dry. CAUTION: The slide will get hot. Let it cool before touching it. If the hot slide breaks, let the pieces cool before carefully discarding them.

F. Use forceps to transfer the root tip to a clean microscope slide. Place another drop of stain on it and put a plastic coverslip on it. Place the slide on a paper towel on a flat, hard surface. Cover it with another paper towel folded over in fourths. Press down over the coverslip with your thumb. NOTE: Be careful not to break the slide when pressing.

G. Check the preparation with your microscope under low power. First identify the square cells, and then look for blue threadlike structures (chromosomes). If the cells are not spread apart, squash them again, adding more liquid along the sides of the coverslip if necessary. When you find cells with blue threads in them, switch to the high-power objective.

H. If the staining is too dark, destain the preparation. To do this, place a drop of distilled water on one edge of the coverslip and use the corner of a paper towel to draw fluid from the opposite edge. Do this until the chromosomes are clearly visible.

I. Seal the coverslip by drying the slide around the coverslip, and then coating the edges of the coverslip with clear nail polish. Make sure you do not allow the nail polish to touch the microscope lens.

J. Scan slide at low power to identify cells in different phases of mitosis.

6. Switch to high power and make a diagram that includes one cell in each phase: **interphase, prophase, metaphase, anaphase,** and **telophase.** Indicate whether they are in early or late stages of each phase. (Use your textbook to help you identify cells in each phase.)

K. Find a cell that is in metaphase and count how many chromosomes it has.

7. How many chromosomes are there?

Part II: Modeling Mitosis

L. Roll out 2 pieces of modeling clay, one of each color, 20 cm long. (The diameter of the clay should be about 0.5 cm.) Place the 2 rods next to each other, and cut them 7.5 cm from each end with a spatula. You now have 6 chromosomes: one 5-cm pair and two 7.5-cm pairs.

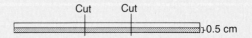

Strategy for Modeling
As you manipulate the clay, keep in mind that everything you are doing is a model of an actual biological event in mitosis. Concentrate on the mitotic events that are occuring as you work with your model.

M. Create a **centromere,** the point on a chromosome to which the spindle attaches during mitosis, by rolling a small piece of clay into a ball. Make 6 such centromeres. Press a centromere into one end of each of the 5-cm rods. Place a centromere at the center of one 7.5-cm rod of each color. On the 2 remaining 7.5-cm rods, put a centromere about one-third along the length of each. The chromosomes with matching centromeres are homologous chromosomes. Each pair should have one rod of each color.

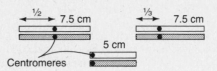

8. What stage of mitosis is this?
9. What do the 2 colors represent?

N. Press the teeth of a small comb lengthwise along each chromosome to create a deep, notched line down the center of each. Try to make your notches all the way through the chromosomes.

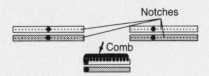

10. What does the notched line represent?

O. Line up the chromosomes end to end along the center of 2 paper towels.

11. What stage of mitosis is this?

P. Use a spatula to divide the centromere of each chromosome lengthwise. Then, place a second spatula right next to the first one.

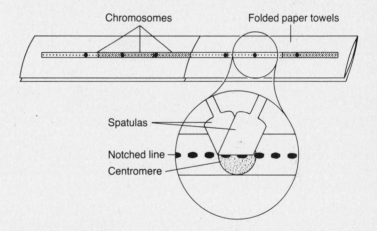

Q. Slowly draw the spatulas apart, separating the chromosomes along the notched lines. (If the chromosome starts to break up instead of separating along the notches, press through the notches again with your comb.)

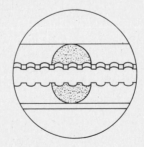

12. What stage of mitosis is this?
13. What mechanism are you modeling in Step Q?
14. As you separate each centromere and pull the chromosomes to opposite poles, record the shapes you observe until the

chromatids (paired chromosomes) are completely apart. Repeat this procedure for each chromosome.

15. What effect does the location of the centromere have on the shape of the dividing chromosome?

R. Complete your separation of the chromatids by grouping them opposite from each other in separate groups.

16. What stage of mitosis is this?
17. What process normally occurs next? What occurs during this process?

Postlab Analysis

18. In your model, where would you expect the centrioles to be?
19. When and where would you observe nuclear membranes?
20. Did the correct sorting of the pairs depend on how the chromosomes lined up before division?
21. In your model is it possible to end up with 2 chromosomes of the same color in a chromosome pair?

Further Investigations

1. The salivary glands of the fruit fly (*Drosophila*) contain large multi-stranded chromosomes. Banded patterns of some of its genes can be identified. Because the fruit fly has only 4 chromosomes, it is easy to tell them apart. If you would like to prepare and observe these chromosomes, ask your teacher for the procedure.
2. A **karyotype** is a common way of viewing an organism's genetic makeup. It is prepared by ordering specially stained metaphase chromosomes according to their sizes and shapes. Photocopy the photomicrograph prepared from the cells of a parasitic worm that causes the disease **schistosomiasis**. Cut out the chromosomes and order them into a karyotype, matching their sizes and shapes.

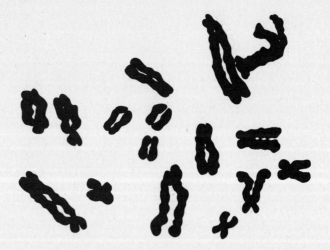

Investigation

9.2 | *Mitosis*

1. _____

2. _____
   _____
   _____
   _____

3. _____

4. _____
   _____

5. _____

6. Make a drawing in the space provided.

7. _____

8. _____

9. _____

10. _____

11. _____

12. _____

13. _____

14.

15. _____

_____

_____

16. _____

17. _____

_____

18. _____

19. _____

_____

_____

20. _____

21. _____

10

Genetic Probability

Learning Objectives

- To use a Punnett square to determine the probable outcome of a genetic cross.
- To relate the number of offspring in a sample to the reliability of genetic predictions.

Process Objectives

- To predict the genotypic and phenotypic ratios of offspring resulting from the random pairing of gametes.
- To model the random pairing of alleles.

Materials

For Group of 4
- 50 White beans
- 50 Colored beans
- 2 Small containers

Sample Punnett Square

	d	d
D	Dd	Dd
D	Dd	Dd

How can the genetic makeup of offspring be predicted?

Introduction

Seedless oranges, thornless roses, and turkeys that fit in smaller ovens are all the result of controlled breeding. Controlled breeding is the careful crossing of parents that show particular desirable traits. People have used this method in agriculture for thousands of years without knowledge of genetics. Today, plant breeders can more accurately predict the traits of offspring by applying their knowledge of genetic mechanisms.

In this investigation, you will model the inheritance of a pair of contrasting traits and predict the outcome of a **genetic cross.**

Prelab Preparation

Refer to Chapter 10 in your textbook to review the terms: **allele, dominant, recessive, homozygous, heterozygous, genotype,** and **phenotype.**

1. Give a definition and an example for each term.

When the genotypes of parents are known, **Punnett squares** can be used to predict the possible genotypes and phenotypes of offspring. Punnett squares also indicate the **probability** of obtaining each possible genotype and phenotype. Refer to Chapter 10 in your textbook to review Punnett squares and their use in predicting the outcomes of crosses. Then study the sample Punnett square.

2. List the genotypes of both parents and each of the possible offspring.

3. Will the offspring be homozygous or heterozygous?

Consider the dominant human trait of dimples. In the sample Punnett square, *D* represents the dominant allele for dimples, and *d* represents the recessive allele for no dimples.

4. Which parent has dimples?

5. Will the offspring have dimples? Explain your answer.

Procedure

Throughout the investigation, work in groups of 4.

A. You will use a blank Punnett square to "cross" a female heterozygous for dimples with a male heterozygous for dimples.

6. Write the possible alleles of the female gametes on the left side of the Punnett square. Write the possible alleles of the male gametes across the top. Complete the Punnett square so that each possible pairing of gametes is shown.

7. List all possible genotypes and phenotypes of the offspring.

8. For each offspring, what is the probability that it will not have dimples? (Examples: 1 out of 4, or 25%; 2 out of 4, or 50%.)

B. You will create a model that shows the random pairing of alleles. White and colored beans will represent the 2 alleles controlling stem length in pea plants. A white bean represents *T*, the dominant allele for tallness; a colored bean represents *t*, the recessive allele for shortness. Get 50 beans of each color.

C. Count 25 beans of each color and place them in a container. Mix these beans and label the container "female gametes." Place the remaining 25 beans of each color in another container and mix them. Label this container "male gametes." Each container of beans represents the gametes produced by one parent.

9. Each parent contributes one allele at random to each offspring. What are the genotype and phenotype of each parent? What genotypes and phenotypes are possible among the offspring?
10. Construct and complete a Punnett square for this cross.
11. Predict the genotypic and phenotypic ratios for the offspring resulting from random pairings of the gametes in your containers.

D. You will model a cross between the 2 parents by performing 10 random pairings of beans from your containers. Close your eyes. Take one bean from each container and place the pair on the table.

12. Ask one of your teammates to record in your data table, Results of Random Pairing of Gametes, the genotype and phenotype of the resulting offspring.

E. Return the 2 beans (alleles) to their respective containers and mix the beans in each container. Repeat the procedure 9 times.

F. Have each of your teammates repeat Steps D through E.

13. Use the data table, Determination of Ratios for Offspring, to determine the genotypic and phenotypic ratios for your offspring.
14. Compare your ratios with those of your teammates.

Postlab Analysis

15. Pool the data collected by your team. Compare the genotypic and phenotypic ratios for your team's offspring with those for other teams' offspring.
16. Pool the data of the whole class. What are the genotypic and phenotypic ratios for the offspring of the whole class?
17. Compare the actual genotypic and phenotypic ratios for your offspring with your predicted ratios. Compare the actual ratios produced by your team and by the whole class with your predicted ratios.
18. Under what circumstances are predicted genotypic and phenotypic ratios for offspring most reliable?

Further Investigations

1. Review the **testcross** in Chapter 10 of your textbook. Use your beans to set up a container of "gametes" produced by an individual displaying a dominant phenotype. The individual may be either heterozygous or homozygous for the trait. Have a teammate perform a "testcross" to determine the genotype of the individual.

2. The model cross that you performed in this investigation was an example of a **monohybrid** cross. Review the **dihybrid cross** in Chapter 10 of your textbook. Use 4 colors of beans or other like objects to design and carry out a model dihybrid cross of 2 heterozygous parents.

Strategy for Predicting
To predict the genotypic and phenotypic ratios for the offspring, refer to the Punnett square constructed for this cross. Consider how predicted ratios are related to probabilities indicated by a Punnett square.

Strategy for Modeling
As you repeat the pairings of the gametes, remember that your gamete selections must be random. How can you ensure random pairings?

10 | *Genetic Probability*

1. _____

_____

_____

_____

_____

_____

_____

_____

_____

2. _____

_____

3. _____

4. _____

5. _____

6. Use Punnet Square 1 to predict the possible genotypes and phenotypes of the offspring.

Punnett Squares

    1   2

7. _____

_____

8. _____

9. _____

10. Use Punnet Square 2 to predict the possible genotypes and phenotypes of the offspring.

11. _____

12.–13. Enter your data in the tables.

Results of Random Pairings of Gametes				Determination of Ratios for Offspring		
Trial	Offspring Genotype	Offspring Phenotype		Genotypes Produced	Totals	Genotypic Ratio
1	_____	_____		_____	_____	_____
2	_____	_____		_____	_____	_____
3	_____	_____				
4	_____	_____		Phenotypes Produced	Totals	Phenotypic Ratio
5	_____	_____		_____	_____	_____
6	_____	_____		_____	_____	
7	_____	_____				
8	_____	_____				
9	_____	_____				
10	_____	_____				

14. _____

15. _____

16. _____

17. _____

_____

18. _____

11 | *Human Genetic Traits*

How common are certain inherited traits?

Learning Objectives
- To determine how many of your classmates display certain inherited traits.
- To discover whether a dominant trait is always found in a majority of a population.

Process Objectives
- To measure the occurrence of 3 pairs of genetic traits.
- To analyze your data by comparing them with data for larger populations.

Materials
For Group of 4
- Hand-held mirror
- Hand-held calculator

Introduction

Can you roll your tongue? Do your fingers have mid-digital hair? Do you have a widow's peak? These 3 traits are known to be inherited in humans.

Every human being has 46 chromosomes. Each chromosome contains thousands of genes. Although we do not know the function of most of these genes, there are a few genes that produce easily identifiable traits. In this lab you will study the 3 inherited traits mentioned above. You will find your own phenotype for each of these traits. Then you will pool your data with other students' data. For one of these traits, you will compare your data with national figures for that trait.

Prelab Preparation

Before doing this lab, reread Section 10.1 and your notes on Mendel's inheritance principles. Review in particular how dominant and recessive genes are inherited and expressed. Make sure you know how to describe the phenotype of a person whose genotype is given to you. You will be studying 3 pairs of human traits in this lab.

The ability to roll your tongue is dominant to the inability to roll your tongue. If you cannot roll your tongue into the shape shown, you are a nonroller.

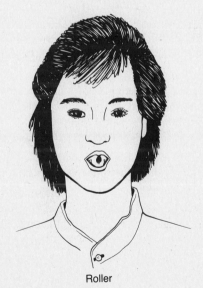

Roller

The presence of mid-digital hair is dominant to the absence of mid-digital hair. Look at your fingers. Notice that each finger has 3 segments. All humans have hair on the backs of their fingers on the segments nearest to the palm. However, some people also have hair on the middle segment. You should check each finger and, if you have hair on the middle segment of your fingers, consider yourself as having mid-digital hair regardless of how many mid-digital hairs you have. (Some people have a lot; others have only a few.) Look at the illustration to determine whether you have mid-digital hair.

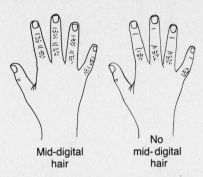

Mid-digital
hair

No
mid-digital
hair

The presence of a widow's peak is dominant to the absence of a widow's peak. If your hair comes to a point in the middle of your forehead, you have a widow's peak. Use a mirror to compare your forehead with the illustration.

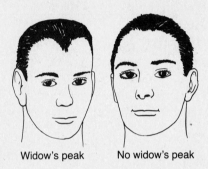

Widow's peak   No widow's peak

Procedure

Part I

A. Try to roll your tongue.

    1. Are you a roller or a nonroller?

    2. Record your phenotype for tongue-rolling in the data table.

B. Now you will gather data for your entire class. Then you will calculate the percentage of students in your class who exhibit each trait.

    3. Count how many people are able to roll their tongues and how many people are not. Enter these numbers in the appropriate columns on the chart for individual and class data.

To calculate the percentage of the class who are rollers, first find the total number of students in the class. Then divide the number of rollers by the total number of students. The answer will be a decimal fraction between 0 and 1. Multiply this answer by 100 to get the percentage. For example, if there are 18 students in the class and 5 are rollers, the percentage of rollers is:

$$\frac{5}{18} \times 100 = 0.28 \times 100 = 28\%$$

    4. Perform this calculation with the tongue-rolling data you collected. Record the percentage of students who are tongue-rollers and the percentage who are not.

    5. Now collect class data for mid-digital hair. Follow the same procedure for calculating the class percentages and entering data on the data table as you did for tongue-rolling.

Strategy for Measuring

Work with your lab partners to confirm your phenotype observations.

6. In the same manner, gather class data for widow's peak. Follow the same procedures you used for recording the preceding traits.

C. Review the data you collected for these 3 traits.

7. Does the same percentage of students always display the dominant trait? Explain your answer.
8. For which pairs of traits do the majority of students display the dominant trait?
9. For which pairs of traits do the majority of students display the recessive trait?
10. Why is it possible that a majority of people may display a recessive trait?

Part II

D. In Part I you examined yourself and members of your class for 3 inherited traits. Now you will concentrate on one of these traits, tongue-rolling. You will obtain more data and do some calculations that will enable you to measure how many genes in a sample population are R (rolling) and how many genes are r (nonrolling).

As a class, obtain the tongue-rolling phenotypes of 100 people. Begin your data collection with the members of your class. Then each class member should test enough additional people (friends and relatives) so that you have a total of 100 different people. Construct a table like the one shown that includes the name and phenotype of everyone in your sample. (Remember to include your class data on the table.)

11. Why do you need the names of the people? Work cooperatively to make sure you have 100 different names.
12. Enter your compiled data in the data chart for 100 subjects.

Nationally, about 70% of the population are rollers, and approximately 30% are nonrollers.

13. How do your data compare with these figures? Discuss possible reasons for any differences between your data and the national data.

E. Now you can calculate the frequency of R and r genes in your sample by using an algebraic equation, the Hardy-Weinberg Principle. According to this principle, the distribution of dominant and recessive alleles in a population remains constant from one generation to the next as long as no external factors such as mutation and migration arise. Further, mating must be random, and the population must be large for the formula to work. There must be no natural selection. Specifically, the equation for a gene with two alleles is:

$$p^2 + 2pq + q^2 = 1$$

where p = frequency of R (dominant gene), expressed as a decimal fraction
q = frequency of r (recessive gene), expressed as a decimal fraction
p^2 = number of RR individuals
$2pq$ = number of Rr individuals
q^2 = number of rr individuals
and $p + q = 1$

The Punnett square on the left illustrates the formula in a more familiar way. Together the 4 squares represent the total population under study. In other words, if you were to assign values to each of the genes in the squares, these values would add up to one, or 100% of the population.

Now return to the algebraic formula. Knowing the percentages of

Tongue Rolling

Name	Phenotype (tongue roller or non-roller)
1	
2	
3	
4	
100	

	R	r
R	RR (pxp) (p²)	Rr (pxq)
r	Rr (pxq)	rr (qxq) (q²)

individuals possessing each trait, rolling and nonrolling, does not in itself tell us the frequency of R and r genes in the population. To determine these frequencies, do the following:

14. Express the percentage of nonrollers as a decimal fraction (e.g., 36% = 0.36)

We know from Hardy-Weinberg that $q^2 = rr$. The square root of q^2 will give us the frequency of r. Using our example, the square root of 0.36 equals 0.6.

15. Calculate q for *your* sample.

Now, since Hardy-Weinberg states that p + q = 1, then p = 1 - q. Therefore, using our example, p = 1 - 0.6 = 0.4.

16. Calculate p for *your* sample.

We can check these calculations by plugging the example values into the Hardy-Weinberg equation.

$$p^2 + 2pq + q^2 = 1$$
$$(0.4)^2 + 2\,(0.4 \times 0.6) + (0.6)^2 = 1$$

17. Check *your* values by plugging them into the Hardy-Weinberg equation.

18. Considering your findings in Questions 15 and 16, what is the frequency of the gene for tongue rolling, *R*? What is the gene frequency of *r*?

Postlab Analysis

19. What percentage of people did you find in your sample of 100 who could roll their tongues?

20. How does this figure compare with the national average of 70%?

21. How might your frequency values change if you increased your sample from 100 to 200 people?

22. Do you think there was any bias in your sample? Explain.

Further Investigations

1. You have just learned that the Hardy-Weinberg Principle works as long as certain external factors do not exist. Using your textbook and library resources, find a "near-perfect" Hardy-Weinberg human population. Prepare a short report for your class that describes this population. What characteristics of this population make it useful to study? Mention any genetically linked traits that may have been traced by researchers of this population.

Strategy for Analyzing
In what ways are the 100 people you sampled similar to the total U.S. population? How are they different?

Investigation

11 | *Heredity of Human Traits*

1. _____

2.–6. Enter your answers on the data chart.

Individual and Class Data

Trait	Dominant Gene	Recessive Gene	My Phenotype	Class Data			
				Number with dominant phenotype	Number with recessive phenotype	Percent having dominant phenotype	Percent having recessive phenotype
Tongue Rolling	Roller	Nonroller					
Mid-digital Hair	Has mid-digital hair	Does not have hair					
Widow's Peak	Has widow's peak	Does not have peak					

7. _____

_____

8. _____

9. _____

10. _____

11. _____

12. Enter your answers on the data chart.

Tongue Rolling Data for 100 Subjects

Number of rollers	
Number of nonrollers	
Percentage of rollers	
Percentage of nonrollers	
Possible genotypes of rollers	
Possible genotypes of nonrollers	

13. _____

_____

14. _____

15. _____

16. _____

17. _____

_____

18. _____

_____

19. _____

20. _____

21. _____

22. _____

_____

12 | *Gene Regulation*

How are genes turned on and off?

Introduction

All cells contain the genes coding for every protein they will ever make. Although thousands of genes are present, only some of these are **expressed** at any given time. This happens because gene expression is **regulated**.

Gene regulation is a complex, and not yet totally understood, process. However, most evidence indicates that gene expression is primarily regulated at the level of **transcription;** whether or not a gene will make its protein product is determined by mechanisms that control mRNA synthesis.

The best understood example of this type of control is that of the *lac* **operon in prokaryotes.** The *lac* operon contains 3 structural genes. These genes code for 3 enzymes which are involved in the metabolism of lactose. When lactose levels are high, the production of these enzymes is **induced.** As lactose levels decline, production of these enzymes is inhibited. Lactose, a sugar, serves as an environmental cue to regulate production of the *lac* genes. (The *lac* operon is discussed in more detail in your textbook.)

Bacteria live in environments that may change suddenly and they use several types of signaling systems to rapidly regulate the expression of essential genes. In addition to chemical agents, light, pH, temperature, and age can all influence gene activity.

In this lab you will manipulate temperature to regulate the production of a red pigment called **prodigiosin** (proe-DIJ-ee-OSE-un) in the bacteria *Serratia marcescens (S. marcescens).*

Prodigiosin

Prelab Preparation

Review Chapter 12 in your textbook.

S. marcescens is predominantly **saprophytic** and is commonly found in soil and on plants. It is also frequently ingested by insects for which it is a parasite. After being eaten, the bacteria enters the bloodstream of its host and causes blood poisoning. This effect is believed to be due to the red pigment, prodigiosin, that *S. marcescens* produces. In addition to its effect on insects, prodigiosin is also toxic to protozoans and fungi.

S. marcescens produces prodigiosin only at certain temperatures. Although prodigiosin is a peptide, it is not encoded by a gene. It is formed by the joining of several amino acids. This joining is controlled by a **regulatory enzyme.** It is the expression of the gene for this enzyme that is regulated by temperature.

Learning Objectives
- To understand regulation of gene expression in a bacterial strain.
- To use aseptic technique to grow bacteria.

Process Objectives
- To experimentally regulate gene expression in bacteria.
- To predict the results of environmental influence on gene expression.
- To infer the purpose for a regulated gene product.

Materials
For Class
- 2 Cultures of *Serratia marcescens,* grown at different temperatures
- 2 Incubators

For Group of 2
- 2 Nutrient agar plates
- Bunsen burner or alcohol lamp
- Wire inoculating loop
- Wax pencil or marker

By an unknown mechanism, the transcription of the mRNA of this enzyme is induced only at certain temperatures. As the levels of this enzyme rise, more molecules of prodigiosin are produced. The accumulation of the prodigiosin pigment in individual bacteria eventually causes the bacterial colonies to appear red.

1. What is the environmental signal that regulates transcription of the *lac* operon genes?
2. How is prodigiosin made?
3. What is the signal that regulates the expression of the gene for the prodigiosin regulatory enzyme?
4. How will you know when the gene for the prodigiosin regulatory enzyme has been induced?

Procedure

Working in groups of 2, obtain 2 agar plates but do not open them until you are ready to inoculate them.

A. Label the bottoms of both plates with your names and the date. Mark one plate A-27°C and the other B-37°C. Draw a line down the center of each bottom, dividing each plate in half. Mark the left half of each plate 27°C and the right half 37°C.

Strategy for Experimenting
By experimenting with bacteria grown originally at 2 different temperatures you can determine whether previous culture conditions affect gene expression.

B. Obtain 2 stock cultures of *S. marcescens* — one that has been grown at 27°C and one at 37°C. Examine their color and other physical characteristics.

5. What color are colonies of *S. marcescens* that were grown at 27°C? At 37°C?

C. You and your partner will each inoculate one plate. Decide who will do plate A-27°C and who will do B-37°C. You will each inoculate the left half of your plate with the stock culture grown at 27°C and the right half with the stock culture grown at 37°C.

D. You will inoculate your plates by the streaking method. Light your bunsen burner or alcohol lamp. CAUTION: Use proper procedures when lighting bunsen burners or working with an open flame. Place the stock culture plates and the plate to be inoculated right side up on the bench, between you and the flame. Flame your loop along the length of the wire until it glows orange. Working quickly but carefully, lift the lid of the 27°C stock culture about 2–3 cm (enough space for you to insert your loop). Remove your loop from the flame and place it carefully onto the agar surface where there are no colonies. (You will hear the loop sizzle as it is cooled by the agar.) Then, touch the loop to one of the colonies to obtain your **inoculum** (sample of bacteria cells to be transferred). Withdraw the loop and replace the lid. Raise the lid of your plate 2–3 cm and inoculate the half of your plate you marked 27°C by lightly moving your loop along the surface in a zig-zag fashion (see the illustration).

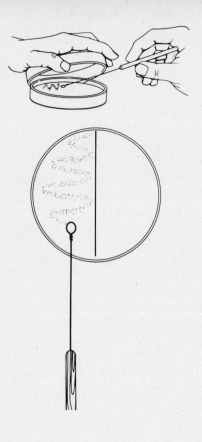

NOTE: Make sure you inoculate the correct half of the plate (remember, it was inverted when you labeled it). Withdraw the loop, replace the lid, and again flame the loop. Repeat the same procedure with the 37°C stock culture. Your partner should perform the identical procedure with your other plate. CAUTION: Wash your hands and the lab bench after transferring bacteria. Flame the loop before putting it away.

E. Incubate the plates for 24 to 48 hours. Plate A will be incubated at 27°C and Plate B will be incubated at 37°C.

You and your partner now have 4 cultures: (1) grown at 27° and recultured at 27°; (2) grown at 37° and recultured at 27°; (3) grown at 27° and recultured at 37°; and (4) grown at 37° and recultured at 37°.

6. After you put your cultures in the incubators, record your predictions for the color of each of your cultures in your data table.

Part II

F. Remove your cultures from the incubators.

7. Examine your cultures and record your observations in the data table.

8. Did your results match your predictions? If there was any difference, how might you account for it?

Postlab Analysis

9. What results would you predict if both cultures were grown at 32°C?

10. What would you expect if the bacteria were grown at 37°C for 8 hours, then at 27°C?

Strategy for Predicting
To make a prediction you must have sufficient background data. Review the material about prodigiosin and *S.marcescens* and identify the relevant information.

Strategy for Inferring

Biological processes usually have a survival advantage for the organism which uses them. Consider why it might be advantageous for *S. marcescens* to make prodigiosin at certain temperatures.

11. If a culture is grown at 37°C for 48 hours then the plate is returned to 27°C for 48 hours, what color will the colonies be?
12. What is the average room temperature in degrees centigrade? What is normal body temperature in degrees centigrade?
13. What would you infer to be the reason for thermal regulation of prodigiosin production?
14. For what purpose would you infer that *S. marcescens* makes prodigiosin?

You may continue to grow your cultures for 2–4 more days if you wish to find out what happens over a longer period of incubation.

Further Investigations

1. In your library, research *S. marcescens* and prodigiosin.
2. Look up in a biochemistry book, or other reference source, how gene expression is regulated in bacteria and animals.
3. Pick a particular animal cell type, such as neurons or muscles. Try to identify the stimuli that might regulate gene expression in these cells.
4. Many researchers are working to learn how to control the expression of particular genes. Read about some of their recent work and some of the medical and industrial applications of this research. (You might also want to visit a nearby college or university where this type of work is being done and speak with one of the scientists conducting such research.)

Investigation

12 | *Gene Regulation*

1. _____

2. _____

3. _____

4. _____

5. _____

6.–7. Enter your answers on the data chart.

Data Chart for *Serratia marcescens*

Experimental Conditions	Experimental Cultures	
	Predicted Results	Actual Results
Originally grown at 27° C and recultured at 27°C		
Originally grown at 37°C and recultured at 27°C		
Originally grown at 27°C and recultured at 37°C		
Originally grown at 37°C and recultured at 37°C		

8. _____

_____

_____

_____

9. _____

10. _____

_____

11. _____

12. _____

_____

13. _____

_____

_____

_____

14. _____

_____

_____

13 *Manipulating DNA*

How is DNA analyzed?

Learning Objectives

- To learn how restriction enzymes are used to manipulate DNA.
- To understand the principle underlying gel electrophoresis

Process Objectives

- To use models to simulate the technique used to cut long DNA chains into smaller sequences.
- To analyze DNA fragments.

Materials

- Pencil and paper

Recombinant DNA Construction

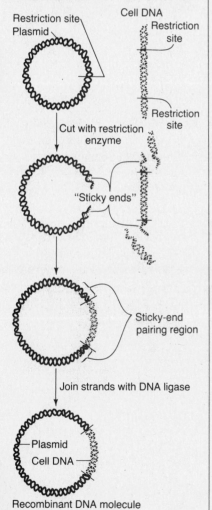

Recombinant DNA molecule

Introduction

Biotechnology is the manipulation of the biological capacity of cells and their components. For thousands of years people have used biotechnology by using yeast to make flour into bread and grapes into wine. Today, we are using biotechnology to study the basic processes of life, diagnose illnesses, and develop new treatments for diseases.

Some of the tools of biotechnology are natural components of cells. **Restriction enzymes** are made by bacteria to protect themselves from viruses. They inactivate the viral DNA by cutting it in specific places. **DNA ligase** is an enzyme that exists in all cells and is responsible for joining together strands of DNA. Scientists use restriction enzymes to cut DNA at specific sequences called **recognition sites**. They then rejoin the cut strands with DNA ligase to make new combinations of genes. **Recombinant DNA** sequences contain genes from two or more organisms.

Using this technique, researchers have gained the ability to diagnose diseases such as sickle cell anemia, cystic fibrosis, and Huntington's chorea early in the course of the disease. Many researchers are also applying the techniques of biotechnology to find new treatments for genetic diseases.

In this lab, you will use paper models to simulate the cutting of DNA; you will also model gel electrophoresis to analyze the DNA fragments produced. You will learn how these techniques are carried out and also some of their applications.

Prelab Preparation

As the illustration shows, the sequences of DNA can be written in many ways. However, the simplest way is to use single-letter codes for the nucleo-

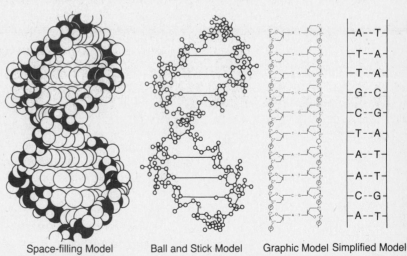

Space-filling Model   Ball and Stick Model   Graphic Model Simplified Model

tide bases: A **(adenine)**, T **(thymine)**, C **(cytosine)** and G **(guanine)**. The DNA sequence in the illustration can be written in its shorthand fashion as:

```
A T T G C T A A C A
T A A C G A T T G T
```

1. What are the complementary bases for the DNA sequence below?

```
T A A G C C G T A G G T T G G A A C T C C
```

2. Write your own double-stranded molecule, 20 base pairs long.

Procedure

Part I: Restriction Enzymes

A. There are now about 200 known restriction enzymes that cut DNA at specific recognition sites. For example, the restriction enzyme *Hin*d II recognizes the base sequence G T C G A C.

3. Copy the sequence below and indicate on it the *Hin*d II recognition site.

```
T A A G C C G T C G A C T C G A A C T C C
```

4. Write out the DNA sequence complementary to the one in question 3. Read the complementary strand in reverse and indicate the *Hin*d II recognition site on it.

B. When the restriction enzyme *Hin*d II recognizes the sequence GTCGAC, it will cut the DNA strand between the cytosine (C) and guanine (G) on both strands. Therefore, it will leave blunt ends on the fragments:

```
      Hind II
        ↓
 - G T C G A C -              - G T C        G A C -
 - C A G C T G -      →       - C A G        C T G -
        ↑                               Blunt ends
      Hind II
```

The restriction enzyme *Eco* RI cuts its recognition site at nonadjacent points on the DNA molecule, leaving "sticky" ends. *Eco* RI recognizes the base sequence G A A T T C and cuts this sequence between the guanine (G) and adenine (A) bases:

```
     Eco RI
       ↓
 - G A A T T C -           - G              A A T T C -
 - C T T A A G -    →      - C T T A A              G -
         ↑                            Sticky ends
     Eco RI
```

Therefore, a molecule of DNA cut at an *Eco* RI site would appear:

```
 - C G T T A T G -                  - A A T T C G T A G -
 - G C A A T A C T T A A -               - G C A T C -
```

Sticky ends can bind to similar sticky ends from other *Eco* RI-digested DNA fragments. After recombining, the ends are joined by DNA ligase, to form a new pattern of bases. By cutting the DNA from 2 different organisms with the same enzyme and recombining with DNA ligase, scientists make recombinant DNA.

5. Copy the sequence given below and complete the strand complementary to it. On both strands indicate the *Eco* RI restriction sites. Remember to read the complementary strand in reverse.

```
G C C T C T A A G A A T T C A G T T C G
```

6. Once the *Eco* RI has cut the above DNA chain, how many fragments of DNA would there be? Would the ends be blunt or sticky? How many bases would there be in each fragment? NOTE: When counting the length of a DNA fragment, count only the number of bases on the upper strand.

C. Below you will see two sequences of DNA—DNA IA and DNA IB. (DNA IB is a mutant variation of DNA IA).

7. What is the difference between the two sequences?

8. Copy the sequences below. Identify the *Eco* RI recognition sites on both sequences and mark the sites where *Eco* RI would cut. (You will use these sequences again in Part II.)

DNA IA

```
T T G C A A G T C A G A A G A A T T C A A C C T A G G A A T T C T A A G C G C
A A C G T T C A G T C T T C T T A A G T T G G A T C C T T A A G A T T C G C G
```

DNA IB

```
T T G C A A G T C A G A A G A A G T C A A C C T A G G A A T T C T A A G C G C
A A C G T T C A G T C T T C T T C A G T T G G A T C C G G A A G A T T C G C G
```

9. How many fragments of DNA were made from each sequence after digestion with *Eco* RI?

10. What are the lengths (in basepairs) of the fragments from the DNA IA and DNA IB digestions?

11. Can you recombine any of the "sticky" ends of DNA IA and DNA IB to make a new sequence of DNA? If so, write out the sequence for at least one such recombination.

Part II: Gel Electrophoresis

D. Scientists identify differences in DNA sequences by measuring the length and number of fragments created by digestion with restriction enzymes. A technique called **gel electrophoresis** is used to separate fragments according to length. DNA fragments (cut with an appropriate restriction enzyme) are placed on one end of a specially-prepared block of agarose called a gel. An electric current is applied across the agarose which causes the strands to migrate through the gel. (Since DNA molecules are negatively charged, they migrate towards the positive electrode.) The agarose is like a sponge with small holes in it. Therefore, the smaller DNA fragments can move through the gel at a faster rate than larger fragments and the larger fragments are found nearer the point of origin. Scientists then use a special stain to make the DNA fragments visible as bands. By counting the number of bands researchers can tell how many fragments exist. By observing the distance each fragment has migrated, they can determine how big each fragment is.

12. How many fragments are produced by cutting DNA X? By cutting DNA Y?

Model of Gel Electrophoresis Migration

A. Restriction Enzyme Digestion

Two different strands of DNA 1,000 base pairs (bp) long, cut with the same restriction enzyme (RE).

DNA X

    RE RE
    ↓   ↓

DNA Y

            RE
            ↓

13. The fragments generated from DNA X and DNA Y in Figure A were then analyzed by gel electrophoresis. From the data in Figures B and C, calculate the sizes of the fragments.

B. Migration Distances of DNA in Gels

The size of the DNA fragments is proportional to the distance traveled.

Fragment Size (base pairs)	Distance From Origin (cm)
600	1
400	3
300	4
200	5
100	6

C. Gel Electrophoresis

Appearance of DNAs X and Y after digestion and separation by gel electrophoresis.

Distance Migrated

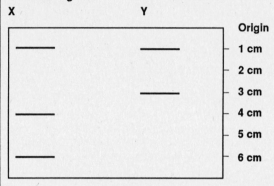

14. On the simulated gel, draw bands corresponding to the positions to which the *Eco* RI digests of DNA IA and DNA IB would migrate. The fragments of each type of DNA should be in separate lanes and at distances from the origin proportional to the numbers of bases in each fragment. (Remember that smaller fragments move farther.) From the following distances, determine how far your fragments will migrate: 40 base pairs will migrate 2.5 cm; 30 bp will migrate 5 cm; 20 bp will migrate 7.5 cm; and 10 bp will migrate 10 cm. Make sure all the DNA is accounted for.

Strategy for Modeling

Use the DNA sequences below question 8, together with the new data in question 14, to model the migration of DNA IA and DNA IB on the simulated gel.

Postlab Analysis

Strategy for Analyzing

In describing the effect of a mutation in the recognition site of a restriction enzyme you will want to analyze your data for DNA IA and its mutant variation DNA IB.

15. If a mutation causes one of the base pairs in a recognition site to change, what effect will it have on the ability of the restriction enzyme to cut at that site?

16. Explain how scientists use restriction analysis to determine that one sequence of DNA is different from another.

Further Investigation

1. Map the recognition sites of several different restriction enzymes on a DNA molecule. Use the DNA sequences in this lab and ask your teacher for the recognition sites for other restriction enzymes. Make simulated gel electrophoresis patterns for the products of each restriction enzyme.

Investigation

13 | *Manipulating DNA*

1. _____

2. _____

3. _____

4. _____

5. _____

6. _____

7. _____

8.

9. _____

10. _____

11. _____

12. _____

13. _____

14. Record your answers on the simulated gel on the next page.

15. _____
_____
_____

16. _____
_____
_____

DNA fragments added here

14

Coacervates

Learning Objectives
- To learn the recipe for making coacervates.
- To decide whether coacervate behavior allows coacervates to be classified as living.

Process Objectives
- To develop an awareness of the dynamic model technique.
- To organize your data according to the characteristics of life.
- To infer whether coacervates are alive.

Materials

For Group of 2
- 4 Medium-sized test tubes
- 4 Rubber stoppers to fit the test tubes
- Test tube rack
- 2 10-mL Graduated cylinders
- 5 mL 1% Gelatin solution
- 4 mL 1% Gum arabic solution
- Glass stirring rod
- pH Paper (with pH scale)
- 2 Medicine droppers
- Microscope slide
- Coverslip
- Compound light microscope
- Drawing paper
- Lab aprons
- Safety goggles
- 5 mL 1% Hydrochloric acid
- Dilute solutions of water-soluble dyes (e.g., Congo red, neutral red, methylene blue, food coloring)

Water molecules

Protein and carbohydrate molecules

Can a mixture of protein and carbohydrate molecules behave as a living organism?

Introduction

How did life on earth begin? What were the first organisms like? Scientists such as Alexander Oparin believe that all life developed gradually in the ancient seas. According to Oparin, prehistoric oceans probably contained a rich mixture of organic chemicals, including proteins and carbohydrates. When protein and carbohydrate molecules are mixed together in liquid water, **coacervate** (koe-AS-ur-VATE) droplets, showing lifelike characteristics, may form. In this lab, you will make coacervates and observe their behavior.

Prelab Preparation

A model is a visual or verbal construction that helps make difficult concepts more easily understood. It often simulates something that is impossible to study firsthand due to size or unavailability. This lab permits observation of coacervates, which scientists believe to be similar to the ancestors of the first living organisms. Proteins and carbohydrates are used to model the probable contents of prehistoric oceans.

Since this lab is an actual "working" model, it is referred to as **dynamic.** A dynamic model differs from a **static** model, such as a replica of the human ear, in that it changes as you observe it. This dynamic model uses a process that occurs in nature—the formation of coacervates—to simulate the possible beginnings of life on earth.

1. List the basic characteristics of life. You may refer back to Investigation 1, "What Is Life?"

Because the pH of your solution will have to be carefully adjusted in order for coacervates to form, review Investigation 3.1 on the concept of pH.

Procedure

A. Get 4 test tubes, 4 rubber stoppers, and a test tube rack. Pour 5 mL of gelatin solution (a protein) into one test tube. Then add 4 mL of gum arabic solution (a carbohydrate). Stopper the tube tightly, and mix the contents gently by turning the tube upside down several times. NOTE: *Do not shake;* it will affect the solution.

B. Uncork the test tube, and dip a glass stirring rod into the mixture. Touch a drop of the mixture onto a piece of pH paper.

2. Is there a color change? Record the pH under Trial 1 in your data chart.

3. Hold the tube up to a light source. Is the mixture clear or cloudy? (If it is cloudy, the mixture may contain coacervates.) Record your results as Trial 1 in your data chart.

C. Using a medicine dropper, put 2 drops of the mixture on a clean slide. Put a coverslip over the drops, and look at them under low power (10×). You may not see coacervates until the pH (acidity) is adjusted, so do not be discouraged. If present, coacervates will look like droplets with tiny bubbles inside. (If the mixture is not cloudy, move on to Step D, and come back to Questions 5–8 later.)

4. Describe the appearance of the coacervates in your data chart.
5. Record the number of coacervates in the microscope field.
6. Draw a coacervate (under high power) on a separate piece of paper.

D. CAUTION: Wear lab apron and safety goggles; hydrochloric acid causes burns. Using a second medicine dropper, add 2 drops of dilute hydrochloric acid to the mixture in the test tube. Cork the tube and turn it upside down once or twice to mix the acid.

E. Repeat Step B. This is Trial 2 on your data chart.

7. What is the pH now? Is the mixture cloudy?

F. Repeat Step C.

8. Does a change in pH cause a change in coacervate structure?

G. Repeat Step D. Continue the procedure (Steps B through D) until coacervates are no longer observed and the mixture is no longer cloudy. When the mixture is no longer cloudy, proceed to Step H. Pay attention to changes in the appearance, number, size, and activity of the coacervates as more and more acid is added.

9. Draw and label coacervates from 3 different trials.

H. When the mixture is no longer cloudy, make a slide, as before, but put 1–2 drops of dilute dye on the mixture before adding the coverslip. Observe the coacervates.

10. What characteristic of coacervates does the dye allow you to observe?
11. Infer, based on coacervate behavior with the dye, whether coacervates are living or nonliving.

Postlab Analysis

Look at your list of characteristics of living organisms for Question 1.

12. In what ways do coacervates seem to be alive?
13. In what ways do they seem not to be living?
14. Describe how the coacervates changed as the pH was altered?
15. What does this tell you about coacervates?

Further Investigations

1. If a live ameba culture is available, make a wet mount from it. Compare amebas and coacervates in terms of appearance, shape, number, grouping, size, and activity. Add dye as you did with coacervates. Notice how the amebas are affected by the dye. How does this compare with the effect of dye on coacervates.

2. Do some further reading on the subject of the origin of life on earth. Prepare a report on this fascinating topic. Important areas to discuss include: Oparin's and Haldane's concurrent hypotheses; conditions necessary for the origin of life on earth; Miller's experiment, simulating the early development of organic material; formation of organic polymers and aggregates; development of life (metabolic) processes.

Strategy for Inferring
Imagine yourself as the first scientist to ever see coacervates. Finish one of these statements: "It makes sense that coacervates are not alive because..." or "It makes sense that they are alive because... ."

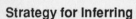

Strategy for Modeling
A model should emulate the environment of the process it is depicting.

Investigation

14 | *Coacervates*

1. _____

_____

_____

2.–5. Enter your answers on the data chart.

Coacervate Data										
Trial	1	2	3	4	5	6	7	8	9	10
pH										
Cloudy/clear										
Microscopic Observations										

6. Make a drawing on a separate sheet of paper.

7.–8. Enter your answers on the data chart.

9.

10. _____

_____

11. _____

_____

12. _____

_____

_____

13. _____

_____

_____

14. _____

15. _____

15

Patterns of Variation

Learning Objectives

- To observe genetic variations in particular traits among members of a population.
- To understand that genetic variation occurs frequently, and that it has an important effect on populations and species.

Process Objectives

- To collect and organize data demonstrating variations in particular traits within a population.
- To infer how individual variation can potentially contribute to the continuing evolution of a species.

Materials

For Group of 2

- 20 Mung bean sprouts
- Metric ruler
- 2 Pieces of Graph paper

How much variation in genetic expression is there within a population?

Introduction

Clasp your hands together so that your fingers interlock. Is your left or right thumb on top? The trait of having either the left or right thumb on top is like left- or right-handedness. Both traits are genetically determined characteristics that occur naturally within the human population and are passed on to children by parents. You are in the majority if your left thumb falls on top.

Many variable traits can be identified within a population, and sometimes these variations, which are mutations, seem to have no apparent survival advantage or disadvantage. It is possible that a mutation that occurs infrequently may eventually prove to be beneficial in response to an environmental change. In this case, the few members of the population that have the new characteristic are more likely to survive and to reproduce, thus passing the gene on to future generations. Eventually, a previously uncommon variation in a population might become the standard.

Prelab Preparation

When you study individual variations of a characteristic, you look at the frequency and the degree to which the variation occurs within a population. You can then measure the **mode,** or the most commonly occurring value of the particular variable.

1. If, in a group of 13 people, 3 people are 5'4" tall, 6 people are 5'7", and 4 people are 5'8", what is the value for the mode? The range of variation for height in this population is from 5'4" to 5'8".

Generally, you can learn much about the variability of a particular characteristic within a population by drawing a **distribution curve,** which is a graph that illustrates the range of measurements for the characteristic (height, for example) and the number of individuals having each measurement.

2. If you measured the heights of several other groups of 13 people, would the mode be the same from group to group? What factors could influence it?

3. When you are measuring variability in a population, would you expect the most common measurements to be at a midpoint in the range of measurements, or at either end of the range?

Procedure

Individuals within a species show variations in their rates of growth even if conditions are identical. You will measure the lengths of 20 mung bean sprouts that have been grown under the same conditions—in a dark, moist area—for several days.

Measure to the nearest millimeter 20 sprouted beans from the point where the sprout grows out of the bean to the tip.

  4. Record the numbers in the space provided.
  5. What is the range of measurements for your sprouts? What is the mode?
  6. What do you think you would see if the sprouts had been measured a day earlier? What about a day later?

B. A distribution curve plots the lengths of sprouts in millimeters against the number of sprouts of each length.

  7. Mark the appropriate increments along the axes of the graph provided and plot your data.
  8. In millimeters, what is the difference between the lowest measurement and the mode? Similarly, what is the difference between the highest measurement and the mode?
  9. What advantages and disadvantages do you think the longest sprouts have for survival?
  10. What factors do you think might have affected the shortest sprouts?

C. Measure the length of the earlobes of 10 people. Measure from the lowest point of the inside to the lowest point of the outside.

  11. As with the bean sprouts, record the earlobe lengths in millimeters in the space provided.
  12. What is the range of earlobe lengths in the population you studied? What is the mode?

D. Draw a distribution curve showing the earlobe lengths you measured versus the number of earlobes of each length, just as you did for the bean sprouts.

Strategy for Organizing Data

Organize your measurement values on a single line from lowest to highest, consecutively. Beneath each value, write the number of sprouts of that length. Transfer this information to your graph.

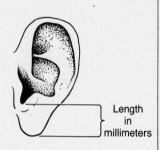

Length
in
millimeters

Postlab Analysis

  13. What shape does a distribution curve generally take?
  14. If you were to measure the earlobe lengths of 100 or 1,000 people, what kind of distribution curve would you expect?
  15. What kind of curve would you get if you plotted the values from males and females separately?
  16. Give 3 examples of genetic variations within an animal population that could be of adaptive advantage to that species.
  17. Similarly, can you think of 3 genetic variations that could be of adaptive value for a plant population?

Strategy for Inferring

Think about characteristics of several animal species that help to meet the animals' basic needs (food, shelter, health, safety).

Further Investigations

1. Measure a variable trait in the human population. Is the second or fourth finger longer? Place your hand on a sheet of ruled paper so that the tip of the fourth finger (counting the thumb as the first) is on a line and the finger is perpendicular to the line. Look to see whether the second finger is above or below that line. Test your partner and a number of other people, making sure to use the same side hand, and testing an equal number of males and females. Record the sex of each person examined. What pattern, if any, is there to the results?

2. Design a study in which you can measure variations within a population. Grow plants from seeds, maintaining the same conditions for all the plants. Take measurements of each plant during the growth period and draw a distribution curve, as you did in class, showing the average rates of growth versus the number of plants. Given the same environmental conditions, might difference in height be attributable to genetic differences?

Investigation

15 | *Patterns of Variation*

1. _____

2. _____

_____

3. _____

_____

4. Enter your answers on the data chart.

Lengths in Millimeters of Mung Bean Sprouts

Sprout Number	1	2	3	4	5	6	7	8	9	10
Length (mm)										
Sprout Number	11	12	13	14	15	16	17	18	19	20
Length (mm)										

5. _____

6. _____

_____

_____

7. Enter your data on the graph.

Distribution Curve Showing Range and Frequency of Mung Bean Sprout Length

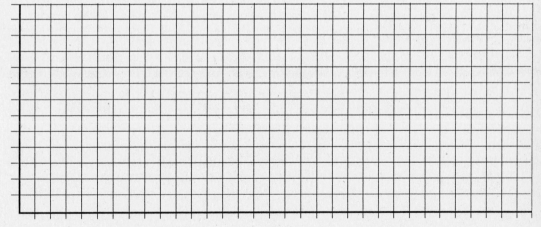

Number of Sprouts

Length of Sprouts in Millimeters

8. _____

_____

_____

9. _____

_____

_____

_____

10. _____

_____

_____

11. Enter your answers on the data chart.

Lengths of Earlobes to Nearest Millimeter										
Earlobes	1	2	3	4	5	6	7	8	9	10
Length (mm)										

12. _____

_____

_____

13. _____

_____

_____

14. _____

_____

_____

15. _____

_____

_____

16. _____

_____

_____

17. _____

_____

_____

16 | *Resistance in Bacteria*

Learning Objectives

- To investigate natural selection by observing bacterial survival during environmental changes.
- To learn basic techniques of microbiology.

Process Objectives

- To experiment with bacteria that carry selectively expressed genetic mutations.
- To analyze data on bacterial growth.

Materials

For Class

- Ultraviolet light source
- Timer or watch with second hand
- Incubator, stabilized at 38°C
- *Escherichia coli* stock culture

For Group of 2
For All Parts

- Glass-marking pen or crayon
- Inoculating loop
- Ultraviolet goggles
- Bunsen burner

Part I

- Nutrient agar plate

Part II

- Incubated plate from Part I
- Nutrient broth tube, with cotton plug or stopper

Part III

- Incubated nutrient broth tube from Part II
- 5 Nutrient agar plates

Part IV

- Culture plates from Part III

How could you use mutant strains of bacteria to observe natural selection?

Introduction

After studying many different live organisms and fossil remains of others, Charles Darwin hypothesized that organisms evolve slowly in response to changes in the environment. He thought that environmental changes favored certain individuals. These individuals survived and passed the adaptive traits to their offspring. Darwin called the survival of certain individuals natural selection. In cases of natural selection, individuals with the genetic makeup that favors survival will pass their genes to their offspring.

Each species of bacteria has variations caused by mutant genes in the gene pool. Although not normally expressed, the mutant genes may help individuals to survive changing environmental conditions. For example, streptomycin is an antibiotic that kills most bacteria. However, some bacteria have variations or mutations that make them streptomycin resistant. These bacteria survive streptomycin treatment to form new colonies. Antibiotic-resistant **pathogenic** (disease-causing) bacteria make medical treatment difficult. Scientists are always searching for new antibiotics that will overcome resistant pathogens.

The presence of mutations that confer resistance presents a quick way of observing natural selection. Since bacteria reproduce in about 20 minutes, growth of resistant colonies may be observed over a 24-hour period.

Prelab Preparation

In this investigation, you will study natural selection in *Escherichia coli* (*E. coli*) bacteria over several days. You will expose these bacteria to ultraviolet light. Most *E. coli* are sensitive to ultraviolet irradiation and will die. However, some contain a mutation that makes them ultraviolet-resistant. These mutant cells will survive exposure to ultraviolet light. You will test colonies grown from these survivors to see if they pass this trait to their progeny.

It is important to plan ahead when working with bacterial cultures. You must carefully follow sterile technique. If working conditions are not sterile, your results could be incorrect due to contamination. Also, nonsterile conditions would be a risk to your health, especially if you were working with pathogenic organisms. Read through the procedure that you will be using in this lab and review the sterile technique used.

1. Why is it important to sterilize the loop and the tube lip both before and after removing a bacterial sample?
2. Why should you never set stoppers and petri covers down while you are performing transfers?

Each colony on a culture plate grows from a single cell. When you transfer organisms from a colony on a culture plate to nutrient broth, you are growing the progeny of an individual bacterial cell. Exposing the progeny to ultraviolet light demonstrates that resistance to the light is genetic and, thus, can be passed from generation to generation. This passing of mutations from parents to progeny is the basis of natural selection.

Controls are essential to any experiment. You will incubate a control plate that has not been inoculated with bacteria and a control plate that has been inoculated but not irradiated with the ultraviolet light. (See Step E.)

3. Why is it necessary to incubate a control plate that has not been inoculated with bacteria?
4. Why would you incubate a plate that has been inoculated with bacteria but has not been exposed to the experimental variable (the ultraviolet light)?

Procedure

Whenever you perform a procedure in this investigation, use the sterile technique described below.

Procedure for Removing a Sample from a Stock Culture

Light your bunsen burner or alcohol lamp, keeping it a safe distance away from your hands and clothing. Sterilize the inoculating loop by moving it through the flame until the wire glows. Allow the loop to cool (but do not wave it around) before touching it to a culture. Remove the stopper from the culture tube, but do not put the stopper down. Hold the culture tube near the bottom and pass the lip (top) through the flame. Insert the loop into the culture to pick up some bacteria. Remove the loop from the culture. Flame the lip of the tube again after removing the sample. Replace the stopper in the tube. CAUTION: Use proper procedures when using a bunsen burner; turn it off when you are not using it.

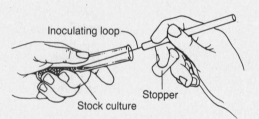

Procedure for Inoculating an Agar Plate

Use an inoculating loop to obtain a sample from the stock culture (as described above). Uncover the plate and lightly run the loop over the surface in a zig-zag pattern. Do not break through the surface of the solid media. Replace the cover of the petri dish immediately. Resterilize the loop. CAUTION: Wash your hands thoroughly after handling the bacterial cultures.

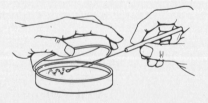

Strategy for Experimenting
When marking petri dishes for bacteriological experiments, mark the bottom of the dish, containing the culture, instead of the cover. Covers can get mixed up. Bacterial plates should be stored and incubated upside down.

Part 1 (Day 1)
A. Label one agar plate on the bottom with your initials and the date. Inoculate it with a sample from the stock *E. coli* culture.
B. Wearing safety goggles to protect your eyes from the ultraviolet (UV) light, place the inoculated plate under the light source and uncover the

plate. Turn the light on for 60 seconds. Turn the light off and replace the cover. Incubate at 38°C for 24 hours. CAUTION: Do not look at the UV light or expose your skin to it; UV light can damage your eyes and skin.

5. What type of *E. coli* do you expect will grow on the inoculated, irradiated plate? Explain.
6. Make a hypothesis about the possible results of exposing *E. coli* to ultraviolet light.

Part II (Day 2)

A bacterial colony appears as a small opaque spot or mound on the surface of the agar.

7. Count and record the number of bacterial colonies on the incubated plates.

C. Use sterile technique to transfer a sample from only one distinct colony to a tube of nutrient broth. Initial your tube. Gently roll the tube between your fingers to mix. Incubate at 38°C for 24 hours.

8. Do you think that all the cells in the colony will be resistant to ultraviolet irradiation? Explain.
9. Other than in laboratory experiments, under what conditions does ultraviolet resistance enable a bacterial cell to survive?

Part III (Day 3)

D. Label with your initials and the date 5 nutrient agar petri plates. Mark one MC for *media control* and do not inoculate. Use sterile technique to inoculate the other 4 plates from the culture you incubated in Part II. Mark one of the inoculated plates CC for *culture control* and do not irradiate. Mark the last plates Plate 1, Plate 2, Plate 3.

E. Follow the procedure in Step B to expose plates 1, 2, and 3 to ultraviolet light: Plate **1** for 1 minute; Plate **2** for 2 minutes; Plate **3** for 3 minutes. Cover and incubate all plates at 38°C for 24 hours.

10. What growth do you expect in each of the 5 plates?
11. Which plate do you think is most likely to have only colonies containing the mutant gene for ultraviolet resistance? Why?

Part IV (Day 4)

F. Examine your plates and count the number of colonies on each plate.

12. Record your results on the Team Chart of Colony Counts.
13. Compare your results with your classmates' using the Class Chart of Colony Counts.

Strategy for Analyzing

Compare the number of colonies in your MC plate with the class average for MC plates. Make similar comparisons between your other plates and the class averages for those plates.

Postlab Analysis

14. Was everyone's media control (MC) plate negative (without any bacterial growth)? Why?
15. Did everyone have growth of colonies on the culture control (CC) plates? If not, explain why.
16. If no colonies grew on the plates that were exposed to ultraviolet light but did grow on the CC plate, what would this mean?
17. Explain what conditions are necessary for natural selection to occur and what conditions affect the speed at which it occurs.
18. How did this experiment allow you to observe natural selection?

Further Investigations

1. Develop a strain of bacteria that is resistant to streptomycin. Be sure to use sterile technique and nonpathogenic bacteria. Outline your steps, and present evidence that your strain is a streptomycin-resistant mutation.

2. Write a report about the resistance of flies, mosquitoes, and other pests to DDT. Include the implications for public health and for agriculture.

Investigation

16 | *Resistance in Bacteria*

1. _____
   _____

2. _____
   _____
   _____

3. _____

4. _____

5. _____

6. _____
   _____
   _____

7. _____

8. _____
   _____
   _____

9. _____

10. _____
    _____
    _____
    _____

11. _____
    _____
    _____

12.–13. Enter your results on the data charts.

Team Chart of Colony Counts

Plate	MC	CC	1	2	3
Exposure Time					
Number of Colonies					

Class Chart of Colony Counts

Initials	Numbers of Colonies				
	MC	CC	1 min	2 min	3 min
Average					

14. _____

15. _____
_____

16. _____

17. _____
_____
_____
_____

18. _____
_____
_____

17

Human Evolution

Learning Objectives
- To identify differences and similarities that exist between modern apes and humans.
- To categorize fossil forms of hominids by examining their structural features.

Process Objectives
- To classify primates based on differences in features of the face, teeth, jaws, and braincase.
- To make inferences about evolutionary changes in the living and fossil anthropoids studied.

Materials
- Accurate drawings of anatomical structures of modern humans, African apes (chimpanzee or gorilla), and fossil hominids
- Life-size casts of anthropoid forms (optional)

For Each Student
- Metric ruler
- Protractor

How is comparative anatomy used to infer relationships between living and fossil human and ape species?

Introduction

You may have wondered how scientists who study fossils and evolution can distinguish and classify the different species they observe. What techniques reveal that a tooth, skull fragment, or footprint is from a particular time period or that one fossil is related to another? Comparative morphology is an effective method in the study of modern and fossil **anthropoids,** the subgroup of primates that includes monkeys, apes, and humans. The subgroup of primates known as **hominids,** including humans and their immediate ancestors, is also studied this way.

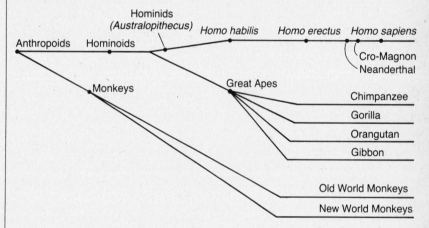

Prelab Preparation

Anthropology is the study of the physical, social, and cultural development of humans. As such, it encompasses the study of non-human primates, prehistoric humans and their cultures, and all present human cultures. When found, fossil skeletons of early humans are incomplete. Usually, just a few teeth and small pieces of a skull or a jawbone are discovered and thus available for anthropological study.

> **1.** Why do you think fossil skeletons of early humans are not complete?

For over a century, paleoanthropologists have used comparative morphology to infer relationships because it is based on accurate examination and description of physical features. For this investigation, imagine yourself at an excavation site where fossil anthropoids have been unearthed. Your task is to determine their identity. You will observe anatomical features such as manual dexterity and the form of the braincase, teeth, and jaws to discover important information about the anthropoids that have been "unearthed" in this investigation.

Procedure

A. You have an opposable thumb that enables you to easily manipulate objects and use a variety of tools. You may recall that the human's increased ability to perform tasks is attributed to a proportionally larger brain. Other primates, such as the gorilla, also have opposable thumbs, but the gorilla's hand is used for both walking and grasping. Study the drawings of the primate hands. Compare the hands of the human and the gorilla.

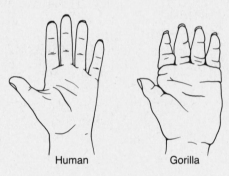

Human   Gorilla

2. Note and record in your data chart for "Comparison of Hands" the observed similarities and differences.

Structurally similar features in different organisms are called **homologues.** They are one category of evidence used for inferring patterns of evolutionary relationships between both living and fossil forms.

B. Refer to the checklist of features for the comparative study of human and ape skulls.

Brain Capacity: The circle drawn on each skull represents the brain capacity of each primate. Measure the *radius* of each circle in centimeters, cube this number, and multiply by 1,000 to *approximate* the lifesize brain capacity in cubic centimeters.
Lower Face Area: Measure *a* to *b* and *c* to *d* in centimeters for each skull. Multiply these 2 numbers together and multiply the product by 40 to *approximate* the life size lower face area in square centimeters.
Brain Area: Measure *e* to *f* and *g* to *h* in centimeters for each skull. Multiply these 2 numbers together and multiply the product by 40 to *approximate* the life size brain area in square centimeters.
Jaw Angle: Note the two lines in the nose area of each skull. Measure the inside angle with your protractor. This measurement will indicate how far the jaw projects outward.
Sagittal Crest: This is the bony ridge on the top of the skull. Note its presence or absence in each skull.
Brow Ridge: This is the bony ridge above the eye socket. Note its presence or absence in each skull.
Teeth: Note the number of teeth in the lower jaw, and the number of each kind of tooth.

Fact Sheet
1. A large brain capacity is characteristic of modern humans.
2. Less face area and more brain area is typical of modern humans.
3. A jaw angle of approximately 90° is a trait of *Homo sapiens*.
4. The sagittal crest is the area of attachment for the muscles used to move the lower jaws. As the lower jaw became smaller in later hominids, the size of this bony crest decreased.
5. Modern humans have lost the characteristic of having a prominent brow ridge.

---

3. Identify and record in your data chart for "Comparison of Apes and Human Skulls" both similarities and differences that

are evident from observations of ape and human braincases, faces, teeth, and jaws.

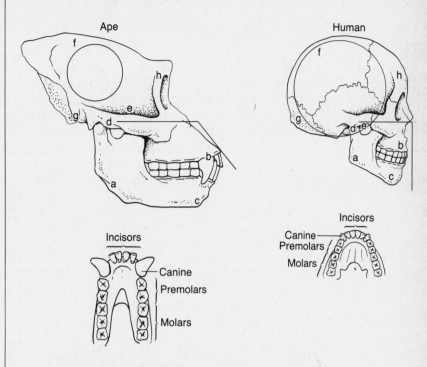

C. In this procedure you are to investigate fossil specimens.

4. Refer again to the checklist of features, and to the data chart for "Comparison of Fossil Hominids." Measure what features you can for the four skulls of fossil hominids. Record your measurements in the data chart.

Strategy for Classifying

Ape-like features of fossil hominids had the same functions as those features in modern apes. Human-like features of fossil hominids had the same functions as those features of modern humans. Consider these as you classify the fossils.

Fossil Hominids

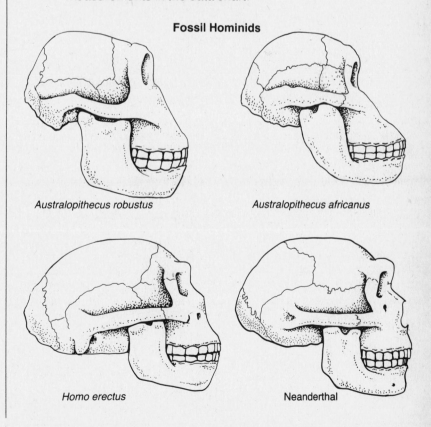

Australopithecus robustus

Australopithecus africanus

Homo erectus

Neanderthal

Strategy for Inferring
Consider where apes and humans
live and the functions of their facial
features as you try to infer their
diets.

5. Classify the features (braincase, face, teeth, jaw, etc.) of these hominids as more ape-like, more human-like, or intermediate on your data chart.

Postlab Analysis

6. From your comparative study of the human and the ape, what insight into their diets can you infer from the teeth and jaws?
7. What is the relationship between the skull features and the brain size of the human as compared to the same features of the ape?
8. Were there similarities among the specimens that you did not expect to find? What were they?
9. Did you have difficulty categorizing the fossil hominids? How might some of the fossils be related to both humans and apes without actually being either?

Further Investigations

1. Continue your study of human evolution by visiting a natural history museum or reading articles in journals or books.
2. Do library research on other hominid fossils for more insight into how humans have evolved.
3. Contact an archaeologist at a local college or university and arrange to visit a site where an excavation of remains of early inhabitants of the area is being conducted.

Investigation

17 | *Human Evolution*

1. _____

_____

2. Enter your answers on the data chart.

Comparison of Hands	
Human	**Ape**
_____	_____
_____	_____
_____	_____
_____	_____
_____	_____
_____	_____

3. Enter your answers on the data chart.

Comparison of Human and Ape Skulls		
	Human	**Ape**
Brain Capacity (cu. cm)	_____	_____
Lower Face Area (sq. cm)	_____	_____
Brain Area (sq. cm)	_____	_____
Jaw Angle (degrees)	_____	_____
Sagittal Crest (absent/present)	_____	_____
Brow Ridge (absent/present)	_____	_____
Teeth (one jaw)	_____	_____
molars	_____	_____
premolars	_____	_____
canines	_____	_____
incisors	_____	_____

4.–5. Enter your answers on the data chart.

Comparison of Fossil Hominids

Fossil Homonid	Brain Capacity (cu. cm)	Lower Face Area (sq. cm)	Brain Area (sq. cm)	Jaw Angle (degrees)	Sagittal Crest (yes/no)	Brow Ridge (yes/no)	Teeth (number of each)
Neanderthal							
Homo erectus							
Australopithecus africanus							
Australopithecus robustus							

H = human-like
A = ape-like
I = intermediate

6. _____

_____

_____

_____

7. _____

_____

_____

_____

8. _____

_____

9. _____

_____

_____

18

Dichotomous Keys

Learning Objectives
- To construct a dichotomous key to classify objects.
- To use an effective classification system to improve communication about organisms or objects.

Process Objectives
- To make careful observations about similar objects based on physical characteristics.
- To classify the objects by group in a dichotomous key.
- To test how well your dichotomous key communicates your intentions by having classmates use your key to classify objects.

Materials
For Group of 2–4
Part I
- 2 Photocopies of geometric shapes
- Scissors
- Small plastic bag or envelope
- Tape or glue

Part II
- Set of cut-out shapes from Part I

How can a classification system help you communicate about organisms or objects?

Introduction

Are you surprised to learn that biologists believe there may be as many as 10 million different species of organisms on earth? Only about 1.5 million of them have been named and described, however. Even though each is different from all others, there are still many similarities among organisms. These similar characteristics have made it possible to place organisms into categories. This categorizing or classifying is very important because it makes it possible to communicate about organisms more efficiently and effectively.

Classification systems are used by nonscientists and scientists alike. For example, when using a library, you locate the materials you need by following the library's system for grouping its collection. **Entomologists** (ent-eh-MOL-eh-jists), scientists who study insects, classify the thousands of species of web-building spiders according to the shapes of the webs they weave.

By classifying organisms according to certain traits, biologists are able to determine which organisms may be related in their evolutionary histories. In taxonomy, the science of identifying and classifying living things, grouping is done on the basis of similarities. But what criteria do you apply to identify and then to classify organisms? In this lab you will learn how to develop a simple system for identifying characteristics, as well as a way to group objects according to the characteristics they have in common.

Prelab Preparation

A simple way to classify objects is by using a **dichotomous** (die-KOT-uh-muhs) **key.** A dichotomous key is a series of steps for classifying an object or organism based upon its particular characteristics. At each step you are given a choice of 2 alternative categories in which to place your organism. The goal is to eliminate one of the 2 categories at each step and arrive at a grouping that specifically defines each object or organism.

1a	Prokaryotic cells — — — — — — — — — — — — —	Kingdom Monera
1b	Eukaryotic cells — — — — — — — — — — — — —	Go to 2
2a	Unicellular organism — — — — — — — — — —	Kingdom Protista
2b	Multicellular organism — — — — — — — —	Go to 3
3a	Heterotroph — — — — — — — — — — — —	Go to 4
3b	Autotroph — — — — — — — — — — — — —	Kingdom Plantae
4a	Absorbs nutrients — — — — — — — — — —	Kingdom Fungi
4b	Ingests food with mouth — — — — — — — —	Kingdom Animalia

1. Use the dichotomous key above to classify the following organisms: mushroom, euglena, moss.

Procedure

Part I

A. Your teacher will provide you with 2 photocopies of drawings of several geometric shapes (see page 110). Use scissors to cut out both sets of shapes. BE SURE TO KEEP THE 2 SETS SEPARATE. Put one set aside in a small plastic bag or in an envelope for later use. Leave the other set on your worktable.

B. Look at the chart that classifies all living things. It was constructed based upon the dichotomous key in the Prelab Preparation. Compare the chart to the key. Note how each step in both the key and the chart contains 2 alternative classifications—1a and 1b, 2a and 2b, etc.

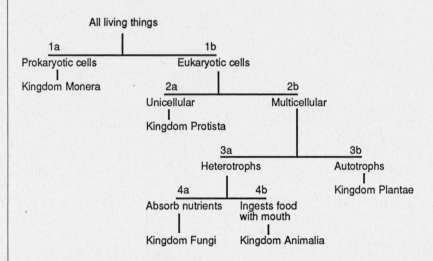

2. How do the 2 alternative classifications at each step differ?

Strategy for Classifying

Use similarities to categorize items into groups. Use differences to separate items as you break down groups further.

C. Now you are going to prepare a chart of your geometric shapes that is similar to the chart classifying all living things. Start at the top of a blank sheet of paper by copying the following :

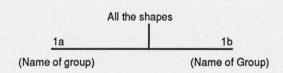

3. How will you proceed in preparing your chart so that it is similar to the one that classifies all living things?

Now examine your set of shapes. As a team, divide the shapes into groups based on their similar characteristics. Give these 2 groups names according to the characteristic they share.

4. Record the grouping names on your chart as the first grouping.

D. Next divide each of the first 2 groups into 2 groups.

5. Record the names of each group on the chart.

E. Continue dividing the groups until each shape has been separated from the others. Record the names of the groups on your chart and tape or glue the shapes onto the chart into their final separate positions.

6. What did you notice about the groups after each division?

Part II

F. Using the chart you have created in Steps C–E, you will next create a dichotomous key for the shapes. Look again at the dichotomous key for all living things. Note once more how each step contains 2 alternative classifications. Further, each step instructs you to proceed to another step, where again you must make another classification. For example, in Step 2, you had to decide between unicellular and multicellular organisms. If your organism was multicellular, you then went to Step 3, where again you had to make a classification—heterotroph or autotroph. If your organism was an autotroph, you proceeded to Step 4 for the final classification. If your organism was a heterotroph, you were able to classify it as a member of the Kingdom Plantae and did not go to Step 4.

To construct your key for the shapes, review the chart you just made. Look at the shapes included in each grouping, and at the group names you assigned. Work as a team to use these names as descriptors in your key. Remember to follow the numbering used in the "all living things" chart and key—1a and 1b for the first grouping, 2a and 2b for the second grouping, etc. Make sure your key includes steps that instruct the user to "go to" the next category of classification, if necessary.

G. Exchange dichotomous keys with another team. Use the key designed by that team to group your second set of shapes into groups. The other team will use your key to divide their second set of shapes into groups.

7. Compare your key with the keys developed by other groups. How are they similar? How are they different?
8. Could another team follow your key correctly to the end? What problems did they have, if any?

Postlab Analysis

9. How does the general diversity of characteristics of the groups compare between the first and third groupings?
10. What advantage does the dichotomous key you developed have over merely writing a description of the geometric figures?
11. Is it possible to use different criteria for classifying the same group of objects? Explain.

Further Investigation

1. Obtain a dichotomous key used by biologists and compare it to the key your group developed. Use the biological key to identify some simple organisms.

Strategy for Communicating
To check if your dichotomous key effectively communicates your classification system, it must be tested by someone other than yourself.

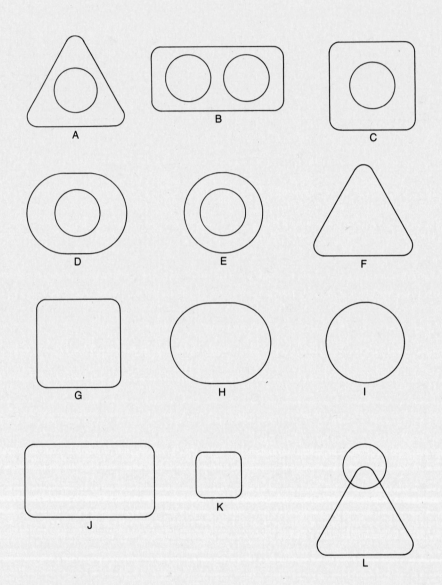

A

B

C

D

E

F

G

H

I

J

K

L

Investigation

18 | *Dichotomous Keys*

1. _____

2. _____

3. _____

_____

4.–5. Record your answers on your chart.

6. _____

7. _____

_____

_____

8. _____

_____

9. _____

10. _____

_____

11. _____

_____

19

Virus Models

Learning Objectives
- To learn that viruses occur in a variety of different shapes.
- To compare the structural properties of 3 different viruses.
- To relate virus structure to viral infection of cells.

Process Objectives
- To model viral structure.
- To communicate similarities and differences in virus models.
- To infer from models how a virus infects a cell.

Materials
Part I: Adenovirus (icosahedral virus)
For Group of 4–6
- Lightweight cardboard or manila folder
- Ruler
- Scissors
- Glue
- 12 Small straws or toothpicks
- 12 Cotton balls (or pieces of sponge)
- 4 Colored pencils or markers
- Compass and protractor

Part II: Tobacco Mosaic Virus (helical virus)
For Group of 4
- Cardboard tube from toilet tissue
- Needle probe
- Modeling clay
- Toothpicks (at least 64)
- Ribbon, about 30 cm

Part III: Influenza Virus (enveloped virus)
For Group of 4
- Sturdy round balloon
- Lightweight cardboard or manila folder
- Scissors
- 2 Colored pencils or markers
- Glue or cellophane tape

How does the structure of a virus make it an effective agent of infection?

Introduction

Viruses are less complex than any cell. All viruses are obligate intracellular parasites: they require host cells in order to reproduce. Viruses never grow— they are assembled directly into the full-sized virus particle.

Virus **capsids,** the protein capsules that contain the viral DNA or RNA, come in a variety of shapes. The capsid structure serves important functions in the process of cell infection by a virus: protecting the virus from the organism's immune system, inserting the virus into the cell, and dispersing new viruses in order to infect more cells.

The capsid is usually made up of a specific number of subunits. Subunits are packed together in precise arrangements. The pieces fit together like parts of a puzzle to give the viruses their exact shapes.

During viral reproduction, the capsid may remain outside the host cell. The tail fibers of the virus attach to the host cell's cell wall or cell membrane. Then, like a hypodermic needle, the virus injects its DNA or RNA into the host cell to infect it. Some viruses enter their host cells intact; the capsids must dissolve before their genetic material can be released and infect the cell.

As you build your models, consider how shape helps the virus function.

Prelab Preparation

Carefully read Chapter 19 on viruses.
1. Name 5 diseases caused by viruses. For each of these, list the shape of the capsid and the type of cell it infects.
2. Using pictures in Chapter 19, draw a simple picture of an icosahedron and a helix.

Procedure

Part I: Adenovirus—Icosahedral virus
A. Cut out your model of adenovirus from lightweight cardboard or a manila folder. You may either prepare 20 triangles, each with one flap on each side of the triangle, for gluing, or a grid as shown in the template for an icosahedron. In this case you will have to score the lines between triangles on the grid. NOTE: Do not forget the flaps on the outer edges needed to glue the icosahedron together. It is very important that your triangles are accurately drawn. Each side of the triangle should be equal in length, with an angle of 60° between adjacent sides.
B. During assembly, 5 triangles must meet at every vertex or corner.
C. After you have constructed your icosahedron, push a small straw or toothpick through each vertex (there are 12 vertices altogether) and glue a piece of cotton or sponge at the external end.

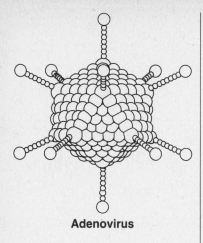

Adenovirus

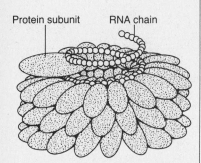

Protein subunit · RNA chain

Tobacco Mosaic Virus (TMV)

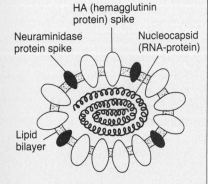

HA (hemagglutinin protein) spike

Neuraminidase protein spike · Nucleocapsid (RNA-protein)

Lipid bilayer

Influenza Virus

D. Draw circular capsomeres on at least five adjacent faces around a vertex of the icosahedron as in the template diagram. Capsomeres are the subunits of the protein capsid. Color these using a color code to show that 4 different protein types make up the face.

    3. Count and record the edges, faces, and vertices in the icosahedron.

    4. What does the word icosahedron mean (use a dictionary)? How does your model agree or disagree with this definition?

E. A DNA molecule bound up with protein is contained within the capsid.

    5. What functions does the capsid serve?

Part II: Tobacco Mosaic Virus—Helical virus

F. Follow the template to construct a helical virus. First, draw a spiral with at least 4 turns around the cardboard tube. Then, use a needle probe to puncture holes into the cardboard tube so that 16 holes are evenly placed in each of the 4 helical turns of the tube.

Template for Helical Virus (TMV)

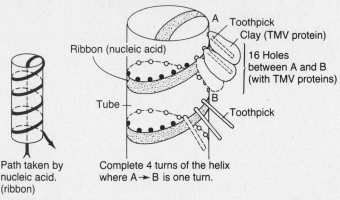

A · Toothpick · Clay (TMV protein)

Ribbon (nucleic acid)

16 Holes between A and B (with TMV proteins)

Tube · B · Toothpick

Path taken by nucleic acid. (ribbon)

Complete 4 turns of the helix where A → B is one turn.

G. Make 16 TMV subunits out of small pieces of modeling clay. Starting at the top of the roll, place a toothpick into each hole and then impale on the toothpick a piece of modeling clay in the shape of the TMV subunit shown in the template diagram. After placing the 16 TMV subunits, continue inserting toothpicks until all holes are filled.

H. Feed about 30 cm of ribbon up through the tube's central hole. Wrap the ribbon around the tube, following the path of the protein units (clay), and pressing it into the clay to hold it in place. The ribbon represents RNA. The external end of the ribbon should be taped or glued down. The template diagram shows the pathway of the nucleic acid and its position relative to the subunits. TMV contains about 2,130 identical protein subunits or capsid proteins. These form 130 turns with 16.33 proteins per turn. The RNA of TMV is 6,390 nucleotides long. Each protein subunit interacts with 3 nucleotides.

    6. Calculate how much ribbon RNA you would need to complete 130 turns in your model.

    7. The rod-shaped virus is 3,000Å (Angstroms) long and 180Å in diameter, a ratio of 17:1. Using this ratio, calculate how long the tube would be if it extended to the full length of 130 turns of the helix.

    8. How might the helical RNA of the TMV become uncoated?

Part III: Influenza virus—Enveloped virus

I. Blow up a round balloon until firm. This represents the lipid bilayer that surrounds the nucleic acid of an influenza virus.

J. The template for the influenza virus uses large spikes to represent the hemagglutinin protein and small spikes to represent neuraminidase. Cut

Template for an Icosahedron

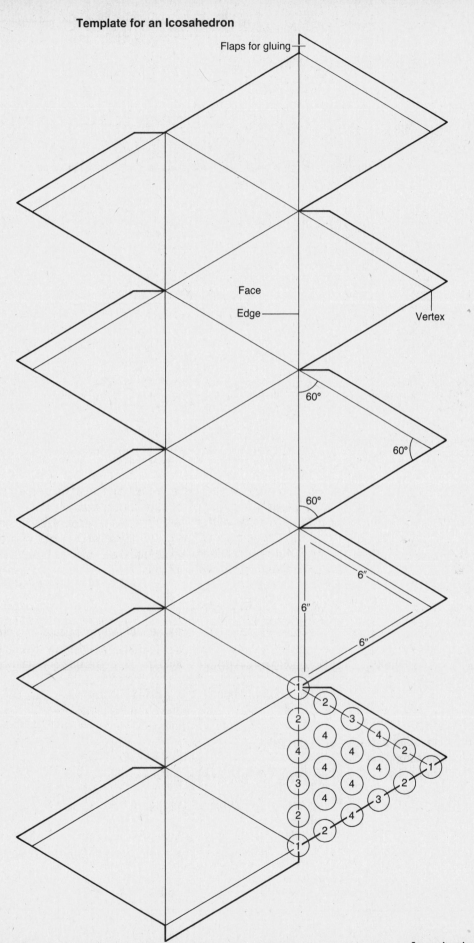

Flaps for gluing

Face

Edge

Vertex

60°

60°

60°

6"

6"

6"

spikes from cardboard and tape or glue them onto the surface of the balloon. Color these spikes differently so that they can be clearly seen.

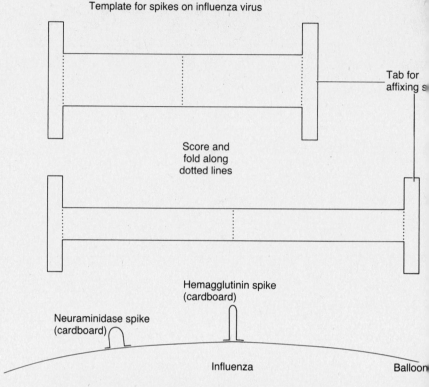

Template for spikes on influenza virus

Tab for affixing s

Score and
fold along
dotted lines

Hemagglutinin spike
(cardboard)

Neuraminidase spike
(cardboard)

Influenza

Balloon

9. What is the advantage of having proteins jutting out?
10. Neuraminic acid is one of the components of cell membranes. What might be the function of neuraminidase (an enzyme) projecting from the virus surface?
11. What part of the influenza virus is recognized by an organism's immune system? What changes in viral structures (mutations) would make a virus better able to escape the immune system?

Postlab Analysis

12. Compare all of the models of adenovirus constructed by the class. How are they the same? Different? What does this tell you about the limits of model making? What specific aspect of viruses does your models not show?
13. Look at each of your models. What inferences may be made about the physical and chemical structure of the nucleic acid?
14. In what way is modeling a virus easier than studying a virus directly?

Further Investigations

1. Discuss as a class the relationship between the structure of a virus and its ability to infect cells.
2. Do library research to find out what viral diseases may be prevented by vaccinations.

Strategy for Inferring
In making inferences, think about what types of evidence would support or not support each inference.

Strategy for Modeling
Consider why model building is particularly useful in visualizing virus particles. How are viruses observed directly?

Investigation

19 | *Modeling Viruses*

1. _____

_____

_____

2.

3. _____

_____

4. _____

_____

5. _____

6. _____

7. _____

8. _____

9. _____

10. _____

11. _____

_____

_____

12. _____

_____

13. _____

_____

14. _____

_____

20

Effectiveness of Bactericides

Learning Objectives

- To work with bacteria under semi-sterile conditions.
- To discover which bactericides best inhibit bacterial growth.
- To compare the susceptibilities of different bacteria to the same bactericides.

Process Objectives

- To organize and share data from your experiment.
- To analyze your data critically.
- To predict which bactericides are more effective in inhibiting bacterial growth.

Materials

Part I
For Class

- Hot water bath (beaker of simmering water)

For Group of 2

- Sterile petri dish
- 2 Masking tape strips (5 cm length)
- Grease pencil
- Test tube containing 20 mL sterile nutrient agar
- Test tube holder (pincer type)
- Test tube rack or small beaker
- Inoculating loop
- Clean paper towel as a work surface
- Alcohol lamp or bunsen burner and matches
- *Sarcina subflava* culture
- *Escherichia coli* culture

Part II
For Class

- Incubator set at 26°C (warm room temperature)

For Group of 2

- Antiseptic sample (from home)
- Disinfectant sample (from home)
- Antibiotic ointment (from home)
- Inoculating loop

What are some effective ways to slow or stop the growth of unwanted bacteria?

Introduction

Most bacteria (and other microorganisms) are harmless. In fact, many bacteria are beneficial. Cheesemaking, decay, and soil building are a few of the important processes that depend on the action of **saprophytic** bacteria, which thrive on decaying organic matter. However, some bacteria are **pathogens** (disease causers). Tuberculosis, tetanus, strep, syphilis, and some pneumonias are a few of the serious bacterial diseases.

Scientists have learned how to use certain chemicals to kill bacteria. Some, such as tincture of iodine and mercurochrome, are **antiseptics** and are used on cuts or wounds to inhibit bacterial infection. Others, like formaldehyde and chlorine bleach, are too concentrated or toxic for use on living tissue. These are called **disinfectants** and are used on clothes, surfaces, or other non-living objects. These agents generally work by either disrupting the cell membrane, causing the bacterium to lyse, or binding to the bacterium's enzymes which inhibits its activity.

Even though antiseptics and disinfectants are very useful in helping to prevent infections, we cannot use them internally to treat an infection. If bacteria enter our bodies, we rely on another class of chemicals called **antibiotics** to kill them. Although their use is now commonplace, antibiotics were only discovered about 50 years ago. Before then, more people died from infections than from all the wars in history combined. The antibiotics were the first of the "miracle drugs" and they have permanently altered the course of history.

Prelab Preparation

Read Chapter 20 in your textbook and review the laboratory procedure.

1. What is the purpose of the alcohol lamp or bunsen burner you will use in this lab?
2. What precautions must you take when working with an open flame?
3. How were the glassware and culture medium sterilized in preparation for this lab?
4. Since only saprophytic bacteria will be used in this lab, what should a suitable nutrient medium contain?
5. What is the actual composition of the agar plates used here?
6. What is agar and why is it used?
7. What are possible sources of contamination in this lab?

You will need to bring in some bactericidal materials for this lab. See the list for examples. Be sure to put them in small plastic or glass jars or bottles with tightly fitting lids. Unless you are given pre-sterilized containers,

- Beaker of lab alcohol (70–95% denatured ethanol, NOT potable)
- Sterile filter paper disks in sterile, covered container
- Alcohol lamp or bunsen burner and matches
- Set metal forceps, preferably with bent tips
- Sterile cotton-tipped applicators
- Sterile distilled water

Part III
- Small metric ruler

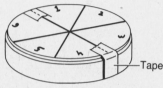

Bottom View of Petri Dish

wash your containers with hot, soapy water (use a dishwasher if available), and make sure all traces of soap are rinsed off with hot water.

Label each container and carefully fill each with one of the following:

Disinfectants—liquid bleach, bathroom bowl cleaner, liquid cleaners such as Lysol, Pine-Sol, Fantastic, Ajax.

Antiseptics—mouthwash, vinegar, rubbing alcohol, tincture of iodine, witch hazel, hydrogen peroxide, dishwasher liquid, mercurochrome.

Antibiotics—small tubes of any over-the-counter antibiotic ointments such as Neosporin, Bacitracin. (Bring in the entire tube.)

Procedure

Part I: Culturing bacteria

It is essential that **aseptic** or sterile technique be used in handling all bacteria. This will avoid contaminating pure cultures with unwanted species and will prevent self-infection. Even "safe" or nonpathogenic bacteria can be harmful in large amounts or in the wrong place, like your eyes.

A. Obtain a sterile petri dish. Use two strips of masking tape to keep the top and bottom halves together. Use a grease pencil to mark off 6 equal segments on the bottom (smaller) half of the dish. Write your names and date on the tape.

B. Place a test tube containing sterile nutrient agar in a boiling water bath until the agar is completely melted.

C. While the medium is melting, obtain a culture tube containing either *Sarcina subflava (S. subflava)* or *Escherichia coli (E. coli)*. Half the class will work with each species. Also, get a rack, paper towel, and inoculating loop.

D. After the agar in your test tube has liquefied, remove it from the hot water bath. Set it in the rack and let it cool until you can hold it comfortably in your hand, but without letting the agar harden.

E. Light the alcohol lamp or burner. CAUTION: Keep it at a safe distance away from clothing and skin . Hold the test tube of agar and your stock culture as shown. Working quickly but carefully, heat the inoculating loop until it glows orange. Have your partner remove the plugs or caps from the tubes (holding on to the plugs and not touching their bottoms which must remain sterile). Pass the lips of both tubes through the flame several times. Touch the loop to the edge of the liquid agar to cool it. Next, run the loop gently over the bacterial growth in the stock culture and immediately submerge the tip of the loop in the nutrient agar. Shake the loop in the liquid. Remove the loop, and quickly flame the lips of the tubes again before replacing the plugs. Flame the loop and put it aside on the paper towel. Turn the flame off.

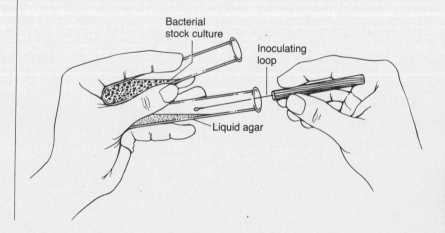

F. Gently roll the tube you inoculated in your hand to mix the bacteria with the medium. Your partner should remove the masking tape from one side of the petri dish and lift that edge just far enough for you to pour the entire contents of your tube into the dish. Replace the lid and tilt or gently swirl the dish to cover the entire surface with agar. Allow the medium to cool and solidify while you clean up and return all equipment. The tube you inoculated has been contaminated with bacteria and should be given to the teacher for special cleaning. CAUTION: Wash your hands and clean your work surface.

G. If you are doing Part II on another day, store your petri dish upside down in the refrigerator until you are ready (no more than 48 hours later).

8. Why do you lift the petri dish lid only part way?
9. What other precautions did you follow to avoid contamination?

Part II: Testing bactericides

To assay bactericidal effectiveness, you will conduct a **sensitivity test.** You will soak small filter paper disks in different bactericides and one disk in sterile water as a control. These disks are placed on the agar plate which already contains bacteria. When you place the agar plate in the incubator, the bacteria will grow uniformly wherever their growth is not effectively inhibited by a bactericide. This uniform growth is called a **bacterial lawn** and the regions where no growth occurs are termed **zones of inhibition.** A large zone of inhibition is created by a bactericide that is more effective at inhibiting that strain's growth than is a bactericide that creates a smaller zone.

H. Retrieve your petri dish and place it right side up on a clean paper towel.

I. Choose which 5 bactericides you wish to test. Cooperate with a group working with the other species of bacteria to enable you to compare the effect of the same bactericides on different bacteria.

10. Enter the names of the bactericides you have chosen in your data chart along with your predictions for the effectiveness of each.

J. Gather all the materials you will need. Carefully light the alcohol lamp or burner. Insert the tips of a pair of forceps in a small beaker of alcohol to sterilize them. Lift out the forceps, shake off any excess, and quickly pass the tips of the forceps through the flame several times. This will remove all traces of alcohol but will not heat the metal. CAUTION: Make sure no alcohol has dripped on your hands!

K. Use the sterile forceps to pick up a disk of sterile filter paper. Put the disk into one of the solutions or use a sterile, cotton-tipped applicator to smear it with an ointment you have selected. Shake off any excess. Open the labeled petri dish slightly and lightly press the disk onto the center of the appropriate segment. Make sure that on segment 1 you place the control disk, soaked in sterile water. Flame the forceps quickly before handling each disk. It is not necessary to immerse the forceps in alcohol again unless they have touched a non-sterile surface between uses.

L. When you are finished, re-tape your petri dish and incubate it at about 26°C for 48 hours or 30–35°C for 24 hours. CAUTION: Wash your hands and clean work surface.

Part III

M. Observe your petri dish after the incubation period. Yellow, glossy areas indicate growth of *S. subflava*. White, cloudy areas indicate growth of *E. coli*. Clear areas are the zones of inhibition. (Hold your dish up to the light to see the zones more clearly.)

Strategy for Organizing

When you are working with another group of students using the same bactericides, number the segments that contain the same antibiotic identically.

Strategy for Predicting

Read the product information that accompanied the bactericides you are using. Think about how the claims they make would apply to the bacteria you are looking at.

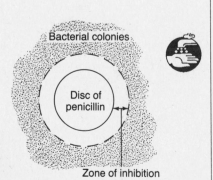

Bacterial colonies

Disc of penicillin

Zone of inhibition

N. Use a metric ruler to measure the distance that the zone of inhibition extends from the edge of each disk. Enter your measurements in the appropriate column of the data table. Enter your cooperating group's measurements for the other bacterial species.

Postlab Analysis

Strategy for Analyzing Data
When you analyze the effectiveness of a bactericide using the zone of inhibition technique you should compare each measurement to the control value.

11. Did the bactericides have the effects you predicted in Question 10? If not, explain.
12. How do your results compare with those of other groups who tried the same bactericides?
13. Note and record results from groups who tried bactericides different from yours.
14. Make a hypothesis about how a zone of inhibition is produced.
15. What do different sizes of zones of inhibition mean for a given strain of bacteria?
16. Why are some bactericides more effective than others?
17. Were both species of bacteria equally sensitive to the same bactericides? On what do you base your answer?
18. What effect might the concentration of the bactericides on the disks have on your results? How might you control variation of this type?
19. Based on your replies to Questions 10 and 11, how do the advertising claims for commercial bactericides compare with their performance?

Further Investigations

1. Use your textbook and the library to investigate the natural sources of antibiotics. Discuss the significance of this.
2. Read about Gram staining of bacteria in your school library. Then, use an inoculating loop and flame to make thin smears of *S. subflava, E. coli,* and other available bacteria on clean microscope slides with a drop of water. Allow the smear to dry thoroughly or pass it through a flame several times to speed the drying. Stain the smear with methylene blue or crystal violet. Rinse off the excess dye very carefully. Place a cover slip on the smear and examine it with a microscope under high power. Draw and measure the different types of cells.
3. Survey your surroundings for the presence of bacteria. Prepare several sterile agar plates by following the same procedure as before but without inoculating the medium. After the plates have hardened, use a loop to inoculate them with samples from your environment (for example, soil, pond water, your skin). Incubate your cultures as you did before. Note the appearances of different bacterial colonies. Examine the colonies under the microscope as described in Further Investigation 2. Look for other, nonbacterial growth, such as yeasts or molds.

Investigation

20 | *Effectiveness of Bactericides*

1. _____

2. _____
_____

3. _____

4. _____

5. _____

6. _____

7. _____

8. _____

9. _____

10. Enter your answer on the data chart.

Effects of Bactericides on *S. subflava* and *E. coli* Bacteria

Petri Dish Segment	Bactericide Used	Type (e.g., antiseptic)	Predicted Results		Actual Results (in mm)	
			S. Subflava	E. coli	S. subflava	E. coli
ex:	Penicillin	Antibiotic	moderate inhibition	strong inhibition	2 mm	4 mm
1	None	Control				
2						
3						
4						
5						
6						

11. _____
_____

12. _____

13. _____

14. _____

_____

15. _____

16. _____

17. _____

18. _____

_____

19. _____

_____

21

Protista

Learning Objectives

- To gain experience with microscope techniques learned in previous labs.
- To compare structure and activity of 3 different groups of protists.

Process Objectives

- To observe living protists under the microscope.
- To infer general characteristics of life from specific observations of protists.

Materials

For Group of 2

- 3 Microscope slides
- Cotton ball
- 5 Medicine droppers or pipettes
- Culture of *Ameba*
- Culture of *Paramecium*
- Culture of *Euglena*
- 3 Coverslips
- Compound microscope
- Distilled water
- Paper towel
- Yeast solution stained with Congo red
- Vinegar
- 10% Methyl cellulose solution (optional)
- Unlined paper

How do the life activities of protists illustrate characteristics that are common to all living organisms?

Introduction

Protists are microscopic, one-celled organisms found in almost every environment on earth. Although the ancestors of all life on earth were one-celled organisms, modern protists have evolved far beyond them. Many protists are extremely complex and perform the same basic life functions as do multicellular organisms. This fact illustrates the basic unity of all life.

In this lab, you will examine representatives of 3 groups of protists: Euglenophyta *(Euglena);* Sarcodina *(Ameba);* and Ciliophora *(Paramecium).* The Division Euglenophyta contains organisms that can function both as photosynthesizing autotrophs and as heterotrophs, depending on light conditions. Although they are classified as algal protists (you will learn more about them in Chapter 22), they are included in this lab because they exhibit many features similar to those of the protozoa. The Phylum Sarcodina and the Phylum Ciliophora contain heterotrophic protists, often called protozoans. The Kingdom Protista contains 4 phyla of protozoans and 6 divisions of algae.

Protists are common in pond water and are easy to raise in a laboratory culture. They are readily observed under the microscope. Comparing these organisms will give you a sense of the structure and activity found throughout the Kingdom Protista and an understanding of their similarity to all living organisms.

Prelab Preparation

1. Review Labs 1 and 14 . Make an updated list of the activities that you think a living organism should perform to qualify as living.

Lists may be discussed in class if time permits. Bring your list to the lab so that you may compare it with the observations and inferences you will make about protists.

Review the text material in Chapter 21 on the different groups of protozoa. *Euglena*, as mentioned above, are discussed in Chapter 22. Go back over the techniques you learned in Lab 2 for successful use of the microscope and for making wet mount slides.

Procedure

You will make 3 wet mounts in this lab. Observe them one at a time under the microscope. Make the slide of *Ameba* first, observe it, and answer the questions. Next, make the slide of *Paramecium* and observe it. Repeat Steps A–G and answer the same questions for each. Then repeat the process for *Euglena.*

CAUTION: A slide is a harsh environment for a unicellular organism. Do not make a wet mount until you are ready to use it.

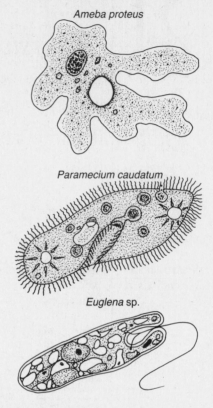

Ameba proteus

Paramecium caudatum

Euglena sp.

A. Make a wet mount of the appropriate culture of protist. Place a few strands pulled from a cotton ball in the center of a clean slide. Use a medicine dropper or pipette to draw in protists from the culture container. *Euglena* and *Paramecium* may be found by skimming the surface. *Ameba* may be found on the bottom. Put a few drops of the culture onto the cotton strands and place a coverslip carefully on the drops. Your teacher will explain when and if to use methyl cellulose in preparing your wet mounts.

Strategy for Observing
One-celled organisms are three dimensional. Continue to check for structural details by changing the fine focus and altering the light conditions.

B. Use the low power of your microscope to find the organisms and to observe their movement; the cotton strands should limit movement and make this easier. Use high power once an organism has been found where it is unable to move out of view; observe fine-level structure and processes. Experiment with different settings of the diaphragm to enhance various aspects of the organism. If the organism moves out of view, move the slide to keep up with it. If you lose it, return to low power to scan a larger field.

    2. Draw each organism on a separate sheet of paper as you observe it under high power. Refer to appropriate diagrams in the text to identify the various parts of each organism. Label each part in your drawing.

    3. Do you observe organisms of different sizes for one species, or are they all the same size? Infer a reason for this.

    4. Do you see any organisms of the same species side by side or attached to one another? Why are they like this?

C. As you observe each organism, fill in the data chart. You will repeat Questions 5–11 for each organism. Use phrases or sentences as is necessary.

    5. What is the general shape of the organism?

6. What is the overall color?
7. If some parts are different colors, what are the parts and what are the colors?
8. Describe the style of movement.
9. What structures are involved in movement? Refer to the descriptions in Chapter 21 for the appropriate terminology.

D. Look closely at the organism's cytoplasm. Protists have one or more **contractile vacuoles,** which collect excess water from the cytoplasm and expel it to the outside. Contractile vacuoles may be observed "pulsing": swelling with collected water and then shrinking as it is expelled.

10. Once you have found a contractile vacuole, count the number of pulses during a certain interval of time and record this in the data chart.(Movement could make this difficult.)

E. Place a drop of distilled water at one edge of the coverslip. Place a piece of paper towel next to the opposite edge of the coverslip. This will draw the distilled water underneath the coverslip.

11. Now observe the pulsing of a contractile vacuole during the same interval of time as before. How has it changed? Describe this in the data chart.
12. Why is the contractile vacuole important for survival in a freshwater environment, such as a pond?

F. Check for the formation of **food vacuoles.** To do this, add a drop of the solution of yeast stained with Congo red dye to one edge of the coverslip. Place a piece of paper towel along the opposite edge of the coverslip. The yeast will be drawn under the coverslip by the flow of water. Observe each organism to see how each reacts to the yeast.

13. What do you observe happening? Pay attention to where the yeast go and what happens to them.
14. Do the yeast change color within the cell? What may be happening to the yeast?

G. Place a drop of vinegar at one edge of the coverslip. Observe each protist.

15. What is the reaction to the vinegar? Does their activity change? If so, how?

Some protists form needlelike **trichocysts** (TRIK-uh-sists) when exposed to acid or to an enemy.

16. Do you notice trichocysts forming along the cell membrane? What is their function?

H. Repeat Steps A–G for each species of protist.

Postlab Analysis

17. *Ameba* and *Paramecium* are among the protists called protozoa, which means first animals. In what way is this accurate? In what way is this misleading?
18. Which style of protistan locomotion seems least specialized? Which style seems most specialized? Explain both of your answers.
19. Which species of protists had contractile vacuoles that were easiest to see? Which seemed to be most active (greatest rate of pulsing) before distilled water was added? Which were most active after the distilled water was added?
20. Did any of the protists give evidence of cell division or reproduction? If so, describe it.
21. Which organism ate the most yeast? Is there any specific

reason why any particular kind of protist would not be expected to eat at all?

22. How does each organism appear to protect itself against danger or enemies?

23. Could you describe one species as the most primitive or the most advanced? Why?

24. Which life activities from your list in Question 1 are shown by one or more of the kinds of protists? Which were not observed? Why not?

Strategy for Inferring
While you make your observation, ask yourself "Why is it like this?" and answer with "It makes sense that is it like this because..."

Further Investigations

1. Stain living protists with various stains and dyes (methylene blue, for example) to see internal structures more clearly.

2. Changes in the water may alter protistan behavior. Observe their responses to concentrated salt or sugar solution (hypertonic). Observe their responses to a range of pH, from acid to base. Observe their responses to a range of temperature.

3. Use the approach of this lab investigation on other kinds of protists (for example, *Euplotes, Stentor, Vorticella*). A great variety of other protozoa may be found occurring naturally in pond water.

4. Mix some powdered activated charcoal in with Congo red-stained yeast, and see if protists that eat the yeast will avoid the grains of charcoal.

Investigation

21 | *Protista*

1. _____

_____

_____

2. Make a drawing on a separate sheet of paper.

3. _____

_____

4. _____

_____

5.–11. Enter your answers on the data chart.

Observations of Three Protists

	Ameba	Paramecium	Euglena
General Shape			
Overall Color			
Color of Parts			
Style of Movement			
Structures for Movement			
Number of Pulses of Contractile Vacuole Per _____ Min.			
Vacuole Reaction to Distilled Water			

12. _____

_____

13. _____

_____

14. _____

_____

15. _____

_____

16. _____

17. _____

_____

18. _____

_____

_____

_____

19. _____

_____

_____

20. _____

21. _____

_____

22. _____

_____

23. _____

_____

_____

24. _____

_____

_____

_____

22 | *Algal Blooms*

What makes a pond die?

Introduction

Have you ever seen a thick green scum that floats on a lake, pond, or other still body of water? This is called an **algal** (AL-juhl) **bloom,** a sudden massive growth of algae. Despite this proliferation of growth—in fact it is because of it—the pond is dying. It is undergoing a process of **eutrophication** (YOU-truh-fuh-KAY-shun). When this happens, bacteria use up all the oxygen in the water, which causes all the animals and plants in it to die.

Eutrophication happens when the algae in a pond are given a good supply of nutrients, causing them to grow fast. Some pollutants, rather than being toxins, are actually good sources of nutrients for algae. Pollutants of this type are fertilizers, which provide nitrates, and laundry detergents, which are rich in phosphates. They are carried into ponds by seepage from household wastewater and runoff from rain.

Once the pond has a rich supply of these nutrients, algae multiply rapidly and a large population quickly forms. The aquatic plants that normally live in the pond are crowded out by the algae. As the decaying algae and plants sink to the bottom of the pond, they reduce its depth and lead to rapid bacterial growth. Finally, the bacteria use up the oxygen in the water which causes the fish to die.

In this lab you will look at the ability of different pollutants (such as detergents and fertilizers) to affect the growth of algae in culture.

Prelab Preparation

Green algae make up the Division Chlorophyta within the Kingdom Protista. Like other protists, they are either single-celled, colonial, or multi-celled. The green algae are considered to have given rise to plants. Like plants, they have chlorophyll, store food as starch, and contain cellulose.

1. What do green algae need in order to grow?
2. What is the major energy source for green algae?
3. What conditions are necessary to culture green algae?

Read the procedures for Part I and Part II. In Part I, you will prepare algal nutrient medium with different concentrations of common pollutants. You will then add cultures of living green algae. In Part II, after these cultures have grown for a week, you will count the number of cells in samples of algae from each culture flask under a microscope. This kind of test is called a **bioassay**. It uses biological results to determine whether an effect has occurred.

4. How will you set up a control for your experiment?

Procedure

Part I: Preparing algal cultures

A. With a 1% solution of one of the pollutants provided, and algal growth medium as the diluent, make up the following five concentrations of pol-

Learning Objectives
- To investigate how algae and other aquatic plants are affected by pollutants.
- To determine how common pollutants contribute to environmental problems.

Process Objectives
- To experiment with the effect of different concentrations of pollutants on the growth of algae.
- To analyze the data you collect from your experiments and interpret your findings.

Materials
Part I
For Group of 5
- Culture of green algae (about 30 mL)
- Freshwater algal growth medium (about 500 mL)
- 1% Solution of a common pollutant (such as laundry detergent or plant food fertilizer)
- 10-mL Pipette
- Pipette bulb
- Five 250-mL Erlenmeyer flasks (or equivalent containers)
- 10-mL Graduated cylinder
- 100-mL Graduated cylinder
- Nonabsorbant cotton roll

Part II
For Group of 5
- Cultures from Part I
- Dropper
- 5 Microscope slides
- 5 Cover slips
- (1-5) Compound microscopes
- Graph paper

lutants in 250-mL Erlenmeyer flasks:

Table of Concentrations

Flask Number	Percent Pollutant Solution	1% Pollutant Solution, mL	Algal Growth Medium, mL
1	Control	None	100
2	0.01%	1	99
3	0.1%	10	90
4	0.5%	50	50
5	1%	100	None

 Measure solutions and water carefully in graduated cylinders or use pipettes to measure small amounts of solutions. CAUTION: Never pipette with your mouth —use pipette bulb.

5. The solutions of pollutants have been incubated with bacteria to break down the chemicals in the pollutants. Why is this important to your experiment?

B. Inoculate each of your five flasks with 5 mL of green algae culture. To control for the amount of algae in each 5 mL, swirl the algae culture before withdrawing each 5 mL. Swirl the flasks to mix the algae with the other solutions. Plug the flasks with cotton. Place the algal cultures about 2 feet from a cool fluorescent light for one week. If you have a light timer, set it for 16 hours on and 8 hours off. If algae are grown in a window, cover the cultures with a piece of tissue paper. This will allow light penetration without overheating. Temperatures should be maintained at 20-28°C. Observe and note any changes you see in the cultures during the week.

6. In which concentration of your solution do you expect to see the most algal growth?

Part II: Analyzing your algal cultures

C. First observe your cultures in their flasks for any apparent differences in growth.

D. Each member of your group should count the algae from one of your 5 flasks. Make sure your slides and coverslips are clean and dry. Swirl each flask to distribute algal cells evenly, then remove a sample from the flask with a clean dropper. Place *only* one drop on your microscope slide and prepare a wet mount.

E. Allow the algae a minute to settle onto the slide. Look at the slide under the low-power objective (100x) of your microscope. Focus until you see the unicellular algae.

7. Describe and draw an algal cell. Identify any cellular structures you can see.

F. Count the number of algal cells in one field. Count four more fields. Move the slide across and down in a uniform pattern so you do not count the same field again.

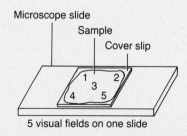
5 visual fields on one slide

Strategy for Observing
Since the green algae are too small to be seen without a microscope, you will look for a greenish tint and, perhaps, cloudiness in your flasks.

8. Record your counts on your data table.
9. Calculate and record the average number of algae cells from the 5 fields of view for each culture.
10. Which solutions had the lowest and the highest counts?
11. You will now plot the number of algal cells vs. the concentration of pollutants on separate graph paper. Use the graph below as a model.

Model Graph

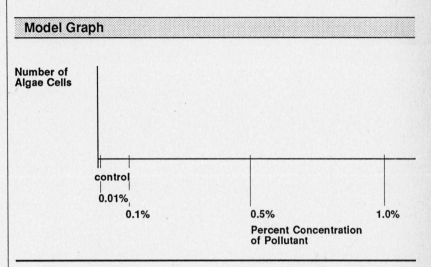

12. Do you find any changes in growth due to more concentrated solutions?
13. If you find the growth differs in populations at higher concentrations of pollutant, what do you hypothesize happens to these?
14. Does your pollutant solution contain a nutrient?

G. Add your team's data to the class graphs on the chalkboard. Use a different color chalk for each type of pollutant.

Strategy for Analyzing
Are there critical values of pollutant concentration above which algal growth declines?

Postlab Analysis

15. Which pollutants supported the least and the greatest growth of algal?
16. What do your graphs tell you about the potential effects of the pollutant on an aquatic ecosystem?
17. How could you modify your procedure to use algal bioassays for routine testing of the quality of pond water.

Further Investigations

1. After you have completed your sample counts, you might like to add phosphates or nitrates to the cultures which showed little growth and grow them for another week. Perhaps the nutrient "boost" will initiate rapid growth.
2. Some micronutrients (for example, metal compounds like iron salts and copper salts) are necessary in low concentrations for algal growth, but are toxic in higher concentrations. How would you design an experiment to test this?
3. Test some common pollutants that you believe might be harmful to algae. These substances could include motor oil, antifreeze, phenolic cleansers, granular road salt, and household chlorine bleach. Design a test to determine at what concentration these substances might be toxic to algae.

4. Repeat your investigation with cultures of other species of algae. Do different species of algae show greater sensitivity or greater tolerance to pollutants in the water?

5. Find out how a water treatment plant removes contaminants from wastewater. What happens to the water once the contaminants are removed?

Investigation

22 | *Algal Blooms*

1. _____

2. _____

3. _____

4. _____
   _____

5. _____
   _____

6. _____

7. _____

8.–9. Enter your answers on the data table.

Table of Algal Cell Counts

Fields	Control	Solutions of (type of pollutant): _____			
		0.01%	0.1%	0.5%	1%
1					
2					
3					
4					
5					
Total					
Average					

10. _____

_____

11. Graph numbers of algae vs. concentration of pollutant on a separate piece of graph paper.

12. _____

13. _____

_____

_____

14. _____

_____

15. _____

_____

16. _____

_____

_____

17. _____

_____

_____

23

Fungi on Food

Learning Objectives
- To locate and recognize fungal growth on food.
- To relate the growth of fungi to environmental conditions.

Process Objectives
- To observe the growth of fungi on food sources.
- To hypothesize about the conditions that favor the growth of fungi and those that inhibit it.

Materials
For Group of 2
Prelab Preparation
- Adhesive labels
- Plastic sandwich bags, with ties
- Moldy foods

Parts I and II
- Dissecting microscope or good quality magnifying lens.

Part I
- 2 Sterile petri dishes of potato dextrose agar
- Grease pencil
- Mold culture (optional)
- Matches or flint striker
- Alcohol lamp or Bunsen burner
- Forceps
- Cotton swab
- 5% Propionic acid solution in a small beaker
- 2 Masking tape strips (5-cm length)

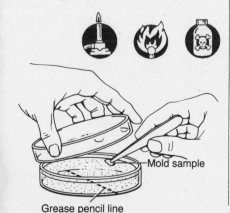

Mold sample

Grease pencil line

Under what conditions do common fungi grow?

Introduction

What moldy, inedible surprises have you found in the back of your refrigerator? Perhaps you have wondered how food becomes moldy and how to prevent it. The term mold refers to several common types of filamentous fungi that create fuzzy growths on moist, organic food sources. Like all fungi, molds reproduce mainly by means of microscopic, airborne or waterborne spores. These dormant spores, produced asexually, are virtually everywhere in our environment, including on our food, waiting for an opportunity to germinate and grow. In this investigation, we will explore the growth requirements of these familiar fungi.

Prelab Preparation

Attach a label to each of several plastic bags and place in each bag a small sample of moldy food from the family bread box or refrigerator.

1. What are some foods that are likely to become moldy?

Seal the bags with ties and mark the labels with the names of the foods. Store the bags at room temperature or in a refrigerator if the lab will be done more than a few days from now.

Procedure

Part I

A. Working with a partner, get 2 sterile petri dishes of potato dextrose agar. Without opening the dishes, turn each one over and draw a line across the middle of the bottom with a grease pencil.

B. Examine your mold samples through a dissecting microscope. Select a dense growth of mold from either a sample you brought from home or one supplied by your teacher.

C. Light an alcohol lamp or Bunsen burner and sterilize a pair of metal forceps by running them through the flame several times. CAUTION: Use extreme care around an open flame. When the forceps have cooled, use them to pinch off a sample of the mold you have selected. Gently touch the sample to the agar in each petri dish in several places on both sides of the grease pencil line. Lift the lids of the dishes as little as possible to do this.

D. Dip a clean cotton swab into the 5% propionic acid solution and lightly swab one half of the agar in each dish, staying to one side of the grease pencil line. CAUTION: Although propionic acid is an approved food additive in low concentrations, it could be toxic if consumed in large amounts. Again, lift the lids of the dishes as little as possible. Immediately place a piece of masking tape at the propionic acid-treated side of each dish so that the tape holds the lid and bottom together. The tape will

Label with name, date, source of mold

Label "Propionic Acid"

Treated side   Untreated side

Grease pencil line

Strategy for Hypothesizing

Remember that a hypothesis should be testable by means of an experiment. List several probable hypotheses or explanations.

Strategy for Observing

Before observing fungal cultures under a microscope, be sure you know what you are looking for. Review Chapter 23 in your textbook.

also provide a surface on which you should write "Propionic Acid." Use a second piece of tape in the same way on the opposite (untreated) side of each dish. Label this tape with your name, date, and source of the mold (food source). You should now have 2 identically prepared cultures.

E. Design an experiment to determine which of 2 opposite environmental conditions are best for fungal growth. Some possible combinations are warmth/cold, light/dark, and moisture/dryness. When you are ready, write one environmental condition on each dish label and incubate the dishes under the conditions you selected. Be sure to enter *your* environmental conditions on the top line in the last 2 columns of the data table.

2. Because of the way the dishes were prepared, what additional factor will you be testing at each location?

3. Form a hypothesis about the conditions that favor fungal growth and those that inhibit it.

Part II

F. After one week, retrieve your mold cultures. Working with your partner, examine each dish briefly through the dissecting microscope.

4. Record your observations in the data table using the symbols given under the table for amount of growth seen. Also record any pigmentation you see.

G. Examine some other students' culture dishes, especially ones prepared from other types of mold.

5. Record your observations in the data table.

6. What, if anything, happened on the side of each dish to which the 5% propionic acid solution was added?

Postlab Analysis

7. Compare the results for the different fungi you observed.

8. What have the results of this investigation shown about the conditions that favor fungus growth and those that inhibit it? Do the results support your hypothesis?

9. Based on your conclusions, if you wanted to keep a nonsterile food product for a long time, under what conditions would you keep it to prevent it from becoming moldy?

Further Investigations

1. If any of the fungi you examined are green, determine which structures contain the green pigment. Do you think the pigment is chlorophyll? Why or why not? How could you find out?

2. On which of the following types of food would fungi be likely to grow sooner:
    a. Additive-free bread or bread containing chemical additives?
    b. Additive-free whole-grain bread or additive-free white bread?
    c. Natural cheese or processed cheese?
    d. Regular strawberry preserves or low-sugar strawberry spread?
    Design a simple experiment to test your hypotheses.

3. Check the labeled ingredients of packaged baked goods (bread, cookies, bread crumbs), processed cheeses (individually wrapped slices, cottage cheese), and other processed foods. How many food additives do they contain? What do you think are the functions of these chemicals?

Investigation

23 | *Fungi on Food*

1. _____

_____

_____

_____

2. _____

_____

_____

3. _____

_____

_____

4.–5. Enter your answers on the data table.

Data Table for Moldy Foods

Species or Type of Mold (Identify by food source)	Condition 1: _____		Condition 2: _____	
	Normal	Propionic Acid	Normal	Propionic Acid
_____	_____	_____	_____	_____
_____	_____	_____	_____	_____
_____	_____	_____	_____	_____
_____	_____	_____	_____	_____
_____	_____	_____	_____	_____

Key: 0 = no growth
       + = poor or little growth
       ++ = good growth
       +++ = excellent growth; signs of reproduction (pigmentation)

6. _____

_____

_____

7. _____

_____

_____

8. _____

_____

_____

_____

9. _____

_____

_____

24.1 *Plants and Energy*

What is the primary energy source in the food chain?

Learning Objectives
- To study the conditions under which starch is made by green plants.
- To understand the role of green plants as food producers for themselves and other organisms.

Process Objectives
- To conduct an experiment that tests for production of starch in green leaves.
- To analyze data on starch production in portions of a leaf exposed to light and in portions kept dark.

Materials
For Group of 2
- 2 Pairs of safety goggles
- 2 Geranium leaves
- 5-mL Iodine solution
- 100-mL 95% Ethyl alcohol
- Hot plate
- 250-mL Beaker
- 1000-mL Beaker
- Tongs
- Petri dish
- Paper towels
- 100-mL Graduated cylinder

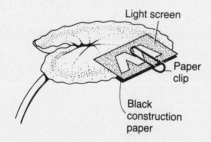

Light screen

Paper clip

Black construction paper

Strategy for Experimenting
Read the procedure carefully for all references marked "CAUTION" before you start working.

Introduction

Why are plants important? From the time of our earliest ancestors, plants have been a source of food, shelter, medicine, and enjoyment. Their most important role has been as food producers for other organisms.

Green plants contain chlorophyll, the pigment that absorbs light energy during photosynthesis. The plant converts some of the sugar it produces into starch, a complex carbohydrate. The starch is stored for future use by the plant. Animals eating the plants will use the starch as their source of energy. This is the beginning of the food chain—the sequence of energy transfer among organisms in an ecosystem.

Prelab Preparation

Your teacher will place geranium plants in a dark place for 3–4 days to keep the plants from performing photosynthesis.

1. What do you think will happen to a green plant kept in the dark for a long time?
2. What do you think will happen in a green leaf if you keep light from reaching a part of it?
3. How will the plants look after being kept in the dark for 3–4 days?

Before the plants are reexposed to light, letter cut outs will be attached to parts of the leaves as screens to block out light. Then the plants will be exposed to light for at least 24 hours.

4. Will there be any change in color after the plants are again exposed to light?
5. Will a leaf have a different appearance in the area where the letter screen blocked the light?

You will test both a light-screened leaf and an unscreened leaf with iodine to determine starch production. Iodine turns blue-black in the presence of starch. First, you will extract the chlorophyll from the leaves.

6. In which area do you predict the iodine will turn blue-black, the area exposed to light or the area blocked from light?
7. Why is it important to remove the chlorophyll before testing the leaf?

Procedure

Work in pairs. CAUTION: Never place the alcohol directly on the hot plate. It can catch fire. CAUTION: *Do not cover the alcohol.* Be careful that the alcohol does not spill and ignite.

A. CAUTION: Wear safety goggles. Fill a 1000-mL beaker about one-third full with water. Place it on a hot plate and bring it to a boil. Place 100 mL of ethyl alcohol into a 250-mL beaker, and then put this beaker into the beaker of water. Bring the alcohol to a boil.

8. Make notes about what parts of the leaf had been covered by the light screen

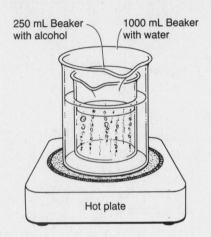

250 mL Beaker with alcohol

1000 mL Beaker with water

Hot plate

B. With tongs, dip one leaf that has been light-screened into the boiling water for 2 minutes. Then dip the leaf into the boiling alcohol for several minutes, until the leaf turns white. Blot the leaf on paper towels and place it in a petri dish. Repeat the procedure with an unscreened leaf as a control. CAUTION: Turn off the hot plate to allow the liquids to cool. Let the beakers cool before moving them.

9. Why do you think the water alone will not remove the chlorophyll?

C. Drop iodine solution over the leaf in the petri dish until it is completely covered. Let the color develop for several minutes. CAUTION: Use care when using iodine to avoid staining hands or clothing.

10. What color are the parts of the leaf that were blocked from the light? What color are the parts that were exposed to light?
11. How do these differences explain the effect of light on photosynthesis?

Postlab Analysis

12. How did the starch pattern of the screened leaf and unscreened leaf compare?
13. Account for the difference in patterns noted in Question 12.
14. Why are plants important?

Strategy for Analyzing Data
Review your hypotheses (answers to Prelab questions 1–6) after you have recorded your observations. Were your hypotheses supported?

Further Investigations

1. How do you think plants exposed to light without red or blue wavelengths would respond to the starch test? Design and execute an experiment to test your hypothesis.
2. Do you think less starch would be made if plants were exposed to lower intensities or shorter durations of light? Design and execute an experiment to test this hypothesis.

Investigation

24.1 | *Plants and Energy*

1. _____

_____

2. _____

_____

3. _____

_____

4. _____

5. _____

_____

_____

6. _____

7. _____

_____

8. _____

_____

_____

9. _____

10. _____

_____

11. _____

_____

12. _____

_____

_____

_____

13. _____

_____

_____

14. _____

_____

_____

_____

24.2 *Plants and Air*

Learning Objectives
- To define substances consumed and produced during photosynthesis.
- To relate the processes of photosynthesis and respiration.

Process Objectives
- To perform an experiment with 2 controls.
- To analyze data related to photosynthesis.

Materials

Part I
For Class
- Light-proof storage area

For Group of 2–4
- 2 Glass funnels
- 6 Small test tubes
- 6 Rubber stoppers for test tubes
- 2 600-mL Beakers
- 0.5% Sodium bicarbonate solution
- Distilled or deionized water
- Bromothymol blue indicator
- Drinking straw
- *Elodea*–a freshwater plant
- Lamp
- 2 Test tube racks

Part II
- Matches
- Wooden splint

How do plants affect the air we breathe?

Introduction

Why are plants necessary for the survival of animals? All living things need the energy stored in chemical bonds to survive. Plants convert the energy in sunlight into chemical energy stored in the form of complex molecules. These molecules are used as food by animals. Energy is released from food molecules, such as glucose, by a process called aerobic respiration. Aerobic respiration requires oxygen for the release of energy. Carbon dioxide and water are by-products of the breakdown of glucose.

Carbon dioxide and water become raw materials for plants to create more glucose by photosynthesis. A by-product of photosynthesis is oxygen. Before the evolution of green plants, there was no oxygen in the earth's air. Now, every animal and plant is dependent on oxygen production in plants.

Prelab Preparation

Review the principles of photosynthesis in Chapter 7. Think of a photosynthesizing plant as a factory. Factories require energy and raw materials to produce a product. Along with its product, a factory also produces wastes. By observing what enters the factory and what leaves, we can begin to understand the process we cannot directly observe.

1. How does knowledge about the materials that enter and leave a factory help you to infer what is happening inside the factory?
2. What is the chemical equation for photosynthesis?
3. What portion of the equation indicates the raw materials used by the plant in this process?
4. What is the source of energy for photosynthesis?
5. How could you turn the photosynthetic process on and off?
6. What are the 2 products of photosynthesis? Why might we consider one to be a waste product?

If CO_2 is added to water, carbonic acid is formed in the solution, which causes the acidity to increase.

$$H_2O + CO_2 \rightleftharpoons \underset{\text{Carbonic Acid}}{H_2CO_3} \rightleftharpoons \underset{\text{Bicarbonate Ion}}{H^+ + HCO_3^-}$$

An indicator can be used to detect changes in the amount of carbon dioxide dissolved in a solution. Bromothymol blue is a substance that changes color according to the pH of a solution. When the solution is neutral (pH 7.0), bromothymol blue appears blue or greenish blue. In acidic solutions (pH < 7.0), the color becomes yellowish, while in basic solutions (pH > 7.0) it becomes deep blue-green.

Fill a test tube about half full with water and add about 1 mL of bromothymol blue. Note the color. Using a drinking straw, gently and slowly blow bubbles through the solution for about 2 minutes. Cover the top of the test tube with fingers or foil to prevent splattering. CAUTION: Do not draw any of the solution into your mouth.

7. What change do you observe in the bromothymol blue?
8. Make an inference as to the cause of this change.
9. How does the presence of carbon dioxide in water affect the pH of the solution?

The following experiment can be used to test the hypothesis that if a plant is conducting photosynthesis, then oxygen is produced and carbon dioxide consumed. Read the procedure. Then check your understanding by answering the following questions.

10. What is the experimental variable in the procedure?
11. What are the controls in this experiment?
12. What results do you predict you will obtain if the hypothesis is correct?
13. What results might you get if the hypothesis is false?

Procedure

Part I

A. Obtain a funnel, small test tube, 3 rubber stoppers, and a 600–mL beaker. Pour 400 mL of sodium bicarbonate solution into the beaker. (Sodium bicarbonate produces a large quantity of carbon dioxide when dissolved in water.)

B. Slowly add bromothymol blue drop by drop until the sodium bicarbonate solution is just distinctly tinted. Record the amount used and repeat that measure when setting up the controls.

14. Describe the tinted color. What does the color indicate?

C. Cut a piece of *Elodea* to fit under the inverted funnel. Arrange the *Elodea*, the funnel, and the 3 rubber stoppers inside the beaker as shown. Be sure that the tip of the funnel stem is completely covered by the sodium bicarbonate solution.

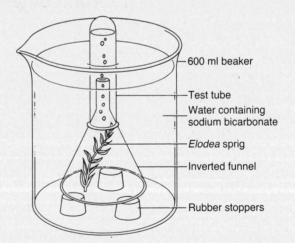

- 600 ml beaker
- Test tube
- Water containing sodium bicarbonate
- *Elodea* sprig
- Inverted funnel
- Rubber stoppers

D. Fill a small test tube with water. Cover the mouth of the test tube with your thumb, invert it, and place it over the stem of the funnel. This must be done quickly to insure that the water remains in the test tube. If an air space appears at the top of the inverted test tube, repeat this step.

E. Place a lamp near the beaker but not touching it. Turn on the lamp and observe the leaf for about 5 minutes.

15. Describe what is happening inside the funnel and test tube.
16. Speculate on the nature and source of the gas collecting at

F. Store your experiment, including the lamp (still turned on), in a safe area for 24 hours.

G. Repeat Steps A through D and place the second experimental setup in a light-proof area for 24 hours.

H. Fill 2 test tubes half full with sodium bicarbonate solution. Add bromothymol blue drop by drop until the solution is just tinted. Place one test tube on a rack next to the lighted beaker and the other test tube next to the unlighted beaker.

    17. For what variable is Step H a control?

Part II

I. Observe the setups from the lighted and unlighted areas as well as their controls.

    18. Record your observations of the colors of the solutions.

    19. What changes have occurred in the small test tubes over the funnels?

J. Read the following procedure entirely before beginning. CAUTION: Be careful when using an open flame. One person should carefully lift the test tube from the funnel kept in the light. Keeping the tube inverted, allow any remaining water to flow back into the beaker. Keep the mouth of the test tube pointing down. Another person should use a match to ignite the end of a wooden splint. Gently blow out the flame, then place the glowing splint up into the test tube. Point the mouth of the test tube away from yourself and others. Hold it at arm's length as the glowing splint is inserted.

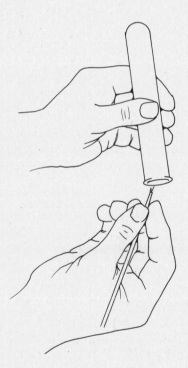

    20. Record your observations.

    21. What do the results of the glowing splint test indicate about the nature of the gas in the test tube?

K. Repeat the procedure described in Step J but use the test tube that was kept in the dark.

    22. Record your observations.

23. Do the results of this test indicate that the gas, if any, in the tube in Step K is the same as that found in Step J? Explain the reasoning for your answer.

Postlab Analysis

24. Do the results of your experiment support the hypothesis that plants consume carbon dioxide during photosynthesis? Summarize the evidence that led to your conclusion.
25. What evidence is there that light alone does not change the color of bromothymol blue?
26. What in your experimental design allows you to answer the previous 2 questions?
27. Do the results obtained in Part II support the hypothesis that plants produce oxygen during photosynthesis? Summarize the evidence that led to your conclusion.
28. Restate the formula for photosynthesis and briefly summarize the evidence you have obtained in this investigation that indicates this formula is true.
29. Is it true that plants could exist on earth without animals but animals could not exist without plants? Explain
30. Close inspection of the beaker that contains bromothymol blue and *Elodea* kept in the dark should reveal that the color has become a lighter yellow than when it was set up. What does this evidence tell you about this plant?

Further Investigations

1. Varying the distance between the *Elodea* setup and a source of light will produce different rates of photosynthesis. Predict the relationship between the distance and the number of bubbles released by the plant.
2. How might you modify this experiment to test the hypothesis that certain colors of light produce a higher rate of photosynthesis than other colors?
3. Design an experiment to test various light sources for their ability to produce photosynthesis. How do incandescent bulbs compare with fluorescent lighting? Would a plant light produce greater rates?

Strategy for Analyzing
The analysis of data must be made with reference to a particular hypothesis. If any of the data does not correlate with predicted results, it stands to challenge the hypothesis.

Investigation

24.2 | *Plants and Air*

1. _____
_____
_____
_____

2. _____

3. _____

4. _____

5. _____

6. _____
_____

7. _____

8. _____

9. _____

10. _____

11. _____
_____

12. _____
_____

13. _____
_____

14. _____

15. _____
_____

16. _____

17. _____
_____

18. _____

_____

_____

_____

19. _____

_____

_____

20. _____

21. _____

22. _____

23. _____

_____

_____

_____

24. _____

_____

25. _____

_____

26. _____

27. _____

_____

28. _____

_____

_____

29. _____

_____

_____

30. _____

25

Plant Diversity

Learning Objectives
- To study the life cycle of a fern.
- To learn how an "alternation of generations" life cycle has been adapted in different divisions of plants during their evolution.

Process Objectives
- To observe fern gametophytes and sporophytes.
- To classify the alternate generations of ferns as sexual or asexual.

Materials
For Group of 4–6
Parts I and II
- Microscope slide
- Coverslip
- Compound microscope
- Dissecting microscope
- Hand lens
- Forceps
- Unlined paper

Part I
- Fresh fern spores
- Living fern sporophyte (with sporangia)
- Living fern gametophyte
- Medicine dropper

Part II
- Living fern frond
- Single-edged razor blade
- Small glass plate

What are ferns and how are they related to other plants?

Introduction

More than 400 million years ago, small green plants became the first organisms to emerge from the sea and make the land their home. These early plants probably evolved from algae, and could not grow very large because they did not have vascular tissue to transport water and vital nutrients to all their parts. They had no roots, and were dependent on a constant supply of water to keep from drying out. The **bryophytes,** (mosses, liverworts, and hornworts) are the only surviving examples of nonvascular plants. From these first plants evolved more complex plants with roots that were able to take advantage of new, drier habitats.

The early vascular plants were very successful, rapidly diversifying and reaching great size in some cases. Like their nonvascular ancestors, they reproduced by means of spores. The most advanced of the spore-producing, vascular plants are the ferns.

Plants that produced seeds, the **gymnosperms**, evolved 350 million years ago, and bore their seeds in cones. Present-day examples of gymnosperms are the palm-like cycads, the ginkgos, and the conifers (pine, fir and redwood). About 100 million years ago, the flowering plants, or **angiosperms,** first arose. Unlike gymnosperms, the flowering plants usually have broad leaves to more efficiently capture the sun's energy, they have more efficient roots and vascular systems, and they reproduce much more quickly. Nearly 75% of all species of plants existing today are angiosperms.

Prelab Preparation

As in all plants, the life cycle of ferns shows an alternation of generations. One part of the cycle is spent as a **gametophyte.** The small gametophyte is made up of cells that each have only one set of chromosomes (haploid, 1N). During the gametophyte generation, eggs are produced in structures called **archegonia,** and sperm in **antheridia.** The egg and sperm unite to produce a zygote, with two sets of chromosomes (diploid, 2N) that grow to become the asexual **sporophyte.** Spores are produced by meiotic division and germinate to produce another generation of haploid gametophytes, thus completing the life cycle. To prepare for this investigation, turn to the data table entitled Parts of a Fern.

1. Using your textbook and the illustration of the Developmental Cycle of Ferns, complete as much of the table as you can. You will finish the table in class as part of your postlab analysis.

Developmental Cycle of Ferns

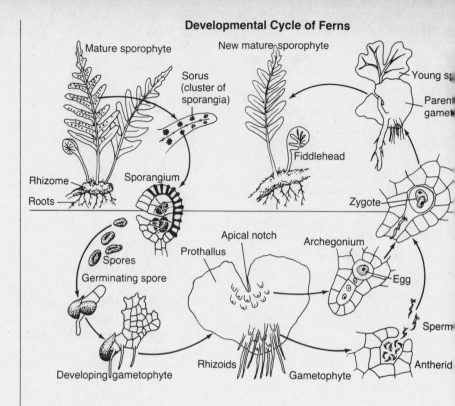

Procedure

Part I

A. As a group, assemble on your work table examples of the stages and structures of the fern life cycle provided by your teacher.

B. Spore culture: Using a hand lens, examine the fern spore culture, looking for spores and signs of the gametophyte generation.

   2. Describe what you observe.

   3. What conditions in nature would favor the germination of spores?

C. Gametophytes: Using forceps, place a sexually mature gametophyte (**prothallus**) on a glass slide so that its lower surface is facing you and prepare a wet mount. Examine the slide under low power and then high power as you answer the following questions.

   4. Describe the size, shape, and color of the prothallus.

D. Locate the **rhizoids,** hairlike outgrowths from cells of the lower surface. Rhizoids help to anchor the plant and, like roots, they absorb water and minerals. Next, find the **apical notch.** This notch is an indentation at the broad end of the plant; it contains small, actively dividing cells.

   5. Are the rhizoids and apical notch together or separate? Why?

E. Look carefully for the antheridia; they are normally located near the rhizoids on gametophytes.

   6. What is their function?

F. Look at or near the apical notch for the archegonia, which may be present. NOTE: Ferns can be unisexual, in which case they would have either antheridia or archegonia but not both.

   7. What is their function?

G. If antheridia are present, tap the cover slip gently to put a little pressure on them. Under low power, watch for the sperm cells that emerge.

Strategy for Observing

Compare each sample with the fern pictures provided here and in your text. How do your samples differ from the pictures? How are they alike?

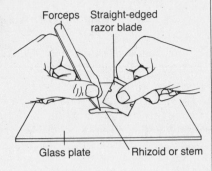

Forceps Straight-edged
razor blade

Glass plate   Rhizoid or stem

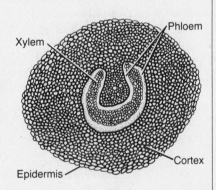

Phloem

Xylem

Cortex

Epidermis

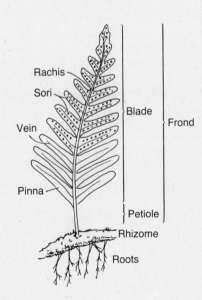

Rachis

Sori

Vein

Blade

Frond

Pinna

Petiole

Rhizome

Roots

8. How does the sperm cell reach the egg?

9. Is there any provision for cross-fertilization in ferns? Explain.

H. Look for young embryos; they are evidenced by bulging, dark-looking masses within the archegonium.

10. Are the cells of the embryo haploid or diploid?

Part II

I. Sporophytes: The sporophyte starts out being dependent on its parent gametophyte, but its stems, roots and leaves soon enable it to be a free-living plant. The stem of a fern sporophyte may be upright (a **stock**) or horizontal (a **rhizome**). Rhizomes grow buried in and horizontal to the ground.

11. On a separate sheet of paper, draw the stem of a fern, indicating the way the roots and leaves are attached.

J. Obtain a small piece of rhizome or stem from your teacher. Lay the rhizome or stem on its side on a glass plate. Use a straight-edged razor blade and forceps to cut a very thin cross-section. CAUTION: Use extreme care to prevent injury. If your cross-section is thin enough, prepare a standard wet mount; if it is too thick, place it on a slide without a coverslip. Observe your preparation with your microscope under both low power and high power. You should be able to find the epidermis and cortex. In addition, you will see **vascular tissue**—**xylem,** for water and mineral conduction, surrounded by **phloem,** for organic molecule transport.

12. How are the vascular tissues arranged in the fern stem?

13. How are vascular tissues important to successful life on land?

K. Examine the fern's roots with a hand lens. Notice that they arise all along the length of the stem.

14. How are roots different from rhizoids?

15. Which are more effective at absorbing water and minerals—roots or rhizoids?

L. Observe a fern leaf, or **frond.** Fronds can be small or large, and simple or complex. Complex blades are divided into sections called **pinnae.** All fronds have a stalk and a blade. The stalk is called a **petiole** below the blade, and a **rachis** as it runs through the blade.

16. Draw and label a frond from your fern on a separate piece of paper.

M. Place a pinna on a glass slide under a dissecting microscope and look at its veins. These veins are actually small bundles of xylem and phloem that branch off the stem to transport water and nutrients to and from parts of the leaves.

17. Draw the vein pattern next to your drawing of a frond (Question 16).

N. Using forceps, remove a **sorus** (cluster of **sporangia**) from the underside of a leaf and transfer it to a clean slide. Examine it dry under low power with the compound microscope. You should be able to see how the sporangia erupt from the leaf surface in clusters. Note the intricate shape of each sporangium.

18. What is the advantage for the fern of producing so many spores?

19. Why, would you infer, are the spores located on the bottom surfaces of the leaves?

Postlab Analysis

20. Now complete the Parts of a Fern table.

Answer the following questions as a class to help you tie together what you have learned about plants.

21. Assume that the gametophyte of plants evolved from their algal ancestors. How has this structure adapted to life on land?
22. How do fern's reproduce asexually?
23. What is the asexual stage of the fern's lifecycle? The sexual stage?
24. What are some of the adaptive advantages of higher plants that ferns lack?

Further Investigation

1. Assuming that plant evolution is continuing, what further changes in the life cycle of a fern might you predict? What kinds of mutations might underlie these changes?
2. Examine living and/or preserved specimens of other plants, including pine branches with pollen and seed cones; winged pine seeds; mature plants of *Lycopodium* (club moss), *Selaginella* (spike moss), and *Equisetum* (horsetail), all with cones; large, fresh flowers of *Lilium* (lily) and *Narcissus* (daffodil). Examine these plants for structures indicating reproductive processes that differ from those of the ferns you have studied. How are the plant specimens alike? How are they different? Create a dichotomous key to help you classify the specimens you have studied. Consult a library source to learn more about each plant—where it grows, its common name, its growth cycle. Prepare a poster illustrating the information you gather.

Strategy for Classifying
Recall that sexual reproduction involves the fusion of haploid cells into a single diploid zygote.

Investigation

25 | *Plant Diversity*

1. Enter your answers on the data chart.

Parts of a Fern

Part	Chromosome Number (1N or 2N)	Definition and Description	Where Found
Spore			
Gametophyte (prothallus)			
Zygote			
Embryo			
Young Sporophyte			
Mature Sporophyte			
Rhizoid			
Antheridium			
Archegonium			
Stem			
Root			
Frond (leaf)			
Sporangium			
Sorus			
Fiddlehead			

2. _____

3. _____

4. _____

5. _____

6. _____

7. _____

8. _____

9. _____

_____

10. _____

11. Make a drawing on a separate sheet of paper.

12. _____

_____

13. _____

_____

14. _____

_____

15. _____

16. Make a drawing on a separate sheet of paper.

17. Make a drawing on a separate sheet of paper.

18. _____

_____

19. _____

20. Enter your answers on the data chart on the previous page.

21. _____

_____

22. _____

23. _____

24. _____

_____

_____

26 | *Leaf Stomata*

Learning Objectives

- To determine the locations and density of stomata on a leaf.
- To describe the effect of an environmental change on stomata.

Process Objectives

- To observe the effect of salt on stomata and guard cells.
- To hypothesize about the response of guard cells to a stimulus.
- To predict how a change in stomata influences photosynthesis.

Materials

For Group of 4

- Distilled water in a small beaker with a dropper
- 5% NaCl solution in a small beaker with a dropper

For Each Student

- 1–3 Microscope slide(s)
- Leaves from a watered, nonwilted plant exposed to light
- Razor blade or pair of forceps
- 1–3 Coverslip(s)
- Compound microscope
- Paper towel

Where are stomata located on a leaf, and how do they work?

Introduction

If asked which plant organ is essential in absorbing compounds for a plant, most people would say it is the root. But roots are not the only plant organs involved in absorbing important compounds. Leaves absorb and release gases, such as oxygen, carbon dioxide, and water vapor. As you might guess, gas exchange in a leaf affects the process of photosynthesis.

Prelab Preparation

Think about the functions of gas exchange between a leaf and the atmosphere. Consider the roles of oxygen, carbon dioxide, and water in photosynthesis and in aerobic respiration within the cells of the leaf.

1. During what part of the 24-hour day would maximum exchange of gases be most likely to occur? Why?
2. Under what environmental conditions would a plant benefit from reducing loss of water vapor through the leaves? Why?
3. What mechanism would you design to allow gas exchange between a leaf and the atmosphere?
4. How could your mechanism control the flow of gases into and out of the leaf, opening to increase the rate of gas exchange, and closing to reduce water loss?

Structures, perhaps similar to the one you would design, are present in the **epidermis** (outer cell layer) of the leaves of many plants. These structures consist of openings called **stomata** (singular, **stoma**), that allow water vapor and gases to enter and exit the leaf. Two guard cells surrounding each stoma regulate its opening and closing. Review the functioning of stomata and guard cells in Section 26.4 of your textbook. In this investigation, you will examine the epidermis of a leaf under a microscope to observe leaf stomata.

Procedure

A. Place a drop of water on a microscope slide. Obtain a healthy, nonwilted leaf from one of the available plants. Holding the bottom surface of the leaf toward you, fold the leaf in half toward you so that the bottom surfaces are together. (See the illustration on page 158.) Unfold the leaf and tear it along the crease by holding the left section of the leaf and pulling the right section down at an angle. A clear, colorless outer layer should be visible along the torn edge. This layer is the lower epidermis.

B. Carefully cut off a small fragment of the transparent epidermal layer with a razor blade or pull it off carefully with forceps. Immediately place the fragment in the drop of water on your microscope slide and position a coverslip over it. Do not allow the fragment to dry out.

Strategy for Observing

Draw the guard cells as soon as you make your observations of the epidermis.

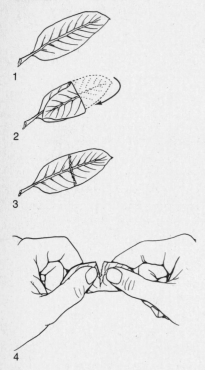

Strategy for Hypothesizing

Your hypothesis should be testable. As you generate hypotheses, think of experiments you could design to test them.

Strategy for Predicting

List several predictions. Select the prediction most likely to be accurate based on your hypothesis (see Question 12).

C. Examine the epidermis through the low-power objective of your microscope. Observe the sizes and shapes of the living cells in the epidermis. The small bean-shaped cells occurring in pairs are guard cells.

   5. Make an outline drawing of a pair of guard cells and the surrounding epidermal cells.

D. Examine the pair of guard cells under the high-power objective. The opening or pore visible between the guard cells is the stoma.

   6. How does the wall of a guard cell vary in thickness? Add this detail to your drawing.

E. Return the microscope objective to low power.

   7. How many stomata are visible in one low power field?
   8. Compare your data with those of other students. What conclusions can you draw about the density of stomata?

F. Repeat Steps A through E with another leaf from the plant you have been using, but this time examine the upper epidermis. Holding the top surface of the leaf toward you, fold the leaf in half toward you so that the top surfaces are together. Then, tear the leaf along the crease, as you did in Step A, to obtain a fragment of the upper epidermis.

   9. How many stomata are visible in your fragment of upper epidermis under low power?
   10. What conclusion can you draw about the locations and densities of stomata on leaves from your species of plant? What advantage would this have for the plant?
   11. Where would you expect to find stomata on a water lily leaf? Why would you expect to find them there?

G. Guard cells control the movement of water vapor and gases between the leaf and the atmosphere by opening and closing the stomata.

   12. State a hypothesis describing guard cell response to water loss and the effect of this response on a stoma.

H. Make a fresh wet mount of the lower epidermis of a leaf. Observe a stoma under high power. Place a drop of 5% NaCl solution at the edge of the coverslip. Take a piece of paper towel and touch it to the opposite edge of the coverslip. The paper towel should draw out the water that bathes the epidermis, and the NaCl solution should replace the water. This will create a concentration gradient between the cells and their environment. Water will move out of the guard cells by osmosis until the salt concentration inside the cells is the same as it is outside. Allow 5 minutes for osmosis to be completed, then observe the guard cells and stoma again.

   13. What has happened to the guard cells and to the stoma?
   14. What property of guard cells permits such a change?
   15. Do your experimental results support your hypothesis?

Postlab Analysis

   16. In what way would the normal pattern of gas exchange between a leaf and the atmosphere be altered by coating both sides of the leaf with petroleum jelly or wax?
   17. Predict what would happen to guard cells and stomata on a hot, dry day. How might this affect photosynthesis? Why?

Further Investigation

1. Compare stomata on leaves from well-watered plants kept in dark with stomata on leaves from well-watered plants kept in light.

Investigation

26 | *Leaf Stomata*

1. _____

_____

2. _____

_____

3. _____

_____

4. _____

_____

5.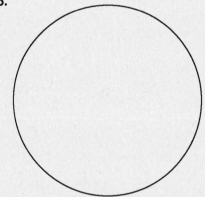

6. _____

7. _____

8. _____

_____

_____

9. _____

10. _____

_____

_____

11. _____

_____

_____

12. _____

_____

13. _____

_____

14. _____

_____

15. _____

16. _____

_____

17. _____

_____

_____

_____

Investigation

27

Flowers and Fruits

How do flowers become fruits?

Learning Objectives

- To investigate the structure of a perfect flower.
- To compare the structures of different types of fruits.
- To learn which parts of the flower become parts of the fruit.

Process Objectives

- To observe carefully the parts of a flower.
- To infer how a flower becomes a fruit.
- To infer the ways in which seeds are dispersed.

Materials

Part I

For Group of 2

- Gladiolus blossom or other fresh flower
- Petri dish
- Dissecting microscope
- Blunt probe
- Unlined paper
- Small glass plate
- Single-edged razor blade

Part II

For Class

- A selection of fruits

Introduction

The shapes and colors of flowers catch our attention and we may wonder why they are so attractive and varied in form. By observing blooming plants, scientists have discovered complex relationships between flowers and their environment. Those plants that depend on the movements of animals for pollination often show adaptations that attract specific animals. The colors, designs, and scents are specific to the sensory range of the pollinator. Even the shape of the flower (such as a snapdragon) may fit its pollinator perfectly, while blocking the entry of other animals.

The pollinator also has adaptations that are specific to flowers it visits so that the pollen is successfully transferred from the pollinator to the stigma. How does successful pollination benefit the pollinator? The transfer of pollen between plants of the same species is the initial step in the process of creating seeds. The structures and behaviors of the pollinator inadvertently increase the pollinator's food supply over time. In this way, successive generations of plants and their pollinators become increasingly dependent on each other.

In this lab, you will study the structure of a flower as related to its pollination and fertilization. You will also have opportunities to observe a variety of fruits and seeds.

Prelab Preparation

Review Chapter 27 in your textbook. Pay particular attention to the structure of a flower, how a flower is fertilized, and how fruits develop. Note methods of pollination and seed dispersal.

1. List and compare the differences between monocots and dicots.
2. What is the function of the pistil? What is the function of the stamen?
3. How is the pollen of plants with colorful flowers usually dispersed? How is the pollen of plants with less-noticeable flowers usually dispersed?
4. How are the seeds of fleshy fruits, such as apples, dispersed? Some fruits, such as the milkweed pod, have feathery structures. How are their seeds dispersed?
5. After a flower is pollinated, which part of the flower becomes the seed? Which part of the flower becomes the fruit?

Procedure

Part I

A. Obtain a gladiolus blossom from your teacher. Observe it carefully. Notice that the sepals are part of the color display, alternating with the petals. Look at the leaves or sepals and notice if the veins are parallel or branched. Count the sepals and the petals. Look at your list of monocot and dicot characteristics from Question 1.

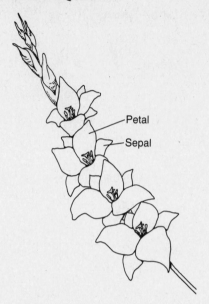

6. Is this flower a monocot or a dicot?
7. Based on your observations, how might the petals function in the process of pollination?

B. Look at the inside of your flower. Arranged around the pistil are the male reproductive structures, the **stamens.**

8. How many stamens are there in your flower?

Observe the top of the stamen, called the **anther.** It is the part of the stamen which contains pollen grains. The anther is supported by a stalk called the **filament.**

9. Draw a diagram of your flower on a separate piece of paper. Label petal, sepal, pistil, stigma, style, stamen, anther, and filament.

C. Remove one stamen from the flower. Place the stamen in the bottom of a petri dish and observe it under a dissecting microscope. Gently touch the anther with a blunt probe. The small particles you see are pollen grains.

10. Draw a diagram of one stamen next to your flower diagram, showing the pollen grains on the anther.
11. Estimate the number of pollen grains. What advantage is there for a plant having so many pollen grains?

D. Look inside your flower again. In the middle is the **pistil,** the female reproductive structure.

12. How many pistils does your flower have?

The top of the pistil is called the **stigma.** Pollination occurs when pollen grains adhere to the stigma. This is where the pollen grains land. Feel the stigma and notice that it is sticky.

13. How is this stickiness an adaptation for obtaining pollen? How is the shape of the stigma an adaptation for collecting pollen from the body of a pollinator?

The long tube below the stigma is called the **style.** The pollen tube grows down the inside of the style on its way to the **ovule.** At the base of the style is the **ovary,** a swollen green globe which contains the ovules.

14. Add the ovary to the diagram you made in Question 9 and label it.

Strategy for Observing
Note the overall structure of the flower before beginning the dissection. Sketching encourages close observation.

E. Remove all the petals, sepals, and anthers from the flower. Be careful not to break the pistil at the base of the style. If you have done this properly, you will be left with the pistil. Hold the ovary firmly on a glass plate. Using a razor blade, carefully make a longitudinal cut through the ovary. CAUTION: Use care with razor blades.

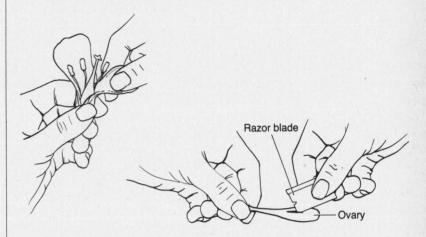

You will see rows of neatly stacked white disks. These white disks are the ovules. Each ovule is one female gamete, the equivalent of an unfertilized egg in an animal. After fertilization by a pollen grain, each ovule will form a seed. After the ovules are fertilized, the ovary will become a fruit containing the seeds. Examine the open ovary under a dissecting microscope.

15. Draw a diagram of the ovary as you see it under the microscope. Label the ovules and show their arrangement in the ovary.

16. Predict the number and relative arrangement of seeds in the fruit of this flower.

Part II

Examine the selection of fruits your teacher has prepared for you.

F. Station 1 will have a peach or plum for your observation.

17. How many seeds are in your fruit? Describe the arrangement of seeds within the fruit.

18. What function do you think the fleshy part of this fruit serves for the flower?

G. Station 2 will have a tomato, apple, cucumber, pumpkin, or green pepper for your observation.

19. Estimate the seeds in your fruit. Describe the arrangement of seeds within the fruit.

20. How many **carpels** (compartments containing seeds) can be seen in the cross section of your fruit?

H. Station 3 will have maple seeds, milkweed seeds in their pods, and/or dandelion seeds for your observation.

21. How are those seeds dispersed? Observe the form and texture of the seeds and describe the adaptations for dispersal.

22. What attributes do you think characterize wind-dispersed seeds? Include how the seed would be attached to a plant and where it might best be located on the plant.

I. Station 4 will have burrs for your observation.

23. How do you think this fruit is dispersed?

Strategy for Inferring

As you observe arrangements of seeds within a fruit, think about their relationship to the arrangement of ovules in the flower. The location and shape of seeds can indicate the method of dispersal.

24. What attributes do you think characterize animal-dispersed seeds?

J. Station 5 will have a coconut for your observation.

25. How do you think this seed is dispersed? How is it adapted for this form of dispersal?

26. What mechanism might the coconut have for detecting that it is on land again and can grow?

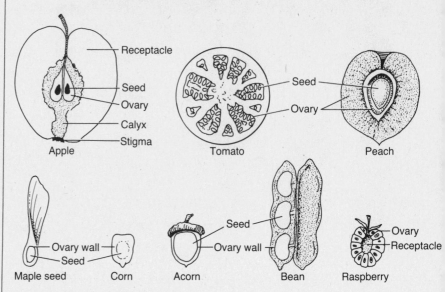

Postlab Analysis

27. Make a list of adaptations of flower structure that increase the efficiency of pollination.

28. Make a list of seed adaptations that increase the efficiency of dispersal.

29. Make a hypothesis concerning the way seeds are dispersed.

30. Summarize the relationship between the structure of a flower and the structure of a fruit.

Further Investigations

1. Use a hand lens to compare a monocot seed, such as corn, to a dicot seed, such as a lima bean. Dissect and sketch the following parts: cotyledon, embryo, endosperm, seed coat, and hilum. Label each part.

2. Collect a variety of flowers. Try to find some from each of the following groups: flowers with only female parts; flowers with only male parts; florets (many tiny flowers on a single stem).

3. An interesting field to explore is **palynology**—the study of fossil pollen. Ancient climates and landscapes can be inferred by identifying pollen from dated samples of soil and artifacts. Investigate how different species are distinguished and how pollen lasts unchanged for thousands of years.

Investigation

27 | *Flowers and Fruits*

1. _____

_____

_____

2. _____

_____

3. _____

_____

4. _____

_____

5. _____

6. _____

7. _____

_____

8. _____

9. Draw flower diagram on separate sheet.

10. Draw stamen diagraom next to flower diagram.

11. _____

12. _____

_____

13. _____

14. Add to your flower diagram.

15. Draw ovary diagram next to flower diagram.

16. _____

17. _____

18. _____

_____

19. _____

_____

20. _____

21. _____

_____

22. _____

23. _____

24. _____

_____

25. _____

26. _____

_____

27. _____

_____

_____

_____

28. _____

_____

_____

29. _____

_____

_____

30. _____

_____

_____

_____

28 *Tropisms in Plants*

Learning Objectives

- To see how plant growth can be a response to external stimuli.
- To learn how light and gravity help regulate plant growth.

Process Objectives

- To observe the effects of light and gravity on seedlings.
- To hypothesize how plant growth is regulated.

Materials

For Group of 2
Phototropism

- 20 Oat seeds
- 4 Plastic pots (5-cm diameter)
- Potting medium (sphagnum or vermiculite)
- 4 Pot markers or labeling tape
- Aluminum foil
- Cotton swabs
- Cardboard box to contain one pot
- Single-edged razor blade or scalpel
- Protractor

Gravitropism

- 20 Oat seeds (presoaked)
- 2 Clean petri dishes
- Grease pencil
- 6 Pieces of filter paper
- Roll of nonabsorbent cotton (enough to fill petri dishes)
- Masking tape or modeling clay
- Shoe box or similar container for several petri dishes

How do plants respond to external stimuli?

Introduction

Have you ever wondered why a potted plant in a sunny window bends toward the window, why the stem and leaves of a plant grow upwards, or why the roots grow down? These are all examples of **tropic** responses of plants—growth in a particular direction or orientation due to some external stimulus.

Tropic responses are caused by plant **hormones.** One hormone called **auxin** is produced by the growing tips of plants. It accumulates on the shaded side of stems and causes the cells on the dark side to grow more than those on the side receiving more light. This response, called positive **phototropism,** results in the plant orienting toward the source of light.

In a similar way, hormones cause roots to grow down and stems to grow up. In this case, however, the stimulus is gravity instead of light. This phenomenon is called positive **gravitropism** (or **geotropism**). Other important tropic responses are **hydrotropism** (a growth response to water) and **thigmotropism** (a growth response to contact with solid objects). All these responses help plants to survive and grow better.

In this lab, you will use oat seeds to observe the processes of phototropism and gravitropism.

Prelab Preparation

This lab consists of 2 experiments, each of which has 4 parts. You will be working in groups of 4. You and your lab partner will study either phototropism or gravitropism, and you will share your data with the 2 members of your group who study the other tropism.

Read about plant tropisms and their hormonal control in Chapter 28 of your text. Study the diagram of the oat seedling and its parts. Each seed is actually a fruit with a complex outer coat.

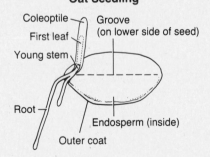

Oat Seedling

1. Which part of the plant (root, stem, or leaf) do you expect to see first as the seeds sprout? Why?

Procedure

Phototropism

Part I (Day 1)

A. Obtain 20 oat seeds from your teacher. Use your thumbnail to remove the outer coat and soak the seeds overnight in water.

Part II (Day 2)

B. Plant 5 presoaked seeds (groove side down) in a row about 1 cm apart in each of 4 plastic pots. Use well-moistened vermiculite or sphagnum moss as a planting medium. Cover the seeds with about 2 mm of moist medium and water gently. Label the pots 1, 2, 3, and 4, and then place them in a warm, dark closet, cupboard, or drawer.

Part III (Day 4)

C. When the seedlings are at least 3–4 mm above the soil, handle the pots as follows:

Pot 1—Place in a well-lighted area where the plants will get light evenly from all directions (a greenhouse is ideal).

Pot 2— Cover each **coleoptile** tip with a 3 mm-high cap of aluminum foil. (A coleoptile is a clear sheath that protects shoots during germination.) Shape the cap with a cotton swab, such as a Q-tip, and make it as lightweight as possible. Place this pot next to pot 1.

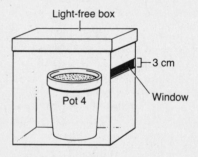

Pot 3—Place this pot in a light-free cupboard.

Pot 4— Cut a narrow, rectangular window in a light-free box with a single-edged razor blade or a scalpel, as shown in the diagram. CAUTION: Handle razor blade and scalpel with care. The bottom of the window should be at the same height as the rim of the pot. Place this pot in the box so that the seedlings line up parallel to the window and will receive light from only one direction. Store the box on a brightly lighted window sill, with the cutout window facing the light.

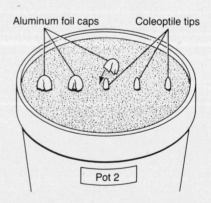

Strategy for Hypothesizing

When you generate a hypothesis, you are making a general statement that can be tested. Think of a broad statement from your textbook that this experiment could be testing. This statement is your hypothesis.

2. Is there a control in your experiment?
3. What hypothesis are you testing here?

Part IV (Day 5)

D. Carefully dig up all the plants. Use a protractor to measure the angle of curvature of each plant as shown in the diagram. (If the plant is straight, the angle is 0 degrees.)

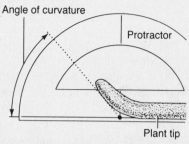

Strategy for Observing

After you measure the 5 angles of plant curvature, you will take an average. This is more statistically reliable and therefore a more accurate observation of what occured.

4. Calculate the average angle of curvature for the plants in each pot and enter your results in the Phototropism Data Table. Include in the table a brief description of the conditions under which the plants in each of the pots were grown.

5. Make drawings of your plants, showing their curvatures.

Gravitropism

Part I (Day 1)

A. Obtain 20 oat seeds from your teacher. Use your thumbnail to remove the outer coat and soak the seeds overnight in water.

Part II (Day 2)

B. Prepare 2 clean petri dishes. On the outer surfaces of the bottom of the dishes use a grease pencil to draw arrows indicating the direction of gravity. Arrange the presoaked oat seeds, with the grooved sides facing the bottom of the dish (as shown in the diagram). Carefully cover the seeds with 3 layers of very wet filter paper. Fill the rest of the dish bottoms with nonabsorbent cotton. NOTE: The cotton must hold the seeds firmly in place when each dish is turned on its side. Place the lid on each dish and seal them with masking tape to prevent evaporation. Mark your names on the masking tape.

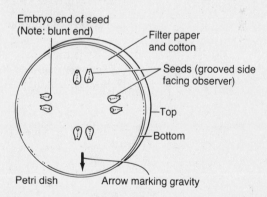

C. Stand the dishes on their edges in a shoe box (together with those of the other groups doing this experiment). Make sure that the arrows you drew are facing down inside the box. Use tape or modeling clay to hold the dishes on their sides. Cover the box and store it in a dark cupboard.

Part III (Day 4)

D. Get your dishes from the cupboard.

6. Make drawings of your seedlings showing the appearance, orientation, and size of all primary roots and shoots (stems and leaves).

E. Label one of your plates "rotated 90°" and return it to the box, oriented on its side as before but with the arrow pointing sideways. Return the other plate to the box in its original orientation.

Part IV (Day 5)

F. Get your dishes from the cupboard.

7. As you did in Part III, make drawings of the primary roots and shoots.
8. Develop a hypothesis to explain the gravitropic effects you have observed.
9. How do you know that phototropism is not a factor in this experiment?

G. Share the data you obtained with the other students in your group and make sure you understand their procedure as well as their results. Make sure you have answered *all* the questions.

Strategy for Observing

When you examine your plants for changes in direction of growth, compare your drawings from day 5 (Part IV) to those you did on day 4 (Part III). By comparing results between your 2 dishes (called an external control) and from the same dish on 2 different dates (known as an internal control), the results are more reliable.

Postlab Analysis

10. How do the shoots grow in relation to the light?
11. Did this experiment clearly demonstrate phototropism? Why or why not?
12. How do the dark-grown and foil-covered-tip plants compare in their growth?
13. Seedlings receiving the same treatment in the phototropism experiment should respond identically. Explain any deviation from the expected results.
14. How do the roots grow in relation to gravity?
15. Which way do shoots grow relative to gravity?
16. How do you explain these responses to the same stimulus?
17. Explain what happened to the seedlings in the dish that was turned 90° from the original direction of gravity. Were they all able to reorient?

Further Investigations

1. To observe thigmotropism, plant presoaked pea seeds in pots of vermiculite or soil. Watch the response of the sprouting seedlings to small wooden stakes placed next to the sprouting seedlings. Include some pots without stakes as controls. Note whether all plants twine in the same direction and, if they do, whether it is clockwise or counterclockwise.
2. In the library, read about Venus's-flytraps *(Dionaea muscipula)* and sensitive plants *(Mimosa pudica)*. Mimosa seeds and Venus's-flytrap can be purchased from supply houses. You can grow and study these interesting species during the school year and watch their rapid responses to stimuli. (These rapid responses are called **nastic movements** and are different from tropisms.)

170   *Investigation 28*

Investigation

28 | *Tropisms in Plants*

1. _____

2. _____

3. _____

_____

_____

4. Enter your answers on the data chart.

Phototropism Data Table

	Average Angle of Curvature	Growth Conditions
Pot 1	_____	_____
Pot 2	_____	_____
Pot 3	_____	_____
Pot 4	_____	_____

5.–7. Make your drawings on a separate sheet of paper.

8. _____

_____

_____

9. _____

10. _____

11. _____

_____

_____

12. _____

_____

13. _____

_____

_____

15. _____

16. _____

_____

_____

17. _____

_____

29

Hydra

How does a sessile animal catch its prey?

Learning Objectives
- To study the way hydras trap and eat their prey.
- To investigate an animal's nervous and muscular systems.

Process Objectives
- To observe how hydras behave when they sense prey.
- To experiment with hydras to determine how they respond to different stimuli.

Materials

For Class
Part I
- Hydra culture
- Beef broth
- Methylene blue
- Vinegar

Part II
- Daphnia culture

For Group of 2
- Medicine dropper
- Culture or petri dishes
- Depression slide
- Dissecting microscope
- Piece of filter paper
- Fine forceps
- Needle probe
- Coverslip
- Compound microscope

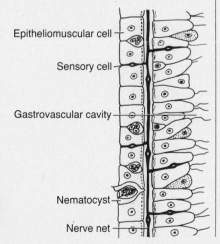

Introduction

You may think of a **predator** as an animal that moves about hunting other animals for food. In many cases you would be correct. Hydras, however, are **sessile** (they do not move about) and are still considered predators. They usually live anchored to twigs or leaves of aquatic plants, preying on organisms that drift to them. Like other members of the Phylum Cnidaria, hydras kill or paralyze their prey with poisoned darts.

The "firing" of these darts is controlled by the hydra's simple nervous system. Sensory cells detect when the prey is within range. Messages are transmitted via a **nerve net.** Coiled stingers, called **nematocysts,** and the glovelike muscle layer of **epitheliomuscular** cells respond to the messages sent through this net.

The nematocysts eject dart-bearing filaments to catch the prey. Poisons in the darts anesthetize or kill the prey. The hydra's tentacles then reach around and pull the prey into its mouth. After digestion occurs in the hydra's gastrovascular cavity, any non-digestible materials are forced out through its mouth.

Prelab Preparation

As you work with your hydra, keep in mind that the hydra's natural environment is a quiet pond. To survive the hydra must catch its food and avoid being eaten. Vibrations and chemicals in the water inform the hydra and trigger its reactions. Handle hydra gently and watch patiently—it may take time for certain types of behavior to occur; other reactions may be instantaneous. At each stage of this investigation, you should observe how long it takes the hydra to act and to complete an activity.

1. Do hydras eat plants, animals, or both?
2. In what ways do the hydra's methods of catching prey affect where the animal lives?
3. Hydras have only a simple digestive system, a gastrovascular cavity. How do they convert food into a usable form?

Procedure

Part I

A. Gently pipette a hydra from the culture dish to a clean depression slide. Allow the hydra to settle. Examine it under a dissecting microscope with high power, making sure the hydra is always in water.

   4. Identify and draw the body stalk, mouth, and tentacles.

B. Experiment to see if the hydra responds to a chemical—in this case a nutrient—stimulus. As a control, first hold a tiny piece of filter paper with forceps and move the filter paper near, but not touching, the hydra's tentacles.

Strategy for Experimenting

When you examine the stimuli that induce feeding behavior in a hydra, try to think of things, other than the stimulus you supply, that might be causing the behaviors you observe.

Strategy for Observing

Study the illustrations of the nematocyst. Make certain you can identify them in your specimen before you introduce prey.

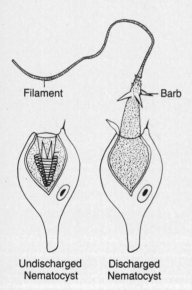

Filament — Barb

Undischarged Nematocyst   Discharged Nematocyst

5. Observe and record the response.

C. Repeat the procedure using the same piece of filter paper, first dipped in beef broth. If the proteins in the beef broth induce feeding behavior, the hydra's stalk should elongate, the mouth disk should expand, and the hydra may sway.

6. Does the hydra demonstrate feeding behavior in response to the beef broth?

D. Investigate whether the hydra responds to a tactile stimulus by showing feeding or protective behavior. Gently touch a tentacle with the needle probe. Observe its response. CAUTION: Be gentle.

7. How does the hydra respond when you touch it with your probe? Try gently touching the hydra with the needle probe on its mouth disk and on its stalk.

8. Does it show protective behavior when touched in either of these parts?

9. Do you think the hydra can distinguish between the touch of a probe and the movement of prey? How might the hydra make the distinction?

Part II

E. Now investigate feeding behavior. Hydras eat small crustaceans, such as daphnia. Pipette several living daphnia onto a depression slide on which a hydra has been placed and allowed to settle (as in Step A). Observe carefully what happens under the dissecting microscope. Watch for the nematocysts to fire and the hydra to catch the daphnia. If the hydra does not respond after a few minutes, it may not be hungry. Try a fresh hydra.

10. Describe the way the hydra captures and ingests daphnia.

11. Eventually, the hydra will digest the daphnia. What do you think happens to the material the hydra does not digest?

Postlab Analysis

12. What might be some adaptive advantages of the cnidarians' unique method of catching prey?

13. How are the cnidarians different from the porifera?

14. How do hydras differ from jellyfish—a more widely known cnidarian?

Further Investigations

1. Hydras can move, although they do not normally do so. If you leave some cultured hydras in filtered pond water a few days without feeding them, you can examine them to see if they have moved in search of food. Devise a system of mapping the layout of the container so that you can observe and record the extent of the movement.

2. Investigate the propagation rate of hydras. Set up 2 equal-sized culture dishes with clean pond water and place the same number of hydra in each dish. Feed the hydras every day. Each time, offer more daphnia to one of the cultures. Count the number of buds in each culture every day. Keep records of the numbers. At the end of a week, graph your results as a function of time.

Investigation

29 | *Hydra*

1. _____

2. _____

_____

_____

3. _____

_____

_____

4.

5. _____

_____

6. _____

7. _____

_____

8. _____

9. _____

_____

_____

10. _____

_____

_____

11. _____

_____

12. _____

_____

13. _____

_____

_____

14. _____

_____

30

Flatworm Behavior

Learning Objectives
- To learn the differences between natural behaviors and "experimental" behaviors.
- To understand that animals may use many senses to perceive their environment.

Process Objectives
- To observe flatworm responses to magnetic fields.
- To perform experiments that will elicit new flatworm behaviors.

Materials
For Group of 4
- Masking tape
- Unlined white paper
- 4 Petri dishes
- Magnetic compass
- Metric ruler
- Dechlorinated water
- 2 Medicine droppers
- 4 Living *Dugesia* (flatworms)
- Stopwatch
- Dissecting microscope or hand lens
- 1–4 Bar magnets

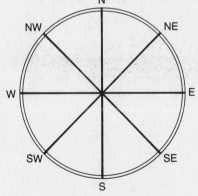

Petri dish outline showing compass points.

Does the common flatworm respond to a magnetic field?

Introduction

We are not conscious of all the information that our brains process. However, everything we perceive must first be recognized by one of our senses. The sensory receptors in our eyes, ears, nose, mouth, and skin **transduce** physical or chemical stimuli into messages that are then transmitted via nerves to our brains. We simply do not become aware of any stimulus that our senses cannot recognize and transduce, such as very high-frequency light or sound.

Other animals, however, may have different senses. Usually, these senses are similar to ours, but some animals have a range of perception that extends beyond ours. In some animals, entirely different sensory systems exist. This is the case with **magnetotaxis**, or movement in response to a magnetic field. In this lab, you will experimentally examine the flatworm's ability to respond to a magnetic field.

Prelab Preparation

It is very difficult for us to understand the sensory input of another species when it does not correspond to one of the human senses. Thus, we know more about how organisms respond to visible light than we do about how they respond to dilute odors or even wavelengths of light outside of our own range of vision. We can, however, design experiments that allow us to investigate an animal's basic response to stimuli.

Read the procedure for this lab.

1. How will this experiment test the hypothesis: Flatworms respond to magnetism?
2. What are the controls in the experiment?

Procedure

Work in groups of 4.

A. Tape a piece of white paper to your desk and trace the outline of a petri dish on it. Use a compass to mark 8 directions on this circle: N, NE, E, SE, S, SW, W, NW. Use a ruler to connect opposite directions, forming an 8-sector "pie." Fill the petri dish 2/3 full with dechlorinated water and, using a dropper, gently transfer a flatworm to the dish. Repeat this procedure with 3 more dishes and worms. NOTE: Each person should observe one worm. Allow the 4 worms 10 minutes to explore their dishes and settle down. CAUTION: Handle the flatworms gently. Do not poke or injure them.

3. Why is it important to allow each worm 10 minutes to settle before recording observations?

Strategy for Observing

As you observe flatworm locomotion, remember that the size of the petri dish limits the flatworm's overall mobility. You should always be mindful of this limitation, taking its effect into account before recording your observations.

Strategy for Experimenting

In performing your experiments, try to examine one variable at a time. You may wish to repeat Steps C and D with the magnet in different places to separate its influence from that of the earth's own magnetic field.

B. Using a stopwatch to keep track of time, observe each worm for 30 seconds, wait 10 seconds, then repeat your observation. Do this for a total of 6 observations per worm.

    4. Record in your data table the direction in which the worm is moving. Using a dissecting microscope or hand lens will make data collecting easier.

    5. Do any of the flatworms prefer only one direction of movement (i.e., N, S, E, or W)?

    6. Do all 4 flatworms show the same tendency?

    7. Why is it better to observe 4 flatworms rather than just 1?

    8. How would placing a magnet next to the dish change the magnetic field?

C. Place the north pole of a magnet against the outside of the dish at the east or west points of the compass that you marked on your paper.

    9. Repeat Step B and record observations.

    10. Is there a tendency for any or all of the flatworms to move toward or away from the magnet?

D. Repeat Step C, placing the south pole of the magnet next to the dish in the same location that you previously placed the north pole.

    11. Record your observations as above in your data table.

    12. Do the effects of the north and south poles differ?

Postlab Analysis

    13. In each of the 3 experiments (no magnet, north pole, south pole), does any one of the 8 compass directions appear to be preferred by a majority of worms?

    14. Does any one worm show a strong directional preference?

    15. How might the natural magnetic field of the earth's north pole thwart your tests using the magnets? How could you control for this?

    16. What metallic element may be accumulating in the tissues of flatworms that allows them to sense magnetism?

    17. How might the flatworms use magnetotaxis in their natural environment?

    18. Do you think that there is an evolutionary advantage for an organism to have the ability to sense magnetic fields? Explain your answer.

Further Investigations

1. Design an experiment that allows you to observe **phototaxis** (response to light) in flatworms. Is their phototaxis a positive response toward light or a negative one?

2. Light sources are often situated above an animal, the direction directly opposite the earth's gravitational pull. Design experiments that separate the influences of gravity and of light on flatworm behavior.

3. Go to the library and read about migration in birds. How do birds know where to go for the winter? What cues do homing pigeons use to return home?

4. Using your own pets (dogs, cats, tropical fish), list as many magnetotaxa, phototaxa, and **chemotaxa** (responses to chemicals) as you can observe. *Do not* intentionally subject your pets to any unpleasant or irritating stimuli. Record your results in a chart labeled with the following column heads: Organism, Magnetotaxis, Phototaxis, and Chemotaxis. Bring your chart to class for discussion.

Investigation

30 | *Flatworm Behavior*

1. _____

_____

_____

2. _____

_____

_____

3. _____

4. Enter your answers on the data chart.

Direction of Flatworm Movement in the Presence of Magnetic Fields

	Observations Without Magnet						Observations With Magnet											
							North Pole						South Pole					
Flatworm	1	2	3	4	5	6	1	2	3	4	5	6	1	2	3	4	5	6
1																		
2																		
3																		
4																		

5. _____

6. _____

7. _____

_____

8. _____

9. Enter your answers on the data chart.

10. _____

11. Enter your answers on the data chart.

12. _____

13. _____

14. _____

15. _____

_____

_____

16. _____

17. _____

_____

18. _____

_____

31

Live Earthworms

How does a live earthworm respond to light, temperature, and moisture?

Learning Objectives
- To learn the external anatomy of an earthworm.
- To study how a live earthworm responds to light, moisture, and changes in temperature.

Process Objectives
- To infer how various responses and characteristics of the earthworm reflect adaptations to its environment.
- To organize and graph data on the effect of temperature on the rate of an earthworm's heartbeat.

Materials
For Group of 2
Parts I and II
- Large live earthworm
- Dissecting microscope

Part I
- 2 Paper towels
- Dissecting pan or large culture dish
- Medicine dropper
- Flashlight
- 2 Pieces of red cellophane
- Rubber band
- 2 Pieces of blue cellophane
- Index card or piece of manila folder
- Hole punch
- Cellophane tape
- Sandpaper

Part II
- 15-cm Petri dish
- Finger bowls or plastic tray
- Warm tap water
- Ice cubes
- Thermometer
- Clock with second hand
- Graph paper

Introduction

Earthworms are classified as annelids, or segmented worms. They have digestive, circulatory, and nervous systems. Gas exchange is through the skin. They live in rich soil, which they eat, digesting the organic matter in it and passing the inorganic dirt particles out of the body. Earthworms are not very mobile creatures. They spend their lives in one small area and as a result do not encounter many other earthworms. Their reproductive strategy is well-suited to this type of life. Because each earthworm is a **hermaphrodite** (has both male and female sex organs), any individual earthworm can cross-fertilize with any other earthworm it encounters.

Prelab Preparation

Review the characteristics of annelids in your textbook before doing this lab. Then answer the following questions.

1. Name 2 major characteristics of annelids.
2. Are earthworms vertebrates or invertebrates?
3. What do you think would happen to an earthworm if it were put in clean sand from a beach? Why?
4. Why do farmers like to have earthworms in their soil?
5. Would you expect an earthworm to be attracted to light?

Learn these 4 terms as they refer to animals:
- ANTERIOR—toward the head end
- POSTERIOR—away from the head end
- DORSAL—toward the back or top side
- VENTRAL—toward the front or underside

6. Why are these terms better than *top, bottom, front,* and *back?*

Procedure

Part I

You will be given a live earthworm. Handle it gently and follow the lab directions carefully.

A. Moisten a paper towel and place it in a clean dissecting pan. Place the worm on it. Watch the worm move and notice which end leads. The worm's leading end is its anterior end. Also identify the worm's posterior end.

B. Look carefully along the dorsal side of the worm. You will see a thick purple line running down the entire length. This is the **dorsal aorta,** the major blood vessel of the earthworm. You will be observing this blood vessel in Part II of this lab.

C. To differentiate between the worm's **dorsal** and **ventral** sides, roll the worm over and observe what happens.

    7. How does the worm respond to being put ventral-side up?

CAUTION: Frequently during the lab, give your worm a "bath" with tap water from a medicine dropper. Otherwise, the worm will dry out and become lethargic.

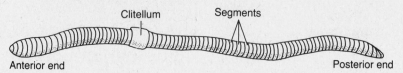

Clitellum   Segments

Anterior end   Posterior end

D. Notice how the body of your earthworm is divided into segments. Starting at about segment #30, there is a thickening in the body of the earthworm. This thickening is called the **clitellum.** The clitellum produces a mucous sac in which the earthworm deposits its eggs.

    8. Why do all earthworms have a clitellum?

E. The worm moves through the soil with the use of **setae.** Setae are bristles located on the worm's ventral surface and sides. They give the worm traction. There are two pairs of setae in every segment. Use a hand lens to locate the setae. You may also use your finger to feel for them.

F. You are now going to see if your earthworm is sensitive to light. Notice that your worm does not have any eyes. Cover the beam of a flashlight with 2 pieces of red cellophane. Fasten them in place with a rubber band. Darken the room and shine the beam on the earthworm.

    9. How does the worm react to the red light?

G. Replace the red cellophane with blue cellophane. Shine the blue beam on the earthworm.

    10. How does the worm react to blue light?

H. Punch a small hole in an index card. Remove the cellophane and tape the card over the flashlight. Shine the light first on the worm's anterior end, then on its middle, and then on its posterior end. Shine the dot of light on one area of the dish. See if the worm will enter the dot of light.

    11. Is any area of the earthworm's body more sensitive to light than other areas? Which one(s)?

    12. How is this sensitivity an adaptation to the earthworm's way of life?

I. You are now going to determine whether an earthworm prefers a moist or dry environment. Place a piece of dry paper towel next to the wet one. Place your worm across the 2 pieces.

    13. Does your worm move toward the wet or dry paper towel?

    14. Why do you think this is so?

J. Place a piece of sandpaper next to the dry paper towel. Place the worm across the 2 surfaces, anterior end on the sandpaper first. Observe the worm's response. Then place the worm across the sandpaper and dry paper towel, this time with the anterior end on the paper towel.

    15. Which surface does the worm prefer?

    16. Why do you think this is so? Are there other factors that might affect which way the worm moves?

Part II

Birds and mammals have complex mechanisms that enable them to maintain a constant body temperature. Earthworms do not. In a warm environment, an earthworm's body temperature will rise. In a cold environment, its body tem-

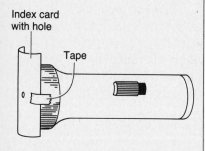

Index card with hole

Tape

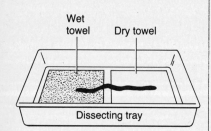

Wet towel   Dry towel

Dissecting tray

perature will drop. As the earthworm's body temperature rises and falls, its heart rate changes. In this part of the lab, you will measure the change in an earthworm's heart rate as its body temperature changes.

K. Add tap water to a large petri dish until the bottom is just barely covered. Place a live, healthy worm into the dish and observe through a dissecting microscope.

L. Find the dorsal aorta, the purple blood vessel running along the dorsal side of the earthworm. Look for rhythmic contractions of this blood vessel. A "ripple" of contraction should appear to be moving from posterior to anterior followed immediately by another contraction. Each "ripple" represents a single "heartbeat."

M. When you are able to count heartbeats (ripples) accurately, your teacher will assign you a temperature point. You will then count the number of ripples per minute at the temperature assigned to you.

N. In order to adjust the temperature, you will "float" your petri dish containing your worm on top of either a warm water bath or an ice bath. Use warm tap water for the warm water bath. Use cold water with ice cubes for an ice bath. Measure the temperature by placing your thermometer next to the worm in the petri dish.

O. Use the following procedure for each temperature assigned to you: Place the petri dish with the worm into the water bath (warm or cold). Watch the temperature until it comes within 1 or 2 degrees of your assigned temperature.

17. Record the temperature.

P. Remove the petri dish and immediately begin counting the heartbeats for exactly 1 minute. It is usually best for 1 student to count while the other student times the minute on a clock or watch. At the end of 1 minute, note the temperature again and calculate the average temperature to the nearest degree.

18. Record this temperature and the number of heartbeats you counted.

Q. You will now pool the data from the entire class into a table showing temperature and heart rate.

19. Copy the table below on the blackboard and enter your data on it.

20. Then graph the heart rate as a function of temperature. Plot the data from the entire class and draw the best possible straight line through the points.

Strategy for Organizing
When graphing the temperature (which is the independent variable) should be on the x-axis. The heart rate (the dependent variable) should be on the y-axis.

Data Chart for Class

Temperature ˚C	Heart Rate

21. Describe the relationship between temperature and heart-beat rate as shown by your table and graph.

Postlab Analysis

22. What is the adaptive advantage of the earthworm's responses to light and moisture?
23. How is gas exchanged between the earthworm's **closed circulatory system** and the environment?
24. What do you think causes the variations of heart rate as temperature is varied? Why would this provide an adaptive advantage for the earthworm?

Further Investigations

1. To learn about the internal anatomy of an earthworm, you might wish to dissect a preserved specimen or work with models, diagrams, filmstrips, slides, films, videotape, videodiscs, or computer software. To do a dissection, ask your teacher if specimens are available or may be obtained. Your teacher can also provide you with instructions to carry out the dissection. To learn about earthworms without dissecting one, ask your teacher to provide you with or direct you to an appropriate alternative such as those mentioned above.
2. What other experiments may test an earthworm's response to light?
3. Design an experiment that would test an earthworm's response to varying levels of moisture.
4. Check a live earthworm's "robin response." Gently grab at the worm's anterior end—as a hunting robin would do. What is the worm's response?

Strategy for Inferring

Think about the relationship between each body structure and its function. Think about what would happen to the earthworm if it did or did not react in a certain way or if it did or did not have a particular body structure.

Investigation

31 *Live Earthworms*

1. _____
2. _____
3. _____
4. _____
5. _____
6. _____
7. _____
8. _____
9. _____
10. _____
11. _____
12. _____
13. _____
14. _____
15. _____
16. _____
17. _____
18. _____

19. Record your data on the blackboard.

20. Make a graph on a separate sheet of paper.

21. _____
22. _____
23. _____
    _____
24. _____
    _____

Investigation

32

Live Crayfish

How is the crayfish's behavior adapted to its environment?

Introduction

Crayfish live on the bottoms of freshwater rivers, streams, ponds, and marshes. They hide during the day and come out for food at night. Their diet includes fish and insects as well as dead organic matter, making them partly scavengers.

Crayfish are members of the Phylum Arthropoda. Arthropods are characterized by their hard external skeletons and jointed appendages. They have complete, or well-developed, digestive, circulatory, respiratory, excretory, and nervous systems. The arthropods include crustaceans (such as crayfish, lobsters, and crabs), insects, centipedes, and millipedes. There are so many species of arthropods that together they make up over 95% of all the known animal species in the world.

In this lab, you will observe the external skeleton and appendages of crayfish. You will also study some of the structures and behaviors that suit it for its particular way of life.

Prelab Preparation

Review the characteristics of arthropods and crustaceans in Chapter 32.

1. What are 2 major characteristics of arthropods?
2. Are crayfish vertebrates or invertebrates?
3. Name the 3 body sections of arthropods.

Learn the following terms:

- anterior—toward the head
- posterior—toward the tail
- dorsal—toward the back
- ventral—toward the underside

4. Using this terminology, describe where the crayfish's eyes are located.

Procedure

Part I: External Anatomy

A. Obtain a live crayfish from your teacher and place it in a transparent plastic box. Fill the box to a level between 2- and 3-cm deep with fresh water. CAUTION: You will be working with a live animal, so it is important to treat it gently and to follow directions carefully.

B. Observe your crayfish for about 5 minutes.

5. Record your observations.

Learning Objectives

- To diagram the external anatomy of the crayfish.
- To observe how the crayfish has evolved as a bottom-dwelling animal.

Process Objectives

- To examine the external anatomy and behavior of a live crayfish.
- To infer what environmental pressures caused the crayfish to evolve as it has.

Materials

For Group of 2

- Live crayfish
- Clear plastic box to hold the crayfish
- Blunt probe
- Food pellets
- Rock
- Flashlight
- Saturated salt solution
- Dechlorinated tap water

Lateral View

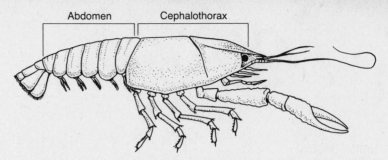

Abdomen   Cephalothorax

C. Now examine the external anatomy of your crayfish. Typically, arthropod bodies are divided into three sections: head, thorax, and abdomen. In the crayfish, the head and thorax are fused into a single **cephalothorax**. Find the cephalothorax and the abdomen of your crayfish. Notice that the crayfish has a hard external skeleton. The covering of the cephalothorax is called the **carapace**.

6. How many segments do you see in the abdomen?
7. What is the relationship between the external skeleton and body size?

D. Locate the body appendages. They are listed in order below, beginning at the head. Notice that they are jointed. This is characteristic of all arthropods.

- antennules (short, double)
- antennae (long, single)
- mandibles (jaws, hard)
- maxillae—2 pairs
- maxillipeds—3 pairs for holding food
- chelipeds—large claws
- walking legs—4 pairs
- swimmerets—5 pairs on the abdomen, one pair on each segment
- uropods—1 pair, used for swimming
- telson—the only non-paired appendage

Dorsal View of Crayfish

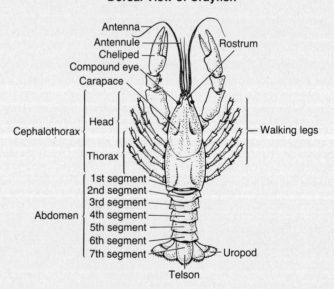

Antenna
Antennule
Cheliped
Compound eye
Carapace
Rostrum

Cephalothorax { Head / Thorax
Walking legs

Abdomen {
1st segment
2nd segment
3rd segment
4th segment
5th segment
6th segment
7th segment

Uropod
Telson

The mandibles, maxillae, and maxillipeds are the mouth parts of the crayfish. They are found on the ventral side of the animal, anterior to the

Crayfish Mouth Parts

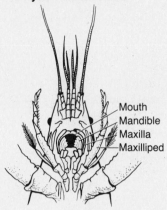

Mouth
Mandible
Maxilla
Maxilliped

chelipeds. It is difficult to see them individually. It helps to gently separate them with a blunt probe. The other appendages should be easy to find.

8. Draw a diagram of the ventral surface, showing all appendages.

E. Determine the sex of your crayfish. To do this, look at the first two pairs of swimmerets, located near the middle of the abdomen. In the male, these are large and stiff and are used for transferring sperm to the female. In the female, they are small, like the other swimmerets. Compare several crayfish so you see the difference between the two sexes.

9. What is the sex of your crayfish?

F. Observe the walking legs of your crayfish. The **gills** are attached to the walking legs and are located under the sides of the carapace.

10. What is the function of the gills?

11. What is the advantage of having gills attached to the walking legs?

G. Observe the large **compound eyes** of the crayfish. They are made up of many photoreceptor units. This makes them especially sensitive to movement. Compound eyes are a characteristic of arthropods.

12. Why is this type of eye well-suited to the crayfish's nocturnal life-style?

H. Finally, observe the **anus** on the ventral side of the abdomen just anterior to the telson. The telson is the single broad, flat, most posterior appendage. After food is digested, the undigested waste passes through the intestine, which runs the length of the abdomen. The intestine passes this waste out the anus.

13. What is the advantage of having the anus at the posterior end?

Part II: Behavior

In this part of the lab, you will observe the behavior of a live crayfish.

I. Place a large rock on the bottom of your aquarium. Observe the behavior of the crayfish for about 5 minutes.

14. How does the crayfish respond to the rock?

15. How does this response reflect the crayfish's way of life?

16. Describe how the crayfish walks.

J. Very gently, touch the antennae with a probe. Observe how the crayfish responds.

Strategy for Observing

When you draw the crayfish, identify all the parts you can observe. Compare your illustration to the listing of the structures in Step D. Try to identify any missing parts and add them to your drawing if you are successful. However, do not include any structures that you could not see on your specimen.

17. Describe your observations.

K. Shine a flashlight on the anterior end of the crayfish. Observe its response. Then shine the flashlight on the posterior end and observe the response.

    18. How did the crayfish respond when you shined light on its anterior end? On its posterior end?

    19. What can you infer from these two responses?

    20. How do these responses illustrate the crayfish's way of life?

Strategy for Inferring

Think of what you have observed experimentally and how this might occur naturally. Relate the behavior you observed in the crayfish to some natural stimulus that would cause it to act like this in its normal habitat.

L. Crumble a piece of food with your fingers and place it in front of the crayfish. Watch your crayfish for several minutes while it eats the food. Notice how it uses its mouth parts. If your partner carefully holds the box up, you can watch the ventral side of the crayfish.

    21. Describe how the crayfish used its mouth parts to eat the food.

M. Put 5 drops of a saturated salt solution in the water near the animal's antennae and observe the results.

    22. How does the crayfish react to the salt?

    23. What is the purpose of the antennae and antennules?

Postlab Analysis

    24. Name at least 4 structural or behavioral characteristics which make the crayfish well-suited to its habitat.

Further Investigations

1. Observe your crayfish over a period of several weeks in an aquarium. Place a 3–inch clay flowerpot on its side in the aquarium. The crayfish will use this for shelter. Note when the crayfish stays in the flower pot most. Is this influenced by the amount of light in the room?

2. Change the color of the gravel on the bottom of the aquarium in which you are keeping the crayfish. Observe any color change in the crayfish over a period of several weeks. Why would this be adaptive?

3. If you would like to learn about the internal anatomy of a crayfish you might wish to dissect a preserved specimen. Ask your teacher how you can do a dissection.

Investigation

32 | *Live Crayfish*

1. _____

   _____

2. _____

3. _____

4. _____

5. _____

   _____

   _____

6. _____

7. _____

   _____

8.

9. _____

10. _____

11. _____

_____

12. _____

_____

13. _____

14. _____

_____

15. _____

_____

16. _____

_____

17. _____

_____

18. _____

_____

19. _____

_____

20. _____

_____

21. _____

_____

22. _____

_____

23. _____

_____

24. _____

_____

_____

33

Caterpillar Behavior

Learning Objectives

- To understand how insect behavior might be influenced by developmental characteristics.
- To learn how insect larvae respond to different stimuli.

Process Objectives

- To observe the behavior of insect larvae when exposed to variable conditions
- To hypothesize about the function of certain larval behaviors.

Materials

For Group of 4

Part I
- 3–5 caterpillars
- 2 pieces of brown wrapping paper, 125 cm x 150 cm
- Chalk or marking pen
- String
- Thumbtack
- Stopwatch or watch with second hand
- Thermometer
- Colored pencils or markers
- Compass (magnetic directional)
- Adhesive tape
- Meter stick

Part II
- Wood board (5cm x 30 cm)
- Stopwatch or watch with second hand
- Rack to support board at 45° angle
- Protractor
- Piece of cardboard
- 3–5 caterpillars and 3–5 mealworms

Strategy for Hypothesizing

Think about the particular locations in which larval insects pupate and compare these with the locations where larvae generally feed.

What environmental and developmental factors influence insect behavior?

Introduction

Beetles, bees and butterflies are examples of insects that live out their life cycles in four distinct stages. The four stages— egg, larva, pupa and adult— make a sequence known as a **complete metamorphosis**. The egg is a structure that protects an organism from unfavorable climatic conditions. The egg hatches into a larva which is specialized for growth. When growth is complete the larva enters an immobile state called a pupa. The covering of the pupa protects the organism as it develops into an adult. The adult usually has wings and is specialized for dispersal and reproduction.

Insects behave differently in each developmental stage. For example, a larva that feeds on tree leaves will be attracted toward light. A larva that pupates in soil will be attracted toward the ground when it is ready to pupate.

Prelab Preparation

When performing behavioral studies on animals, keep in mind the following classifications of behavior. Consider how they relate to the insect species you are studying. **Reflex** actions are simple behaviors, automatic responses to a stimulus. **Instinctive behavior** is a complex series of genetically determined behaviors. **Learned behavior** is instinctive behavior that an animal has modified based on past experience.

Insects in a larval stage are in a period of rapid growth. Their behavioral responses are directed toward obtaining large amounts of food quickly.

1. In terms of behavior, how would you classify the movement toward light of a foliage-feeding larval insect?
2. Why might a larval insect respond differently when it is ready to pupate?

Procedure

Part I

A. Prepare an orientation chart by taping together two 125 cm x 75 cm pieces of brown wrapping paper to made a 125 cm x 150 cm piece. Draw a circle with a radius of 60 cm by using a chalk or marking pen tied to a 60 cm length of string. Secure the end of the string with a thumbtack to the center of the paper and make a circle with the marker, holding the string taut. Using the same technique, draw a circle with a radius of 30 cm within the larger circle. Divide the area of the combined circles into quadrants. Bisect each quadrant and label according to the illustration.

Take the orientation chart and the caterpillars outside. Orient the chart horizontally so that the line marked N on your chart is pointing North according to the compass. Release caterpillars, a few at a time, in the center of the circle. Mark where each caterpillar crosses the circle.

Orientation Chart

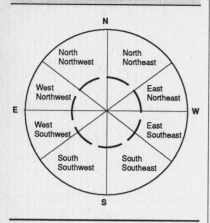

Strategy for Observing

Observe the relative sizes and colorations of the different caterpillars tested by your team. Note whether the behaviors seem dependent on differences between animals, such as different stages of larval development, rather than on the external variables.

3. Record the numbers of caterpillars found in each quadrant. In your data chart for caterpillar orientation also record the conditions: temperature, amount of light (full sun, hazy, shade), direction of the sun.

B. Take the orientation chart and the caterpillars to a site where light conditions are different and repeat the experiment. Use a different color to mark where the caterpillars cross the circles.

4. Record the quadrant and conditions data in the chart for caterpillar orientation.
5. In which direction did most of the caterpillars move? Did the direction vary depending on the amount of sunlight?
6. What was the main factor that influenced the direction chosen by the caterpillars? Form a hypothesis about how this behavior may provide the caterpillars with a survival advantage.

C. Repeat Step A; however, this time test the speed at which each caterpillar reaches the outer circle. Repeat the test under different conditions.

7. Record caterpillar speeds in the appropriate data chart.
8. Did external conditions have an effect on the speed of caterpillar movement? What other factors might influence the speed at which they move?

Part II

D. The response of an animal to gravity is called **geotaxis** (jee-o-taks-iss). When an animal responds to gravity by moving downward, it is called positive geotaxis. When an animal moves upward in response to gravity, it is called negative geotaxis. Compare the geotactic responses of two kinds of insect larvae by using a smooth wooden board approximately 5 cm × 30 cm, placed at a 45° angle. Place 5 to 10 caterpillars or mealworms at the center of the board.

9. Record in your data chart for geotactic responses how many larvae are positively geotactic and how many are negatively geotactic. Record the temperature and amount of light. NOTE: Make sure that light is even on whole surface of the board.

E. Repeat Step D, this time by setting up the experiment in the dark (in a closet or under a cover.) Allow at least 10 minutes for larvae to move.

10. Record the results in your data chart.
11. Do caterpillars respond positively or negatively to gravity?
12. Did lack of light influence their movements? Explain.
13. Do mealworms respond positively or negatively to gravity?
14. Did lack of light influence their movements? Explain.

Postlab Analysis

15. How do the behaviors you observed relate to a particular insect's means of feeding?
16. How might the most frequently observed behaviors favor survival of the insect at that developmental stage?

Further Investigation

1. Design a T-maze and develop experiments that show whether insect larvae can learn. Use rewards, such as darkness, light, or a preferred food, to see if you can induce learned behavior. Can one larva learn from another larva already trained to respond to a T-maze? Before you begin, research the habits and food preferences of the species you will use.

Investigation

33 | *Caterpillar Behavior*

1. _____

2. _____

   _____

3.–4. Enter your answers on the data chart.

Orientation of Caterpillars

	Temp. ___	Temp. ___	Temp. ___
	Light ___	Light ___	Light ___
	Number of Caterpillars	Number of Caterpillars	Number of Caterpillars
NNE	___	___	___
ENE	___	___	___
ESE	___	___	___
SSE	___	___	___
SSW	___	___	___
WSW	___	___	___
WNW	___	___	___
NNW	___	___	___

Timing Movement

Record conditions at time of trial. Record time in seconds for each caterpillar to move from center to outer circle.

	Temp. ___	Temp. ___	Temp. ___
	Light ___	Light ___	Light ___
Caterpillars	Time in Seconds	Time in Seconds	Time in Seconds
1	___	___	___
2	___	___	___
3	___	___	___
4	___	___	___
5	___	___	___

5. _____

   _____

   _____

6. _____

   _____

7. Enter your answers on the data chart above.

   _____

8. _____

   _____

9.–10. Enter your answers on the data chart.

Geotactic Responses of Two Kinds of Insect Larvae

Record U for upward movement and D for downward movement.

Caterpillars	Temp. _____ Light	Temp. _____ Dark	Mealworms	Temp. _____ Light	Temp. _____ Dark
1			1		
2			2		
3			3		
4			4		
5			5		

11. _____

12. _____

13. _____

14. _____

15. _____

_____

16. _____

34 | *Live Sea Stars*

How do sea stars move, feed, and defend themselves?

Learning Objectives
- To relate the structure of the sea star's water-vascular system to its means of locomotion.
- To learn about echinoderm feeding structures and defensive adaptations.

Process Objectives
- To observe how tube feet function in locomotion, feeding, and defense of sea stars.
- To experiment with the effects of various stimuli on the behaviors of sea stars.

Materials
For Class
- Marine (salt water) aquarium
- Live bivalves or fresh extract of clam, oyster, or other bivalve
- Household bleach (10% solution)

For Group of 2
- Compound microscope
- Slide of sea star arm (cross section)
- Live sea star
- Specimen dish
- 2 Pipettes
- Paper towels
- Dissecting microscope
- Blunt probe
- Slide of sea star pedicellariae (whole mount)

Introduction

If you have ever observed sea stars in tidal pools clinging to rocks as if rooted, you may be surprised to find that they have hundreds of tiny **tube feet.** These tube feet help them eat, assist them in repelling undesirable substances, and enable them to move.

Echinoderms use a highly modified part of their coelom, the **water-vascular system,** in locomotion. The water-vascular system exerts water pressure on the sea star's tube feet to make them move. In this investigation, you will learn how this system works.

Prelab Preparation

An echinoderm uses cilia to carry water into its water-vascular system through a porous opening called the **sieve plate.** Water then enters the internal canal system. It flows through the **stone canal** into the **ring canal** which encircles the mouth region. The water flows from the ring canal into the **radial canals**— one in each arm. Many lateral canals branch off from the radial canals and carry water to the hundreds of paired tube feet.

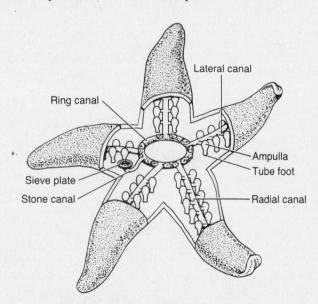

1. Draw an illustration of the water-vascular system and use arrows to show the path of water through it.

Note the illustration on the next page of the tube foot. Water flows into the muscular **ampulla** through the lateral canal. As the ampulla contracts, the valve in the lateral canal closes, preventing backflow into the canal system.

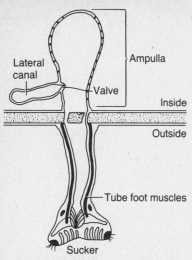

Lateral
canal

Ampulla

Valve

Inside

Outside

Tube foot muscles

Sucker

Tube Foot Cross Section

Strategy for Observing

Look at several individual tube feet. Then, observe simultaneously all the tube feet on a single arm. Notice any patterns of movement that are present.

Strategy for Experimenting

Remember that most animals respond to a stimulus in a manner that would normally enhance their ability to survive.

2. If the ampulla contracts and the valve closes, where does the water go? What happens to the tube foot?
3. As the ampulla relaxes, the muscles in the walls of the tube foot contract. Where does the water go now?

Procedure

A. Examine the cross section of a sea star arm. Using the compound microscope at 100×, compare what you see with the illustration.

4. Draw a diagram from the prepared cross section. Label the coelom, radial canal, and ampulla.

B. Observe a sea star crawling on the side of the aquarium.

5. How do the tube feet move? Are they coordinated?

C. Place a sea star from the aquarium in a specimen dish containing sea water. Using a long pipette, introduce 2 to 3 drops of fresh clam or oyster extract into the water near one sea star arm.

6. What is the response of the tube feet?

D. With another pipette, place 2 to 3 drops of diluted household bleach near a different arm. CAUTION: Immediately after observing response, rinse the sea star with tap water to remove bleach, then place the animal back in the dish with clean (aquarium) salt water. Discard bleach-contaminated salt water.

7. What is the response of the tube feet?

E. Remove the sea star and place it upside down on several wet paper towels.

8. What is the response of the arms? Do the tube feet participate in righting the organism?

F. **Pedicellariae** are small pincers made of calcium on the **aboral surface** of a sea star. They are used to help defend the sea star and to help keep its surface clean of settling organisms. Place your sea star back in the specimen dish. Place the dish under a dissecting microscope and focus on some pedicellariae, which you will find clustered around spines. Using the probe, gently touch some spines in the area of these pedicellariae. Gently touch the pedicellariae outside and within the jaws.

9. How do the pedicellariae respond? What happens when you touch the pedicellariae outside the jaws? Inside the jaws?

Return the sea star to the aquarium.

G. Examine the pedicellaria whole mount under the dissecting microscope.

10. How many plates are present on each pedicellaria?

Postlab Analysis

11. How do tube feet assist sea stars in feeding? What is the advantage of having many tube feet for feeding?
12. List as many structures as you can think of that are found on the aboral surface of an echinoderm. How might these structures be affected by other organisms settling on the sea star?

Further Investigations

1. Consider the properties of a fluid that make it useful for the water-vascular system. Compare the water-vascular system of the sea star to the hydrostatic skeleton of nematodes.
2. Observe the eye spots in the sea stars. Design an experiment in which you could make and test a hypothesis about the function of the eye spots.

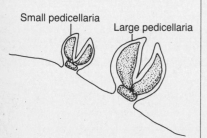

Small pedicellaria

Large pedicellaria

Investigation

34 | *Live Sea Stars*

1.

2. _____

3. _____

4.

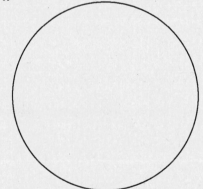

5. _____

_____

_____

6. _____

_____

7. _____

_____

8. _____

_____

9. _____

_____

10. _____

11. _____

_____

_____

12. _____

_____

_____

35 | *Comparing Invertebrates*

What are the major evolutionary trends among the invertebrates?

Learning Objectives
- To learn about similarities and differences among invertebrates.
- To relate structural adaptations to the evolution of invertebrates.

Process Objectives
- To classify invertebrates according to their body plans.
- To analyze your data for trends.

Materials
For Group of 4–6
Parts I and II
- Dissecting microscope
- Specimen dishes
- Dropper bottle of water, or plant spray bottle
- Unlined paper

Part I
Living Specimens:
- *Hydra*
- *Metridium* (sea anemone)
- *Dugesia* (flatworm)

Preserved Specimens:
- *Hydra* cross section
- *Metridium* cross section
- *Dugesia* cross section

Part II
Living Specimens:
- *Anguillula acetiglutinus* (vinegar eels)
- *Lumbricus* (earthworm)

Preserved specimens:
- *Ascaris* (round worm) cross section
- Clam
- Snail
- Squid
- *Nereis* (marine worm)
- *Romalea* (grasshopper)

Introduction

The trend in evolutionary adaptations of invertebrates is from simplicity to increasing complexity.
- The increase in the number of germ layers from 2 to 3 allows for the formation of skeleton, muscles, and blood.
- The presence of bilateral symmetry allows for a more motile life-style than radial symmetry.
- The increasing specialization of segments leads toward the fusing of segments into larger functional units.
- The advance from a simple to a complete digestive system provides the animal with more readily available energy.
- A complex respiratory system rather then gas exchange through the body covering allows a greater body size.
- The increase in cephalization enables the development of more complex sense organs.
- The development of a closed circulatory system increases the efficiency of circulation by providing oxygen and nutrients to the cells more quickly.
- As excretory systems advance from simple diffusion to specialized filters, wastes are excreted more efficiently and water can be reabsorbed.

Prelab Preparation

Most invertebrates are **motile**, moving about freely to capture food. Other invertebrates are **sessile**, or attached to one spot. They feed by capturing food that passes nearby. Body structures and life-style are closely related.

Sessile invertebrates lack appendages for walking or swimming. Instead, they must have a means for anchoring themselves in place. *Motile* invertebrates often have appendages for walking or swimming. These animals have many separate muscles that move the body segments and appendages. Motile invertebrates without appendages have layers of muscles whose contractions and relaxations serve to move the animal about.

Sessile invertebrates have a nerve net of diffuse nerve cells but without a control ganglion. Because *motile* animals must control their body movements and coordinate the movements of appendages, they have a brain or ganglion. Animals that move about also have sense organs concentrated anteriorally to sense the oncoming environment.

1. What shape do you think would be advantageous to slithering and/or swimming movements? On a separate sheet of paper, draw the shape you would expect to find for these types of locomotion.
2. Where would you expect sense organs to be? Near the tail? the head? All around? Explain your choice.

3. Where would you expect the appendages to be? All on one side? On one end? Underneath? Explain your choice.

Procedure

Part I

A. Under your dissecting microscope, examine a live *Hydra*. Observe its movements. Now look at a live *Metridium*, a type of sea anemone. CAUTION: Be sure to keep all live animals moist by applying water with either a medicine dropper or a plant spray bottle.

4. Besides the writhing of its tentacles, does the *Hydra* move about much?

5. Do you see anything that looks like a head on the *Hydra*? On the *Metridium*?

6. In what pattern are tentacles arranged on each organism?

B. Using the dissecting microscope, look at the cross sections of *Hydra* and *Metridium*. Use the figure to help you find the outer cell layer (**ectoderm**) and the inner cell layer lining the gastovascular cavity (**endoderm**). The material between these layers, a solid mass called mesoglea, is largely without cells (**acellular**).

7. What shapes are the *Hydra* and *Metridium*?

8. How is the *Metridium*'s anatomy more complex than that of *Hydra*? What differences might we see, then, in the *Metridium*'s functions?

C. Under your dissecting microscope examine live *Dugesia*, Phylum Platyhelminthes (flatworms), comparing them with the *Hydra* and *Metridium*.

9. Do the *Dugesia* show more movement than the *Hydra*?

Cephalization is the concentration of sensory and nervous elements at the anterior end of an organism.

10. Can you identify a head in the *Dugesia*? What is the most conspicuous feature of the anterior end?

D. Under the dissecting microscope, examine the cross section of the *Dugesia*. Now look at the illustration of the *Dugesia*. Note its outer and inner cell layers. Also note how the volume between its gut and body wall is filled with living cells—**parenchyma**. Recall how similar layers in the *Hydra* and the *Metridium* are filled with acellular mesoglea.

11. Does the *Dugesia*'s shape differ from that of the *Hydra* or the *Metridium*?

12. Do you see any evidence of lungs or gills in the *Dugesia*? If respiratory and circulatory systems are absent, how might this influence the organism's shape?

Part II

E. Under the dissecting microscope, watch the *Anguillula acetiglutinus*, or vinegar eels, as they swim. These organisms belong to the Phylum Nematoda (roundworms).

13. How do the vinegar eels' movements compare with those of the *Dugesia*?

F. Under the dissecting microscope, examine the cross section of another nematode, the *Ascaris*. Notice that there is a large space (**pseudocoel**) between the intestine and the body wall. This space is fluid-filled when the *Ascaris* is alive. In the illustration, you can see the *Ascaris* muscles in cross section. Note how these muscles are longitudinal, or run lengthwise along the organism.

Strategy for Classifying

As you examine each organism in this lab, think about how it has adapted to its environment. How does each animal perform the basic life processes—food gathering, digestion, reproduction, etc.?

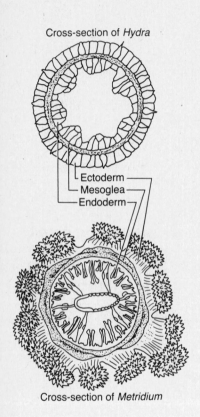

Cross-section of *Hydra*

├ Ectoderm
├ Mesoglea
└ Endoderm

Cross-section of *Metridium*

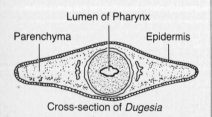

Lumen of Pharynx

Parenchyma   Epidermis

Cross-section of *Dugesia*

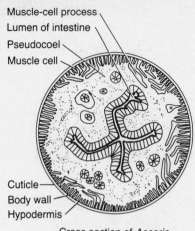

Muscle-cell process
Lumen of intestine
Pseudocoel
Muscle cell

Cuticle
Body wall
Hypodermis

Cross section of *Ascaris*

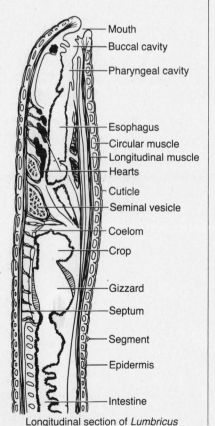

Mouth
Buccal cavity
Pharyngeal cavity

Esophagus
Circular muscle
Longitudinal muscle
Hearts

Cuticle
Seminal vesicle

Coelom

Crop

Gizzard

Septum

Segment

Epidermis

Intestine

Longitudinal section of *Lumbricus*

14. Compare the *Ascaris*'s muscles with any you might have noticed in the *Dugesia*. In which phylum—Nematoda or Platyhelminthes—are the muscles more developed? How does this structure relate to the different movements of each?

Recall your observations of the vinegar eels' movements.

15. What happens in the bands of longitudinal muscles when the body is flexing? Do all the bands of muscles contract simultaneously?

16. Are nematodes shaped differently from flatworms? Explain your answer. What do you think functions in place of a circulatory system in a nematode?

G. Under the dissecting microscope, examine the 3 preserved mollusks your teacher has provided—the clam, the squid, and the snail. When alive, the clam burrows into the sand or mud or attaches to a rock. The snail crawls on surfaces in search of food. The squid swims in search of prey.

17. Of these mollusks, which 2 forms are most streamlined? Why? Which form is most highly cephalized? What might be the advantages and disadvantages of having a shell in each form?

H. Under the dissecting microscope, examine the preserved *Nereis*, a marine annelid or sea worm. Compare it with the live *Lumbricus*, a type of earthworm. Observe how the earthworm moves.

18. What do the appendages (**parapodia**) tell you about the sea worm's mobility? Which worm has the more conspicuous head—the sea worm or the earthworm? How might the structure of a head or the presence of parapodia reflect adaptations to a burrowing lifestyle?

19. How are the sea worm and the earthworm similar?

20. How does the earthworm use its segments when it crawls? How does this differ from the movement of the nematodes you studied before—the vinegar eels?

I. Examine the *Lumbricus* section shown in the illustration.

21. Are the external segments reflected in the internal structure? What effect does this have on the body cavity? How might this affect locomotion?

J. Now examine a preserved insect, the grasshopper (*Romalea*). Notice its head, body segments, and appendages. Much of the grasshopper's internal anatomy is similar to that of the annelid you studied—the *Nereis*—particularly the nervous systems.

22. What external similarities do insects (arthropods) and annelids share?

23. How do the body coverings of insects and annelids differ? What effect might this have on their respiratory systems?

24. Is there any evidence of regional specialization in the body covering of the insect? In the appendages of the insect? How do arthropod appendages compare with those of annelids and with those of the organisms you imagined in your prelab exercise?

Postlab Analysis

25. Summarize in the data chart the characteristics of the animals you have compared. Rate each characteristic, using 0, +, ++, and +++, to indicate the extent of that characteristic's development in a given animal.

Strategy for Analyzing Data
As you compare the animals you have just examined, think about the "starting materials" involved. How might natural selection have acted to produce each unique result? What limitations are involved?

26. How do these characteristics appear to be related to one another?
27. Think about each animal you examined. How might the lack of a respiratory system limit an animal's structure? How might it influence body covering or size?

Further Investigations

1. Study and make a list of some invertebrates that burrow and live underground. What features do they lack, compared with similar invertebrates that move above ground? What features are enhanced? To what extent may adaptations to different environments account for variations within a phylum or other taxonomic group?

2. Make a list of environments that invertebrates inhabit. Make another list of ways that invertebrates obtain food (e.g., herbivory, sit-and-wait predation, prey-chasing, parasitism). Randomly choose one environment and one way of obtaining food. Now "design" an invertebrate that would fit such a description. Remember to consider shape, body covering, appendages, and maintenance systems (e.g., respiration, digestion, circulation). How does your imaginary invertebrate compare with one that actually copes with such an environment? Now add a third constraint (e.g., a certain kind of body cavity) that might be imposed by the ancestor of this organism. Does this alter your design?

Investigation

35 | *Comparing Invertebrates*

1. Make a drawing on a separate sheet of paper.

2. _____

3. _____

4. _____

5. _____

6. _____

7. _____

8. _____

_____

_____

9. _____

10. _____

11. _____

12. _____

_____

13. _____

14. _____

_____

15. _____

_____

16. _____

_____

17. _____

_____

_____

18. _____

19. _____

20. _____

    _____

21. _____

    _____

22. _____

23. _____

    _____

24. _____

    _____

    _____

25. Enter your answers on the data chart.

Comparison of Invertebrate Body Plans

	Cephalization	Locomotion	Body Cavity	Segmentation	Appendages	Bilateral Symmetry
Hydra/Metridium						
Dugesia						
Vinegar Eels/*Ascaris*						
Clam						
Squid						
Snail						
Nereis						
Lumbricus						
Grasshopper						

26. _____

27. _____

    _____

36

How Fish Respire

What factors change the respiration rate of a fish?

Learning Objectives
- To learn how water moves across fish gills during respiration.
- To study how a fish's respiration rate varies in response to temperature change or to active swimming.

Process Objectives
- To organize data for respiration rates under different conditions.
- To analyze data by graphing and extrapolating.

Materials

For Group of 4
- Goldfish
- 500-mL Beaker
- 1-L Beaker
- 250 mL Dechlorinated tap or pond water
- 1 Thermometer
- 1 L Warm water (about 80°C)
- Container of crushed ice
- Plastic stirring rod
- Watch or clock with second hand
- Graph paper

Introduction

The gills of a fish function in water like human lungs function in air. When you inhale, you draw oxygen-rich air into your lungs. Your alveoli, or air sacs, provide a large surface area that allows oxygen from the air to be absorbed into the many capillaries of the alveoli. When a fish gulps water, the water is forced through the gills. The gills are covered with fine filaments that provide a large surface area, richly supplied with capillaries, to extract oxygen efficiently from the water.

However, air contains a concentration of oxygen at least 20 times greater than oxygen-saturated water. It is not surprising that air-breathing mammals spend less than 2% of their energy budget to support respiration while fishes spend about 20% of their energy on respiration.

Different requirements for oxygen can be compared by looking at the rate of oxygen use in similar units of weight. A man at rest needs 200 mL of oxygen per kg body weight per hour. Fish require 100–350 mL of oxygen per kg body weight per hour. However, the amount of oxygen dissolved in water is directly related to temperature. Fresh water can contain 9 mL of oxygen per liter at 5°C, while only 5 mL of oxygen will remain dissolved in water at 35°C.

Prelab Preparation

Observe a goldfish at close range, either in the class aquarium during prelab preparation or at a pet store. Watch the rhythm of the mouth and operculum, the flap of skin which covers the gill, opening and closing. As the mouth opens, water is drawn into the pharynx. The mouth closes and the pharynx constricts, forcing the water out the gill slits. The operculum opens, helping to suck out the water. Compare your observations of the living fish with the illustration of a fish gill.

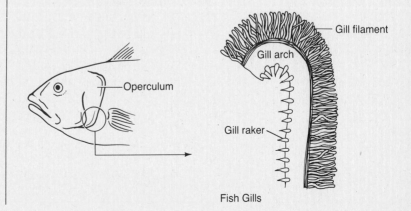

Fish Gills

1. Using information given in the introduction, calculate how many liters of water at 5°C a fish weighing 0.5 kg must move across its gills in an hour. Assume an average oxygen consumption rate of 200 mL O₂/kg/hr.

Read through the laboratory procedure so that you understand what you will be doing in the experiment. Make a note of the data you will need to record. Your teacher will instruct you in the proper handling of the goldfish. Never exceed a temperature of 30°C in the test beaker.

Procedure

Part I: Effect of Temperature on Respiration Rate

A. Put 250 mL of dechlorinated water in the 500-mL beaker. This beaker is your tank for performing tests on the fish. Wet the net before you attempt to catch the goldfish, and make a "clean catch"; do not press the fish against the side of the aquarium as you scoop it up. It is best to put the net under the water, move slowly to corner the fish, then quickly lift the net straight up. Invert the net over the test beaker. Be sure that there is enough water to cover the dorsal fin and to allow the fish to swim freely. Place a thermometer inside the beaker. CAUTION: Handle the fish carefully.

B. You can determine when the fish calms down from the excitement because the rate of mouth openings steadies. Then, find the respiration rate per minute by counting the openings for 2 minutes and dividing by 2. Measure the water temperature of the test beaker.

2. Record the respiration rate and the temperature on your data chart.

3. Why will counting for 2 minutes rather than 15 seconds increase accuracy of the data?

4. Why is it important to wait until the fish calms down before you calculate the respiration rate?

C. Add a few grams of crushed ice to the 1-L beaker and nest your test beaker inside. Add more ice to the larger beaker until the temperature of the water in the test beaker decreases to 5°C.

5. Record the fish's respiration rate, using the same method as above.

D. Slowly add some warm water (about 80°C) to the outer beaker. Monitor the rise in temperature inside the test beaker, adding more warm water to the outside beaker as needed. Allow time for the two water temperatures to reach equilibrium. Pour water from the outer beaker as necessary until the test beaker temperature stabilizes at 15°C.

6. Record the respiration rate of the fish.

E. Continue to raise the temperature in 5° increments up to 30°C.

7. Record the respiration rates of the fish at each new temperature. If the temperature of your water is not exactly 5°, 15°, 20°, and so on, correct the *water temperature* in your data chart.

8. Why is it better to raise the temperature rather than starting with warm water and chilling it?

F. When you have collected all of your readings, gradually decrease water temperature to that of the main aquarium.

Part II: Effect of Activity on Respiration Rate

G. Place a dark cover over the test beaker for 5 minutes. Do not disturb the fish with movement or noise. Remove the cover quickly. The startled fish swims rapidly. Observe any changes in the way the fish respires.

9. Calculate the fish's respiration rate, beginning your count within 30 seconds of this burst of activity.
10. How might this behavior improve the chances of survival of the fish?
11. Do you see any change in the depth of respiration (the distance the operculum is opened)?
12. How would this affect the amount of oxygen available?
13. How does this compare to the respiration rate of a human athlete?

Use care in returning the goldfish to the main aquarium. Gently pour the fish directly from the beaker into the aquarium.

Postlab Analysis

Graph the respiration rate of your goldfish at various temperatures.

14. Using the sample graph below as a model, make a graph of your data on separate graph paper.

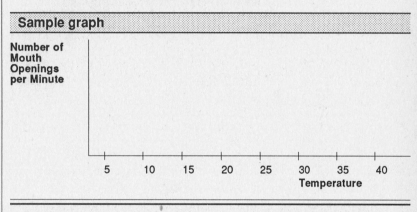

Sample graph

Number of Mouth Openings per Minute

5   10   15   20   25   30   35   40

Temperature

15. Compare your graph to those of other teams in the class. What trend do you observe in the graphs?
16. What kinds of variables might explain differences in the graphs?
17. Using your graph to make extrapolations, what could you expect the fish's respiration rate to be at 35°C? at 40°C? at 45°C?
18. Raising the water temperature beyond a certain point—the thermal lethal point—would kill the fish. Why is there a thermal lethal point? Consider the effects of water temperature on dissolved oxygen in the water and on the enzymes in the fish's tissues.

Further Investigations

1. The water used to cool the reactor of a nuclear power plant is usually returned to a river or the ocean at a temperature considerably warmer than the ambient temperature. Collect data on the temperature of effluent from the nuclear power plant nearest your home. Study the temperature requirements of the plants and animals that inhabit the discharge basin, in terms of thermal lethal point and preferred range.
2. Investigate the effect of light on the respiration rate of a fish. Hypothesize whether the fish's respiration rate will change according to intensity and wavelength of light. Measure respiration rate under different light conditions: a dark room with only enough filtered light to see the fish; in low

Strategy for Analyzing
When interpreting your graph, ask yourself how you may have varied your tests in ways that would have given you different results.

intensity light; and in bright sunlight. Experiment with different colored light filters. Remember to allow the fish to settle down before determining a resting respiration rate.

3. Visit an aquarium and observe at least one of each of the three major classes of fishes: jawless fishes, cartilagenous fishes, and bony fishes. Observe differences in respiratory structures. Show how these differences relate to ways that each kind of fish has adapted to its habitat.

Investigation

36 | How Fish Respire

1. _____

2. Enter your answers on the data chart.

Record of Mouth Openings Per Minute in Response to Temperature

| | Water Temperature | | | | |
	5°C	15°C	20°C	25°C	30°C
Your Team					
Team A					
Team B					
Team C					
Team D					
Team E					

3. _____

4. _____

5.–7. Enter your answers on the data chart.

8. _____
_____

9. _____

10. _____
_____

11. _____

12. _____
_____

13. _____
_____

14. Make a graph on a separate sheet of paper.

15. _____

_____

16. _____

_____

_____

17. _____

_____

18. _____

_____

37

Frogs

Learning Objectives

- To understand the physical adaptations that allow amphibians to live both in the water and on the land.
- To compare the frog's anatomy to that of a human being.

Process Objectives

- To observe internal and external structures of frog anatomy.
- To infer what features of the frog's anatomy and morphology lead us to classify it as an amphibian.

Materials

For Class
Part I
- Live frogs
- Terrarium
- Aquarium
- Bell jar (optional)
- Live insects (crickets or mealworms)

Part II
For Group of 5–6
- Preserved frog
- Dissecting pan
- Scissors
- Forceps
- Blunt probe
- Dissecting pins (about 6)

In what ways do amphibians represent an evolutionary bridge between aquatic and terrestrial vertebrates?

Introduction

Imagine a fish swimming in a pond. One day, the fish develops hind legs with webbed toes. Then front legs appear, lungs develop, and the tail is resorbed. The streamlined shape of the fish thickens, and the head becomes broad and flattened. The "fish" now breathes air and crawls out of the water.

The frog goes through the above transformation as it gradually assumes its adult form. Like all amphibians, it hatches from an egg laid in or near fresh water, emerging in a larval form that resembles a fish.

This larva, commonly known as a tadpole, feeds on aquatic plants as it grows larger. At a certain time in its life, a hormone stimulates it to develop into an adult frog through a process known as **metamorphosis**.

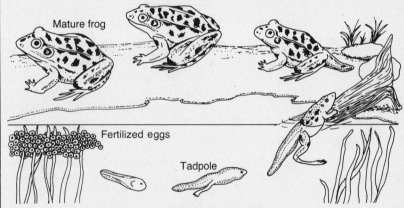

Life Cycle of the Frog

Amphibians can exist in many different habitats. Most frogs and toads spend much of their adult life on land; some even live in trees. However, all must return to water to breed. There are some salamanders that remain aquatic throughout their lives; they breathe through large external gills. This retention of juvenile features in an adult is called **neoteny**.

Prelab Preparation

As you study the frog, a representative amphibian, consider the adaptations that allow it to live both in water and on land. Consider the organ systems that are more fishlike—that is, necessary in an aquatic environment. Think also of the amphibian features that better enable the frog to live on land.

In Part I of this investigation, you will observe a live frog in water and on land. By interacting with it, you can observe its behavior and the way it responds to several stimuli. Do not be afraid to handle it, as frogs do not bite.

CAUTION: Keep in mind, however, that the frog is a living creature that deserves the best of care. Handle it gently, and do not allow it to injure itself by jumping from the laboratory bench onto the hard floor.

In Part II, which is optional, you will dissect a preserved frog in order to study the structures of its mouth and brain, as well as its digestive, circulatory, respiratory, and urogenital systems.

1. Write 2–3 questions about frogs that you would like to have answered. As you work on the lab, see if you can answer these questions as well as the ones in the lab.

2. What do you expect to learn when studying a live frog as compared to a preserved frog?

Procedure

Part I: Observing a Live Frog

A. Observe a live frog on land. Place a frog in a terrarium, or under an inverted bell jar, so that it will not jump away.

3. How is a frog adapted to move on land?

B. Watch the movements as the frog breathes. The frog draws in air by movements of membranes in the throat and forces it into the lungs.

4. How is the way a frog breathes different from the way a human breathes?

C. Introduce a live insect, such as a cricket or mealworm, into the terrarium. Observe how the frog reacts to the presence of the insect.

5. How does the frog catch and eat the insect?

D. Look closely at the frog's eyelids. The frog's eyes have upper and lower eyelids, as well as a clear **nictitating membrane** attached to each lower lid. The nictitating membrane is also found in many reptiles and birds, but in mammals there is only a clear vestige of the nictitating membrane.

6. Generate one or more hypotheses as to the function of the nictitating membrane.

E. Behind, and to the side of, the eyes are the flat, circular **tympanic membranes**. They are analogous to human eardrums. The internal ear structures are sensitive to sound waves and to physical vibrations.

7. Design a simple experiment to determine if a frog can hear.

F. Place the frog in a freshwater aquarium. As you do, examine and feel the skin. The skin is kept moist by mucus produced by mucous glands. The skin acts as a gas exchange (breathing) organ and also helps regulate the levels of salts and fluids (osmoregulation) in the frog's body. Observe the frog as it floats at the surface and swims in the aquarium.

8. Describe how the frog swims. How do you think this differs from the way other amphibians swim?

9. What adaptive advantage does the frog have in being dark-colored on its dorsal surface and light-colored on its ventral surface?

10. Observe the eyes of the frog when it submerges. Was your hypothesis regarding the function of the nictitating membrane correct?

11. How do the front and hind legs differ in terms of length, muscle development, and degree of webbing? Now that you have observed a live frog "in action," what can you infer about the specialized function of each type of leg?

Part II: Dissecting a Frog (Optional)

G. Examine the preserved frog's external features. Look for the **external nares** (nostrils) and tympanic membranes.

Strategy for Observing

As you observe the frog, think first about its major organ systems. Then, observe each organ in turn.

12. How do the positions of these structures help the frog when it swims?

H. Look at the front legs. If there is a lump near the "thumb," the frog is a male. This pad helps the male frog cling to the female frog during **amplexus**, or fertilization of the eggs. Observe several frogs to make sure you see the difference between males and females.

13. What sex is your frog?

I. Observe the **cloaca**, an opening located at the posterior end of the frog. The cloaca is similar to the anus in humans, and is the opening into which urine, feces, and eggs or sperm pass.

J. Mouth: Cut slits in the jaw hinges to open the mouth wide. Look at the attachment and length of the tongue. Look and feel with your finger for teeth. Two **vomerine teeth** extend from the roof of the mouth. (The vomer is the small bone separating the nostrils.) By running your finger along the edge of the upper jaw, you will feel the small **maxillary teeth**.

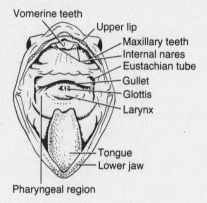

Anatomy of the Frog's Mouth

14. Since a frog cannot chew its food, what do the size and placement of the teeth tell you about their function during feeding?

K. Holes at the back sides of the mouth in the upper jaw are openings for the **Eustachian tubes**. Insert a probe into the Eustachian tubes and then into the **internal nares** on either side of the vomerine teeth.

15. Where do the Eustachian tubes lead, and what is their function?

16. Where do the internal nares lead?

L. Below the Eustachian tubes in the male are the openings to the vocal sacs. Find the **glottis**, a small, slitlike opening in the center rear of the mouth, beneath the **gullet**. The gullet is a tube with a wide, muscular opening. Insert your probe into both of these openings.

17. What are the functions of the glottis and the gullet? (Check your hypothesis when the frog's body cavity has been opened.)

M. Internal Organs: Lay the frog on its dorsal surface in the dissecting pan, and pin the legs to hold it in place. Lift the abdominal skin with the forceps and make an incision with the tips of the scissors up the center of the abdomen, from the cloaca to the lower jaw. Always be careful to cut only one layer at a time so that you do not damage underlying organs. Cut slits laterally from this first incision at the base of the abdomen and at the shoulder. The skin may now be pinned back. Cut through the underlying muscles and the breast bone to open up the chest and abdominal cavity. If the abdominal cavity is filled with dark-colored eggs, your frog is a female. You may remove the eggs to see the underlying organs. You may notice striking yellow, flamelike fat bodies. These store energy, in the form of fat, for use during hibernation.

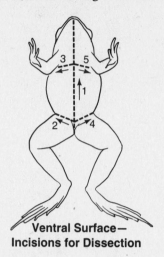

**Ventral Surface—
Incisions for Dissection**

N. Digestive System: The large, three-lobed **liver** will take up a sizable portion of the upper abdomen. Find the greenish **gall bladder** beneath the liver. Trace the muscular opening in the throat with a probe through the **esophagus**. Follow along the esophagus until you see the white **stomach**. The stomach curves around the liver and narrows at the **pylorus**, the opening into the small intestine. The **duodenum**, the short first section, and the **ileum**, the coiled main part of the small intestine, lead into the large intestine. Spread the coils of the small intestine apart. Look within the **mesentery**, the connective tissue, for the **spleen**, a small, spherical organ.

18. Label all the structures in the digestive system on the diagram of the internal organs of the frog.
19. The mesentery is richly supplied with blood vessels. What is their function?
20. Which system is the spleen a part of?

O. Note that under the stomach and connected to the small intestine is the **pancreas**. The pancreas produces many enzymes to aid in digestion.

21. Label these structures on the illustration of the internal anatomy of the frog.

P. Respiratory and Circulatory Systems: Carefully cut away the digestive structures that conceal the heart and lungs. The small lungs lie on either side of the heart. Notice that, unlike humans, the frog has no respiratory diaphragm.

22. What structure in a frog serves the function of the diaphragm in a human?

Q. To see the heart cut through the pericardium, the membranous envelope surrounding the heart. Trace blood vessels that pass between the heart and lungs. Identify the 3 chambers of the heart: the right atrium, the left atrium, and the ventricle. The left atrium receives oxygenated blood from the lungs. The right atrium receives deoxygenated blood from the body. Both atria empty the blood into the ventricle. The blood is then pumped through the Y-shaped **conus arteriosus** to arteries in the body.

23. Why do you think the tadpole has a two-chambered heart, whereas the adult frog has a three-chambered heart?

R. Urogenital System: Lift the intestines and liver gently with the probe and

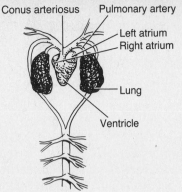

The Frog Heart—Ventral View

carefully remove them. Beneath these you will see the structures of the urogenital system. Remove the **peritoneal membrane** to expose the kidneys— elongated bean-shaped organs lying on either side of the spine— with the fat bodies attached anteriorly to them. Find the thin, twisting **ureters** that drain each kidney. The ureters lead to the **urinary bladder**, located posteriorly, which empties into the cloaca. In the male frog, the **testes** lie below the fat bodies. The testes produce sperm, which pass out the cloaca to fertilize externally the eggs produced in the female frog's **ovaries**. Mature eggs are passed out of the female's body through the cloaca via the coiled oviducts.

24. What organ besides the kidneys regulates salt and water balance in amphibians?

S. Brain: Turn the frog onto its ventral side. Be sure to make very shallow cuts so that you do not damage delicate structures. Remove the skin on the head. Scrape away or pick off the top of the skull, the cranium, until the brain lies exposed. Identify the **olfactory bulb**, **cerebral hemispheres**, **optic lobe**, **cerebellum**, and **medulla**. Look for large nerve bundles (**ganglia**) leading from the brain. Lift up the brain and look for the **pituitary gland**, located beneath the cerebellum at the base of the optic lobe.

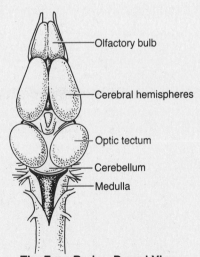

The Frog Brain—Dorsal View

Postlab Analysis

25. What major physical changes occur in the metamorphosis from the tadpole stage to the adult stage that enable the adult frog to survive on land?
26. Amphibians cannot regulate their body temperature. How do frogs survive cold winters? Droughts?
27. Amphibians live in a variety of habitats. Generate hypotheses as to why this is true. What feature of the frog's physiology prevents its successful adaptation to a marine habitat?
28. Compare the structure and function of the frog's and the human's respiratory and circulatory systems.
29. Compare the relative size of the major structures in the frog brain to the relative size of the homologous structures in the human brain. What does this comparison hint at about the frog's senses?

Further Investigations

1. You can study the anatomy of a frog without actually dissecting one. Ask your teacher to provide you with alternatives that are available in your school. Some of these alternatives include models, filmstrips, videotapes, and computer simulations. Alternatively, you may want to assemble a clay model of the internal organs mentioned in Part II. You may also draw the various organs on a series of clear plastic overlays. After performing alternative activities to the frog dissection, answer the questions in Part II of the lab.
2. Some salamanders, such as axolotls and mud puppies, never physically mature or mature only under certain conditions. However, they are able to reproduce. This condition, retaining juvenile characteristics in the adult, is called neoteny. Investigate some salamanders that show neoteny. Find out how they live, how they breed, and what organs are involved. Discuss under what circumstances they can be induced to mature, if they can mature at all.
3. Investigate different kinds of amphibians. Learn about the different kinds of environments in which amphibians live. Find out what they eat, how they breed, and how they escape predators.

Investigation

37 | *Frogs*

1. _____

2. _____

3. _____

4. _____

5. _____

6. _____

7. _____

8. _____

9. _____

10. _____

11. _____

12. _____
_____
_____

13. _____

14. _____
_____
_____

15. _____
_____

16. _____
_____

17. _____
_____

18. Label the diagram of the internal anatomy of a frog.

19. _____
_____

20. _____

21. Label the diagram of the internal anatomy of a frog.

22. _____

23. _____
    _____
    _____

24. _____

25. _____

26. _____
    _____

27. _____
    _____
    _____

28. _____
    _____
    _____
    _____
    _____

29. _____
    _____
    _____
    _____

38

Live Reptiles

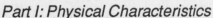

How have reptiles adapted to life on land?

Learning Objectives
- To compare physical and behavioral characteristics of several species of reptiles.
- To examine characteristics that relate reptiles to other vertebrate classes.

Process Objectives
- To observe distinguishing physical characteristics of different species of reptiles.
- To conduct behavioral experiments on different species of reptiles.

Materials
For Class
Parts I and II
- At least 3 species of reptiles
- Fresh reptile food (appropriate for species)

Part I
- Thin wooden dowel

Part II
- Large sheet of fine-grain sandpaper
- Sheet of plexiglass (or other smooth surface), about half the size of the sandpaper

Introduction

Although they appear to be very different, snakes, turtles, and lizards all belong to the same class of animals. Examine their behaviors and physical characteristics and you will see that they are alike in many ways.

For example, all reptiles are **ectothermic**—they cannot regulate their body temperature from within. Reptiles regulate their body temperature by seeking warm or cool locations. In addition to similar thermoregulation mechanisms, many reptiles have evolved similar defense mechanisms. For example, most reptiles have protective coloration to hide from predators. A number of species of lizards and snakes are poisonous.

The Class Reptilia includes many extinct species, such as dinosaurs. The study of petrified dinosaur skeletons reveals an even greater diversity of characteristics than is found in living reptiles—such as the ability to fly.

In this lab, you will observe physical characteristics of the different reptiles. You will compare the reptiles with each other. You will also compare the reptilian structures with those of other vertebrates. Then, in Part II, you will conduct behavioral studies of the animals. Throughout both parts, think about how the characteristics of reptiles help them to survive and thrive on land.

Prelab Preparation

After reviewing Chapter 38 in your textbook, look around your neighborhood for any reptiles that live near you.

1. Where might you find reptiles in your area? If you have found any, what were they and what was their natural habitat?
2. What specific characteristics do you expect to observe that will distinguish lizards, snakes, and turtles from non-reptiles?
3. Do you expect reptiles to be more like amphibians or birds? Why?

Procedure

Part I: Physical Characteristics
Observe at least 2 or 3 species, or more if they are available. Compare the following characteristics to other classes of vertebrates and note similarities and differences among the different kinds of reptiles. CAUTION: Reptiles, turtles in particular, can carry bacteria such as *Salmonella* that can make you sick. You must be very careful to wash your hands well any time you touch the animals or objects that have been near them. Do not keep food you might eat near the animals. CAUTION: Handle the animals gently. Remember, they have sensitive nervous systems. Most reptiles will respond to gentle handling by becoming less afraid of you.

A. Movement: Each species of reptile has characteristic movements or has some specific adaptation for movement.

4. Describe the way each animal moves. How does its movement increase the animal's chances of surviving in its natural environment?
5. Is their movement more like that of amphibians or birds?

B. The Skin: Lightly feel the skin of the reptiles. Scales are formed from **keratinized** skin cells. The large, horny scales located on the belly of snakes are called **scutes.**

6. Do all the species of reptiles have the same type of skin?
7. Is their skin more closely related to amphibians or birds?

C. Sensory Organs: Observe the different reptiles as they investigate their environment.

8. List the specific sense organs of each species of reptile. How do the sense organs vary among the different reptiles?
9. How do their sense organs compare to those of amphibians and birds?

D. Protection: Each species of reptile has structural characteristics that protects it. One example is the turtle's shell, which is a bony outgrowth of skin.

10. List each animal's protective structures and note which species of reptiles have similar adaptations.
11. Are the reptiles' protective devices more similar to those of fish and amphibians or to those of birds?

E. Mouth: Use a thin wooden dowel to observe the inside of the mouths of the different reptiles. CAUTION: Gently hold the reptile's mouth open with the dowel.

12. Do all the reptiles have teeth?
13. How are the mouths specially adapted for different habitats and food gathering?
14. Are the reptilian mouth structures more like those of amphibians or birds?

Part II: Behavioral Characteristics

You will find instructions below for some simple behavioral studies. Also, you will be given general instructions for developing your own behavioral experiments. Do not perform any experiments that might cause harm or pain to the animals. Be sure to note the results of your observations for each species of reptile studied.

F. Study how the animals move on different surfaces. Place a large piece of sandpaper in a box or terrarium. Cover half of the sandpaper with a piece

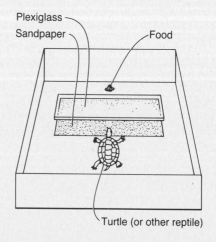

Plexiglass
Sandpaper
Food
Turtle (or other reptile)

of clear glass, plexiglass, or some other similar substance to provide a smooth surface. Observe how the animal moves on each surface. If the animal does not seem to want to move at all, try placing some food on the opposite side of the box. Make certain that you conduct enough trials for your observations to be reliable.

15. Did the surface quality affect how the animal moved? In what ways?

G. Design an experiment to test the reptile's ability to smell.

16. Record your design and results.
17. How can you set up your experiment so that the animal responds only to the smell and not the sight of the smell's source?
18. How could you test whether your animal responds to sounds? To other forms of vibrations?

H. Set up an obstacle course and see how each species of reptile moves through it. After you have seen how a snake moves past an obstacle, repeat the trial but remove the obstacle as soon as the snake's head has passed it.

19. How does the snake move past the obstacle if you remove the obstacle after its head passes the object?
20. How do the reptiles differ in the ways they move through the obstacle course?

Postlab Analysis

21. From the observations you have made, what can you infer about how reptiles have adapted to a land environment?

Further Investigations

1. Design an experiment to condition a reptile to respond to you when you feed it. (This might take several days or weeks since some reptiles eat only every few days.)
2. Using your text or library resources, research how different species of reptiles reproduce, how their reproduction is different from amphibians, and how reptilian reproduction represents adaptation to a land environment.
3. Investigate how the reptiles respond to a visually perceived threat.

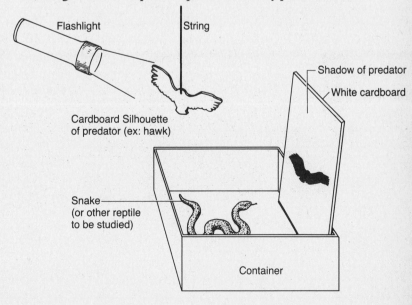

Flashlight
String
Shadow of predator
White cardboard
Cardboard Silhouette of predator (ex: hawk)
Snake (or other reptile to be studied)
Container

Cut a silhouette of an animal that is a natural predator of your reptile from a piece of cardboard. (Owls and hawks are examples of predators of small reptiles.) Place a piece of white cardboard in the animal's container. Hold a flashlight so that a beam of light shines on the cardboard screen. Hold the cardboard shape at different distances so that shadows appear on the screen. Place the animal so that it can see the shadow. Keep the shadow still, then move it around. Change the size of the shadow to make it appear closer to your animal.

Investigation

38 | *Live Reptiles*

1. _____

_____

2. _____

_____

3. _____

_____

4. _____

_____

_____

_____

5. _____

_____

6. _____

_____

7. _____

8. _____

_____

_____

_____

9. _____

_____

_____

10. _____

_____

_____

_____

11. _____

_____

_____

12. _____

_____

13. _____

_____

_____

14. _____

_____

15. _____

_____

16. _____

_____

_____

_____

_____

17. _____

_____

18. _____

_____

_____

19. _____

20. _____

_____

21. _____

_____

_____

39

Bird Beaks and Feet

Learning Objectives
• To compare the beaks and feet of diverse species of birds.
• To relate characteristics of bird beaks and feet to their different functions.

Process Objectives
• To organize data to relate diversity of bird structure to function and habitat.
• To classify birds according to the environment in which they would most likely survive.

Materials
For Class
• Taxidermic specimens, charts, or drawings of a variety of bird species.
• Field guides on birds
• Vertebrate zoology references

How have birds evolved to survive in various environments?

Introduction

Traits that enable separated populations to exploit different foods and habitats become fixed as a result of reproductive isolation. In time, these populations become genetically distinct and can no longer interbreed with each other. By definition they have become different species.

During his exploration of the Galápagos Islands, Darwin found 5 species of ground finches and at least 4 closely related genera. The birds were distinguishable by body size and beak shape. The birds with short, stout beaks were adept at cracking and eating hard seeds. The other birds, with narrow, elongated beaks, were able to dig and spear grubs. The former type occupied the areas of low elevation, which supported shrub growth with seeds; the latter type inhabited the humid upland forests, where grubs were abundant. Darwin inferred that all the birds had evolved from a common mainland species whose offspring, over many generations, were most successful when their food-gathering abilities in various habitats maximized their nutrition.

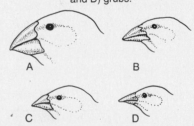

Darwin's finches, adapted to eating
A) seeds, B) insects, C) flower parts,
and D) grubs.

In this lab, you will observe and compare the beaks and feet of several species of birds in order to understand the modifications that have enabled them to survive in different habitats.

Prelab Preparation

Using reference books on vertebrate animals, as well as your textbook, research the traits that are common to all birds. Think about the characteristics that unite the nearly 9,000 existing species of birds.

1. In the chart Characteristics of Class Aves, list at least 6 of the distinguishing features of birds and cite the adaptive function of each.

Procedure

The beak of a bird consists of toothless jaws and a mouth opening that is usually elongated into a pointed shape. It is covered with a lightweight, horny epidermis and is used in the way that you would use a pair of forceps. The base of the beak lacks feathers and may be somewhat scaly.

A. Observe bird specimens, charts, or the drawings in this lab. Use the guide below to answer the questions that follow.

Structure and Function of Bird Beaks

Beak Type	Function
short and stout	eating small seeds
spear-shaped and stout	spearing fish
chisel-shaped	drilling for insects
hooked	catching and tearing prey
tubular	sucking nectar
stout and long	scooping up fish
flat, slight hook	straining algae and small organisms
short	multipurpose

Strategy for Organizing Data
Arranging observations in chart form often makes it easier for you to read and compare a lot of information at one time.

2. Describe in the data chart the structure of the beak of each bird.
3. Use this information to predict the food source that the bird is most adapted to utilize. Add these data to your chart.

The legs of birds are functionally similar to our legs. On the ground or on a perch, the legs of a bird support the entire weight of its body. The leg is constructed like a lever to combine walking and running ability with

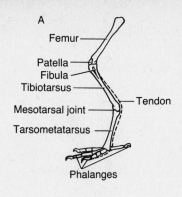

A

Femur

Patella
Fibula
Tibiotarsus

Mesotarsal joint — Tendon

Tarsometatarsus

Phalanges

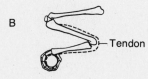

B

Tendon

Strategy for Classifying
Skim the information for similarities and differences in habitat requirements. Think of categories of environments, and identify the birds suited to each.

"take off" capability. A standing bird has its knees directed forward. When a bird perches, its leg bones jack-knife, pulling the ankle tendons taut and clinching the toes around the perch. The bird is thus "locked" on to the perch and can sleep without falling off. It must raise its whole body to straighten the legs and loosen the ankle tendons before it can unlock its feet.

Bird legs generally lack feathers; instead, they are covered with scales similar to those of reptiles. Bird feet generally have 4 toes, which end in claws.

B. Observe bird specimens, charts, or the drawings in this lab. Note the arrangement of the toes on the feet.

    4. Describe in the data chart the leg adaptations of each bird and predict the functions that each adaptation makes possible (see guide).

Structure and Function of Bird Feet

Bird Feet	Function
2 toes in front; 2 behind	
sharp, slender claws	climbing and clinging
3 toes in front; 1 behind	
long legs	wading
short legs, blunt, stout claws	scratching
medium length legs, rounded toes	perching
webbed feet	swimming
large, hook-like claws (talons)	grasping prey
very short, weak legs	hovering or sitting close to branches, flowers

An animal's habitat includes its food, water supply, shelter, and territorial requirements. Sometimes an animal finds shelter in one type of environment and food in another. For example, a hawk may find shelter in wooded areas and hunt rodents in a neighboring meadow. Its ideal habitat could be described as a combination of wooded area and meadow.

    5. Classify the birds you have been studying by the type of habitat to which each would be ideally suited. Enter these data into your chart. Habitats might include wooded area, meadow, lake, shoreline, or some combination of these areas.

    6. Use a field guide on birds to check your data chart and your predictions.

Postlab Analysis

    7. Think about the hummingbird's adaptations for food acquisition. What is the beak's function? How is the rest of the hummingbird's body adapted to this unusual feeding pattern?

    8. What insects have this feeding habit? Describe the similarity of the adaptations of these insects and those of the hummingbird.

    9. Imagine an ideal flying predator. What type of beak and feet would it have? Describe the body structures it might have to increase its hunting efficiency.

   10. What living bird(s) resemble(s) this creature?

Different birds may have similar beak and structures adaptations. Loons, herons, and kingfishers all have long, sharp-pointed beaks for spearing fish. Their feet, however, are quite different. This difference indicates that their methods of movement and of finding and pursuing fish are distinctive.

11. Compare how each of these birds catches fish.

Loons, herons, and kingfishers frequently live together at the same lake. All 3 eat fish. There is a principle of ecology that states no 2 species can occupy the same **niche** (way of life) in any one location or habitat.

12. How do these 3 species of birds conform to this principle?

13. Study the data you have collected about hawks and owls. Apply the above principle of ecology to hawks and owls. When does each of these birds hunt?

14. Different birds may have similar leg and foot adaptations. Choose several examples of this from your data chart and compare their beaks.

15. Describe how their feeding patterns in similar habitats differ so that they do not compete with each other.

The peregrine falcon, bald eagle, eastern bluebird, Kirtland's warbler, and California condor are considered to be threatened or endangered species.

16. If birds are so well adapted to their habitats, why are many of them close to extinction?

Further Investigations

1. Saw a chicken leg bone (femur) in half crosswise. Describe the structural adaptations you see and explain how they relate to standing and flying. How does a bird's bone compare with a mammal's bone?

2. Collect, observe, and draw different kinds of bird feathers. Explain the structure and function of each kind. Research how their colors are produced.

3. Make a trip to the bird section of your local natural history museum. Observe additional birds and add information about their beaks, feet, and habitats to your data table. Be aware of unique adaptations to different types of environments.

Investigation

39 | *Bird Beaks and Feet*

1. Enter your answers on the data chart.

Characteristics of Class Aves

Characteristic	Adaptive Function

2.–6. Enter your answers on the data chart.

Beak and Feet Adaptations

Name of Bird	Habitat Classification	Beak Adaptation		Feet Adaptation	
		Description	Possible Food	Description	Possible Function
Woodpecker					
Quail					
Thrush					
Hawk					
Owl					
Loon					
Pelican					
Hummingbird					
Heron					
Dove					
Duck					
Kingfisher					

7. _____

_____

8. _____

_____

_____

9. _____

_____

_____

_____

10. _____

11. _____

_____

_____

12. _____

_____

13. _____

_____

14. _____

_____

_____

15. _____

_____

_____

16. _____

_____

Investigation

40

Mammalian Behavior

Learning Objectives
- To learn how different species of mammals, and people of different cultures, communicate by sets of nonverbal signals.
- To describe mammalian behavior that occurs in response to changes in the environment.

Process Objectives
- To infer the nature of social interactions from the behavior that occurs among small mammals and among people.
- To perform behavioral experiments about the ways mammals explore their environment and respond to changes in their environment.

Materials
Part I
For Each Student
- Watch with second hand

Part II
For Group of 2–4
- Leather or thick gloves
- 2 Rodents
- Small cage or box
- Clay flowerpot, 10 cm high
- Stopwatch or watch with second hand

In what ways do people, and other mammals, communicate nonverbally?

Introduction

Behavior is the way in which animals respond to changes in their environments. Although scientists can observe a behavior, they cannot be sure about the cause of that behavior. After much study and observation, behavioral experts can make predictions as to the cause of certain behaviors. For example, animals that have very limited ability to communicate vocally have developed patterns of behavior that communicate nonverbal messages. For example, dogs put their tails between their legs to show that they are frightened. Even humans, with highly developed spoken language skills, communicate nonverbally.

Nonverbal communications are learned as part of social interaction while an individual is growing up. Most people are not aware of the many nonverbal messages they send. In fact, messages vary greatly from one culture to another. For example, most Americans shy away from body contact with strangers. In other cultures, people are comfortable in crowds where there is frequent body contact with strangers. Perceptions also govern how animals interpret their environment. For example, when a cat is frightened by another animal, it arches its back and fluffs up its fur, appearing larger in an attempt to scare away the other animal. Similarly, people's perceptions about someone's overall appearance may govern their behavior toward that person.

Prelab Preparation

Most of the exercises in Part I of this investigation must be done outside of the classroom. However, after completing these exercises, you will meet in small groups in class to discuss and analyze your observations. Part II of this investigation will be performed in class.

It is important to record carefully all observations and each behavior seen. The more thorough your notes are in each exercise, the more easily you may infer the cause of the behavioral patterns you are observing.

1. What kind of nonverbal communication tells you that another person is friendly and would like to approach you?
2. What nonverbal communication tells you that another person is unfriendly and does not want to be approached?
3. What kinds of behavior have you observed in pets or other mammals that indicate territorial delineation or protection?

Procedure

Part I

A. Observe the behavior of 2 mammals meeting one another. You may observe pets, other domestic mammals, or wild mammals. The animals may belong to the same or different species. Look for signs of aggression or hostility. Watch which animal demonstrates strength and dominance. Observe which animal shows fear or submission. Take notes on the actions and reactions of both animals. Time the interaction period.

4. Briefly describe the 2 mammals and their behaviors during this interaction.
5. Do you think that the interaction was friendly or hostile? Why?
6. How was the interaction resolved?

B. Observe from a distance 2 people having a conversation. Watch them as they meet. Note whether the pair is male/female, female/female, or male/male. Observe how close they are standing to one another. Time the interaction and record how many times the 2 people shift weight from foot to foot and move closer together or farther apart. Observe their parting gestures. However, do not eavesdrop on their conversation.

7. Record all of your observations.

Use the Distance of Human Interaction table to answer the following question: What is the physical distance on the chart that is closest to the distance estimated between the 2 people that you observed.

Distances of Human Interaction

Characterization of Distance	Physical Distance	Voice Shifts for Zone
Very Close	3 – 6 inches	top secret (soft whisper)
Close	8 – 12 inches	very confidential (audible whisper)
Near	12 – 20 inches	confidential (indoors, soft; outdoors, full voice)
Neutral	20 – 36 inches	personal (soft voice)
Neutral	4.5 – 5.0 feet	nonpersonal (full voice)
Public	5.5 – 8 feet	for others to hear
Across Room	8 – 20 feet	talking to group (loud voice)
Stretching Limits	20 – 24 feet inside 100 feet outside	hailing distance and departures

8. What does the distance between your 2 subjects tell you about their relationship and the nature of their conversation?
9. As the conversation proceeded, did the position of their heads, hands, legs, or the expression on their faces mirror each other's? If so, describe your observations.

C. Small groups should pool observations of several conversations.

10. Based on the several conversations observed by members of

Base your inferences about the mammals' interactions on the behaviors that you observed. Use what you know about the behavior of other mammal species, including humans.

the group, when in the conversation did subjects shift their weight from foot to foot more often?

11. How did the speakers' stances change when they prepared to depart?
12. Did your group's results show any differences in departure signals related to sex? If so, describe these differences in behavior.

Part II

This part of your investigation will be done in class. You will conduct some behavioral studies on small mammals—rodents. Rodents have a highly developed nervous system. You will investigate the animal's responses to a strange animal and you will investigate escape behavior. Handle the animals carefully and gently, without frightening them. Mice and rats may be picked up gently by their tails. CAUTION: Do not pick gerbils up by the tail because the tips of their tails will break off.

D. Place a small mammal into a cage or box containing another mammal unknown to it, but of the same species. Time the interaction and take notes on the 2 mammals' behaviors. CAUTION: Separate the animals immediately if one attacks the other. Wear leather or thick gloves when handling the animals.

13. How did the animals react to one another?
14. Do you think the animals showed aggressive or submissive behavior, territoriality or flight? Explain why.

E. Place a small mammal in an empty clay flowerpot. Record the time it takes for the animal to explore the flowerpot, seek a way out, and escape. Also record the number of times that it attempts to escape before succeeding. Wait for about 10 minutes, then put the same animal back into the flowerpot and again time its escape and record the number of attempts. If an animal does not try to escape, ask your teacher for a different specimen.

15. Does it take the animal more or less time to escape on the second try? More or fewer attempts? Explain what kind of behavioral process this shows.

Behavioral study on small mammals.

Postlab Analysis

16. How do behavioral adaptations protect the individual?
17. How can differences in people's behavior affect the ways in which people from different cultures interact with each other?
18. List some of the ways mammals communicate nonverbally.
19. Why are behavioral studies a part of the study of biology?

Further Investigations

1. Investigate classical and operant conditioning in guinea pigs: Raise guinea pigs. Feed them fresh lettuce or other fresh vegetables every day, but ring a bell just before placing the food in the cage. Observe the behavior they exhibit when they know the food is there. See if, after a week of performing this exercise, the same behavior is repeated when the bell is rung, but the food is not placed in the cage right away. Be sure that you also always feed them an appropriate food.

2. Research the different methods that mammals use to mark a territory. Discuss why different animals mark territories. For example, does an animal mark a territory to scare off other animals or just to indicate where its home is?

3. Break into groups and have 2 of the group members role play a simple conflict situation. As a group, discuss how body language was used to communicate the feelings of the 2 participants. Think up several different dilemmas and act these out, keeping in mind the importance of body language in communicating feelings.

4. Observe the play behavior of different mammals. A local animal shelter, pet shop, zoo, aquarium with marine mammals, or wildlife rehabilitation center are some of the places you might visit to observe play behavior. What type of play did the mammals you observed engage in? Did they play together or by themselves? What objects, if any, did the mammals play with or in? What are the functions of play behavior?

Investigation

40 | *Mammalian Behavior*

1. _____

2. _____

3. _____

_____

4. _____

_____

5. _____

6. _____

7. _____

_____

_____

_____

_____

_____

8. _____

_____

9. _____

_____

10. _____

11. _____

12. _____

_____

_____

13. _____

14. _____

_____

15. _____

_____

16. _____

_____

17. _____

_____

18. _____

_____

19. _____

_____

41

Vertebrate Skeletons

Learning Objectives
- To compare the skeletons of representative animals of 5 vertebrate classes.
- To look for evidence of homologous structures.

Process Objectives
- To observe homologous structures in vertebrate skeletons.
- To infer the evolutionary relationship of vertebrates based on homologies.
- To classify unknown specimens based on their similarities to and differences from known vertebrates.

Materials
For Class
- Fish skeleton
- Frog skeleton
- Turtle skeleton
- Pigeon skeleton
- Cat skeleton
- Human skeleton
- Mystery bones

How does your skeleton compare with the skeletons of other vertebrates?

Introduction

The absence or presence of a backbone is a characteristic used to classify animals. Those animals that have flexible backbones or vertebral columns are called vertebrates. Fishes, frogs, turtles, pigeons, cats, and humans are among the diverse organisms that belong to this group.

Examination of other similar skeletal features in vertebrates indicates important links among members of this widely varied group. Biologists consider these skeletal similarities to be evidence that similar bones in different animals evolved from the same bone structures of a common ancestor.

Prelab Preparation

Consider the structure and function of your hand and arm. Can you detect any similarities between the structures and functions of your hand and arm and those of a pigeon's wing or a cat's paw and leg? Not only do the skeletal systems of most vertebrates perform the same kinds of functions, but they also contain basically the same kinds of bones.

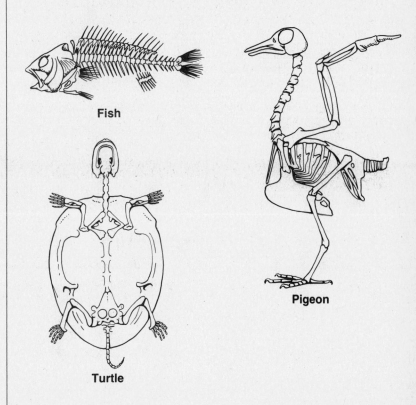

Fish

Turtle

Pigeon

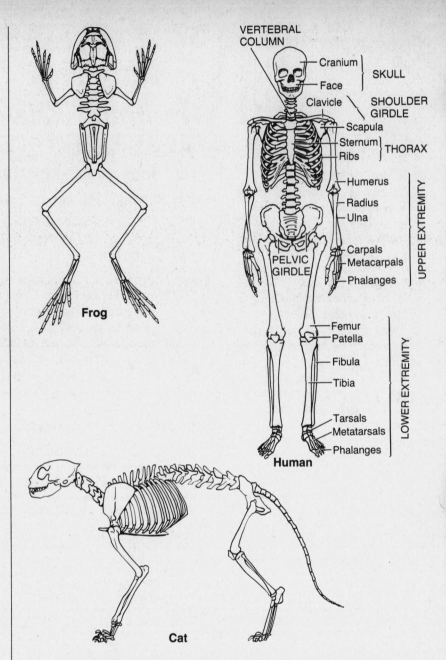

Frog

VERTEBRAL
COLUMN

Cranium — SKULL
Face

Clavicle — SHOULDER
GIRDLE

Scapula

Sternum — THORAX
Ribs

Humerus

Radius — UPPER EXTREMITY
Ulna

PELVIC
GIRDLE

Carpals
Metacarpals
Phalanges

Femur
Patella

Fibula

Tibia — LOWER EXTREMITY

Tarsals
Metatarsals
Phalanges

Human

Cat

1. What are homologous structures?
2. What type of evidence would indicate that your hand, pigeon's wing, and a cat's paw are homologous?
3. How could you collect the evidence that you described in Question 2?

Procedure

A. Examine the models or the illustrations of vertebrate skeletons.

4. List 4 similarities that are shared by all the skeletons.
5. List 3 differences that exist among the skeletons.

B. Examine closely the backbone of each of the skeletons.

6. How are the backbones similar?
7. How are the backbones different?

Strategy for Observing
Compare one portion of the skeleton at a time. Compare heads, then backbones and rib cages, followed by forelimbs and hind limbs.

8. Which animals have backbones that are most alike? On what observations do you base your conclusion?

9. Which 2 animals have backbones that are least alike? What observations led you to this conclusion?

C. Study the upper limb of the human skeleton. The upper limb includes the shoulder, upper arm, forearm, wrist, and hand.

10. How many bones make up the upper arm and the forearm?

11. What is the name of each of these bones and where is each located?

12. How does the length of the upper arm compare with that of the forearm?

13. Describe the bones of the wrist and the hand.

D. Compare the human upper limb with the forelimbs of the cat, the pigeon, the turtle, and the frog.

14. How does the upper portion of each limb compare with the others in the group?

15. How does the forearm of each limb compare in number of bones?

16. How does the length of the upper arm compare with the length of the forearm for each of the limbs?

17. Compare the wrist and hand portions of the limbs.

18. Which animal's forelimb is most like the human upper limb? Describe the observations that led you to this conclusion.

19. Which animal's forelimb is least like the human upper limb? Describe the observations that led you to this conclusion.

E. Look at the hind limbs of each specimen.

20. How are they similar?

21. How are they different?

22. Which two are most alike? Summarize the observations that led you to this conclusion.

F. Very often, bones of unknown origin are discovered. Such unknown bones are compared with known specimens to determine their possible origin. Inspect the set of "mystery bones."

23. Describe the bones in this collection. What parts of the skeleton are represented in the collection?

24. Which vertebrate class is represented by these bones? On what evidence do you base this conclusion?

G. Examine the bones of Unknown Limb A and Unknown Limb B in the illustration. Animal A is known to subsist on a diet of fruit, which it reaches by flying up into the trees. Animal B is known to subsist on a diet of plankton, which it easily obtains in its marine habitat.

25. Describe the limbs of Animal A and Animal B.

26. What evidence is there that the limbs of these animals are homologous to the limbs of the other vertebrates you have been studying?

27. How is the structure of each of these limbs related to its function?

Postlab Analysis

28. What evidence have you obtained to support the inference that vertebrates evolved from a common ancestor?

29. Based on your observations of the skeletons, why should human beings and cats be considered more closely related

Strategy for Observing
Count the number of bones; observe and compare their relative lengths and their arrangement.

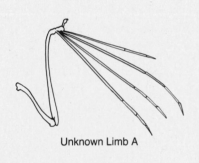

Unknown Limb A

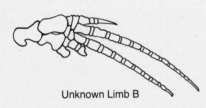

Unknown Limb B

Strategy for Classifying
Look for features similar to those of the skeletons you studied; organisms that share common features are likely to be closely related.

Strategy for Inferring
An inference is the most likely explanation for evidence such as the observation of homologous structures in different kinds of organisms.

than humans and any of the other vertebrates observed in this lab? How is the classification of humans and cats consistent with this conclusion?

30. Which vertebrate class is represented by the animals possessing Unknown Limb A and Unknown Limb B? On what evidence do you base this conclusion? Why is a similarity in physical features a more dependable indication of relationship than a similarity of diet, habitat, or method of locomotion?

Further Investigations

1. Birds are thought to have evolved from ancient reptiles. Make a more detailed study of the pigeon skeleton. Describe the most striking differences between the skeletons of birds and those of other vertebrates. Also, use library resources to find out more about the evolution of birds and reptiles.

2. Find evidence in the skeletons of the different vertebrates you have studied that shows adaptations for each animal's way of life. Use library resources to confirm your findings.

Investigation

41 | *Vertebrate Skeletons*

1. _____

_____

2. _____

3. _____

4. _____

_____

5. _____

_____

6. _____

7. _____

8. _____

9. _____

10. _____

11. _____

12. _____

13. _____

_____

14. _____

_____

15. _____

_____

16. _____

_____

17. _____

_____

_____

18. _____

_____

19. _____

_____

20. _____

21. _____

22. _____

_____

23. _____

_____

24. _____

_____

25. _____

_____

26. _____

_____

27. _____

_____

28. _____

_____

29. _____

_____

_____

30. _____

_____

_____

42 *Muscle Function*

Learning Objectives

- To describe how skeletal muscles work during physical activity.
- To learn how the cardiovascular system and muscle function are related.

Process Objectives

- To hypothesize how various stimuli will affect skeletal muscle functions.
- To analyze and graph personal and class data on muscle fatigue.

Materials

For Group of 2

- Clock or watch with second hand
- Laboratory tubing
- Weights (2 books)
- Ice cubes
- 250-mL beaker
- Graph paper

Strategy for Hypothesizing

Think about a physical activity that you have done that caused muscle fatigue. Base your hypotheses on your memory of fatigue from that activity.

What is the relationship between stimuli and fatigue in skeletal muscles?

Introduction

Do you ever exercise to a point where you experience fatigue? Within muscle cells, fatigue is a temporary loss of ability to respond to stimuli. Fatigue occurs when the energy supply to the muscle cells has been depleted and waste products have accumulated.

During moderate exercise, the blood supplies enough oxygen from inhaled air to provide the muscles with energy. This is known as **aerobic** exercise. Well-conditioned marathon runners pace themselves so that their bodies' need for oxygen is about equivalent to the amount of air they inhale. Athletes whose sports require intense, **anaerobic** exertion, such as weight lifters, draw on sources of energy that do not depend on inhaled oxygen.

Proper exercise improves the aerobic capacity of the cardiovascular system and decreases the chance of developing cardiovascular disease.

Prelab Preparation

Read the procedure that follows and make a hypothesis about what will happen to the major muscles in your upper arm — the **biceps** and the **triceps**. The biceps is the muscle on the front of your arm; the triceps is the muscle on the back side of your arm.

Stretch one arm in front of you and open your hand. Now place your other hand around the upper half of your extended arm. Make a fist and tightly bend (flex) your extended arm.

1. Describe the changes you feel as you bend your arm.
2. Generate several hypotheses about what conditions would make your arm muscles tire quickly.

Procedure

As you do each part of the investigation, your partner will be the timekeeper and recorder. Then you will reverse roles and become the timekeeper and recorder for your partner.

A. Continual Muscle Stimulation: To gather data about muscle fatigue, hold your arm straight in front of you at a 90° angle. With palm turned up, open and close your hand into a fist (flexing) as many times as possible in 20 seconds. Keep your arm out. Without resting between trials, repeat this procedure 9 times. (See drawing on next page.)

3. Your partner will record your counts on your data chart. Average the counts from your 10 trials and enter your average on your data chart.
4. How does the feeling in your arm and hand change as you progress through your trials?

Record data while your partner performs Step A.

B. **Restricted Blood Flow:** To investigate how blood supply affects muscle function, have your partner tie a piece of laboratory tubing snugly around your upper arm. NOTE: Be sure not to tie it too tightly; you should be able to fit one finger under the rubber hose. You will perform 10 trials of the hand-flexing exercise for 20 seconds, as in Step A.

    5. Record the number of times for each trial on the data chart and then enter your average.

Record data while your partner performs Step B.

C. **Weight Lifting:** To investigate the effect of weight on muscle function, you will test how long you can hold weights before your arm becomes tired.

    6. Predict which arm will tire first. Explain your choice.

Hold one weight (book) in each hand. Keep one arm straight by your side and extend the other arm as in Step A. Hold this position as long as you can. Repeat this action extending your other arm.

    7. How long does it take for each arm to become tired? Explain your results.

D. **Varying Temperature:** You will examine how temperature influences muscle activity. Write your full name 3 times on a piece of paper.

    8. Do you think your ability to write your name will be affected by how warm or cold your hand is? Give reasons for your answer.

Now tightly hold several ice cubes in your hand for one minute. Quickly drop the ice cubes into the beaker. Immediately write your full name 3 times. Do not dry your hand before writing. NOTE: Use pencil; the water will make ink run.

Warm your hands by rubbing them together. Write your full name again 3 times.

    9. Look at all of your signatures. How did the different temperature treatments affect your ability to write?

E. Pool the class data from Step A: Continual Muscle Stimulation on the board. Prepare a bar graph using your data.

Strategy for Analyzing
Compare the signature from your first trial with your last trial. Does the legibility increase or decrease?

Postlab Analysis

    10. Did you experience muscle fatigue? Under what conditions?
    11. How did the restriction of blood flow affect your counts?
    12. What can you infer about the relationship between the cardiovascular system and muscle fatigue?
    13. Study the graph you prepared in Step E. How do your results compare with the class data? Suggest variables that might have influenced the class results. Would you consider your own results average, above average, or below average?

Further Investigations

1. Repeat the investigation experiments with subjects of different ages and compare the results with your results.
2. Contact groups that study cardiovascular or muscular function and disease for information about their specific research. Ask how physical activity deters later health problems. Also ask how activities such as musical training can benefit an individual.

Investigation

42 | *Muscle Function*

1. _____

2. _____

3. Enter your answers in the data chart.

Hand Flex Data											
Exercise	**Trial**										**Your Average**
	1	2	3	4	5	6	7	8	9	10	
Hand flexing (Step A)											
Hand flexing with tubing (Step B)											

4. _____

_____

5. Enter your answers on the data chart.

6. _____

_____

_____

7. _____

_____

_____

8. _____

_____

_____

9. _____

_____

10. _____

_____

11. _____

_____

12. _____

_____

_____

13. _____

_____

_____

Investigation

43.1 | *Daphnia Heartbeat*

How do chemicals affect daphnia heart rate?

Learning Objectives
- To compare the effects of different chemicals on the heartbeat of a daphnia.
- To relate changes in chemical concentration to the heart rate of a daphnia.

Process Objectives
- To experiment with a living animal under the microscope.
- To analyze team and class data.

Materials
For Group of 4
- Daphnia
- 2 Medicine droppers
- Depression slide
- Cotton fibers or thread
- Coverslip
- Compound microscope
- Stopwatch or clock with second hand
- Paper towel
- Dilute solutions of adrenaline (0.001%, 0.0001%, 0.00001%), ethyl alcohol (2%, 5%, 10%), coffee and/or cola drink (10%, 20%, 30%), or aspirin (0.5%, 1%, 2%)

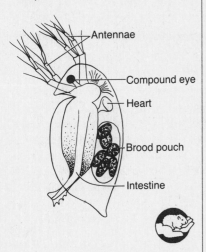
Antennae
Compound eye
Heart
Brood pouch
Intestine

Introduction
How do stimulants and depressants affect heart rate? When chemicals are ingested, they are carried through the body by the blood. While such chemicals sometimes act on the brain centers that control your heartbeat, they also affect the heart directly.

Daphnia, or water fleas, are small enough to be studied under a microscope and so transparent that we can see the heart beating. While daphnia are invertebrates, arthropods living in freshwater ponds, their hearts respond to chemical substances much as our own hearts do.

Prelab Preparation
Work in groups of 4 throughout this investigation.
What is your normal heart rate? How would you measure it?
1. Describe a simple technique for measuring your heart rate and record your measurement as "beats per minute."
2. How many times should you repeat this measurement? Why? Average your repeat measurements and record.

Look at the diagram of the water flea to familiarize yourself with the location of the internal organs. The water flea's heart rate is very fast (several times faster than yours). Do you think you could use the same technique for determining the heart rate of daphnia as you used to determine your heart rate? If not, what modification of your technique or what new technique would you propose?
3. Discuss your ideas with your lab team and record the technique that you have designed. Let your teacher check your technique before you begin. Remember that you want to keep your water flea alive and uninjured while you examine it and experiment with different chemical substances.

Procedure
A. Place the rubber bulb of a medicine dropper over the small end of the dropper as shown on the next page. Transfer a water flea from the culture jar to the center of a depression slide. Add 3 or 4 cotton fibers or bits of thread to the drop of water and cover with a coverslip. Examine the water flea under the low power of your microscope. Even though the heart is transparent, you will find it easily; it is constantly beating. CAUTION: Be careful not to crush the water flea with the coverslip or let the water on your slide dry up.
4. Follow the technique you designed in the Prelab Preparation and record in your data table the number of heartbeats per

Strategy for Experimenting

In designing your experiment, you can include everyone in your team by assigning tasks for observers, timers, and recorders.

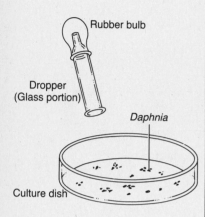

Rubber bulb

Dropper
(Glass portion)

Daphnia

Culture dish

Strategy for Analyzing

As you analyze the data, check whether your data support your hypotheses or not. If necessary, make a new hypothesis.

minute. Repeat your measurements (rotating tasks among your team) and calculate the average.

B. Obtain a solution of one of the chemicals in the list of materials.

  5. Based on your general knowledge of body systems, make a hypothesis as to how this chemical will affect the water flea's heart.

  Remove the slide from the microscope and use a clean medicine dropper to add one drop of the solution to the edge of the coverslip. Draw the solution under the coverslip by holding a piece of paper towel at the opposite edge. Replace the slide on the microscope. Determine the effect of this solution on daphnia by repeating your heartbeat-counting procedure. CAUTION: Do not ingest any substance used in the lab.

  6. Record your team's data in your data table, including the name of the chemical you used and the concentration.
  7. Which part of your experiment was your control? Explain your answer.

C. If you have time, repeat your procedure with a more dilute or more concentrated solution of your chemical. Use a new water flea for each solution and do not forget your control.

  8. Record your data.
  9. What do you think would happen if you used a full-strength solution of your chemical?

D. Add your team's data to the class table on the board.

Postlab Analysis

  10. Which chemicals stimulated the heartbeat and which slowed it down?
  11. Did the chemicals used affect the water flea's heart in the ways you predicted? Which results were different?
  12. Generate as many hypotheses as you can to explain why a known stimulant might depress the heart rate or a known tranquilizer speed it up.
  13. Does each water flea have the same heart rate under control conditions? What factors could account for these differences?
  14. Identify possible sources of error in your technique.
  15. How would you design an experiment to test for complicating factors in the solution: low pH, salt concentration, presence of unknown chemicals?

Further Investigations

1. How would variations in temperature affect the heartbeat of the water flea? How could you test your hypothesis?
2. Using several concentrations of one of the chemicals available, collect sufficient data to make a graph that would show how heartbeat relates to concentration. Is there a threshold level below which a solution has no effect? Is there a maximum level that would be fatal to the water flea or has an effect opposite to what you expected? How could the water flea be used by a professional lab to detect the presence of hormones or drugs, or to determine their concentration in blood or urine? NOTE: You should not experiment with lethal concentrations of the chemicals. Check with your teacher before proceeding.

Investigation

43.1 | *Daphnia Heartbeat*

1. _____
_____

2. _____
_____

3. _____
_____
_____

4. Enter your answer on the data chart.

Daphnia Heartbeat Rate

Normal

	1	2	3	Average
Number of Beats Per Minute				

Experimental

Chemical _____   **Concentration**

	1	2	3	Average
Number of Beats per Minute				

5. _____
_____

6. Enter your answer on the data chart.

7. _____

8. Enter your answer on the data chart.

9. _____

10. _____
_____

11. _____
_____
_____

12. _____

_____

_____

_____

13. _____

_____

14. _____

_____

15. _____

_____

_____

43.2

Pulse and Breathing Rate

Learning Objectives
- To perform a series of tests to see how they affect your pulse and breathing rates.
- To learn how the circulatory and respiratory systems work together.

Process Objectives
- To organize your data by pooling results with the rest of the class.
- To infer the effects of certain conditions on the circulatory and respiratory systems by analyzing data.

Materials
For Group of 2
- Stopwatch, or watch with second hand
- Paper bag, lunch size

How does exercise affect your heartbeat and respiratory rate?

Introduction

When you throw a ball, you first make a conscious decision, then your brain sends a message to your arm muscles to throw the ball. However, you do not make a conscious decision for your heart to beat or for your lungs to take in air. Your cardiovascular and respiratory systems are controlled by a part of the brain called the **medulla.** The medulla acts in response to signals received from the body when oxygen is depleted or carbon dioxide is too concentrated. **Homeostatic** mechanisms operate to maintain the body in a balanced state. For example, a high level of carbon dioxide in your blood triggers the respiratory center in the medulla to increase the rate and depth of your breathing.

How then, can you affect the way your heart beats and how rapidly you breathe? One way is exercise.

Prelab Preparation

This investigation requires you to take your partner's pulse. Practice now to enable you to take your partner's pulse quickly and accurately during the lab.

Sit down. Turn one hand palm upward. With the fingertips of your other hand, press gently about 5 cm up from the wrist. Press near the outer edge of your arm, where you feel a soft spot. You should feel a slight throbbing. Now try taking someone else's pulse.

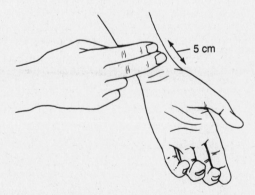

1. What is a pulse?
2. Is your pulse the same in both wrists? What might account for any difference?

Before you begin, read the procedure carefully. After you have completed each test and filled in your chart, switch roles with your partner.

If you feel dizzy during any of the tests, sit immediately and call your teacher. (Watch your partner for any symptoms of faintness.)

After each trial, rest and allow your respiratory and pulse rates to return to normal even if it is not required as part of the trial. As you are waiting between trials, answer the questions and enter your results on your data chart.

As you go through your experimental trials, you and your partner will record the data you collect on your data charts.

Procedure

A. Resting Rates: After being at rest for a few minutes, have your partner count your pulse while you watch the clock. Count the pulse for 15 seconds, and then multiply by 4 to get the pulse rate per minute. Do this 2 or 3 times.

    3. Calculate your average resting pulse rate and record it in the chart.

B. Have your partner count the number of breaths you take in 15 seconds and multiply by 4 to get your breathing rate per minute. Again, repeat this 2 or 3 times and calculate your average resting breath rate.

    4. Record your resting breath rate in the chart.

C. Have your partner time how long you can hold your breath.

    5. Record how long you can hold your breath in the chart.

    6. How do you expect your breathing rate to change during exercise?

    7. How do expect your pulse rate to change during exercise? Why?

    8. Do you think you will be able to hold your breath for a longer or shorter time after you exercise?

D. Walk normally for 2 minutes. Then, have your partner count your breathing rate per minute while you walk in place. Immediately afterward, have your partner time how long you can comfortably hold your breath.

    9. What is your breath-holding time and breathing rate? Record them in the chart.

E. Walk normally for 2 minutes. Then, have your partner take your pulse rate while you are still moving. Immediately afterward, have your partner time how long it takes for your pulse rate to return to its resting rate (which you measured in Step A). Take measurements every 30 seconds until your resting rate is reached.

    10. What was your pulse rate after exercise? How long did it take to return to normal? Record these data in your chart.

    11. What difference in time do you expect for your pulse to return to its resting rate after strenuous exercise compared to mild exercise?

    12. How do you think your respiratory rate after 2 minutes of walking will compare with your respiratory rate after 2 minutes of running?

F. Run in place for 2 minutes. Have your partner time how long you can hold your breath and then have your partner count your breathing rate.

    13. Record your breath-holding time and your breathing rate in the data chart.

    14. What effect does vigorous exercise have on your breathing rate?

G. Run in place for 2 minutes. Then, have your partner determine your pulse rate while still running in place. Next, have your partner time how long it takes for your pulse rate to return to its resting rate. Take several measurements at 30–60 second intervals.

15. What is your pulse rate? How long does it take to return to normal? Record these data in the data chart.
16. In Step H, you will breathe into a paper bag. What effect do you expect this to have on your oxygen–carbon dioxide balance?
17. Do you expect your breathing rate following paper-bag breathing to increase or decrease? Do you expect your breaths to be deeper or shallower?

H. Breathe in and out of a paper bag for 2 minutes. Immediately after, have your partner time how long you can hold your breath and determine your breathing rate and pulse rates.

18. Record your breath-holding time, breathing rate, and pulse rate in the chart.

I. Sit down for this trial. NOTE: Stop and call your teacher if you feel dizzy. Hyperventilate (breathe as deeply and fast as you can) for one minute. Have your partner time your rate of breathing during hyperventilation. Immediately following hyperventilation, have your partner time how long you can hold your breath. Then, have your partner time your breathing rate after holding your breath.

19. Record your results from Step I in the chart.
20. After hyperventilating, was your breathing deeper or shallower than usual?
21. Record your data from all of the tests on the class chart. What do the data tell you about the average resting breathing rate and pulse rate?

Strategy for Analyzing

Make a bar graph of the class results for at least one set of tests, comparing them to the resting results from Steps A and B.

Postlab Analysis

22. What do the data tell you about the average breathing and pulse rates during exercise?
23. Compare the differences in time it takes for the pulse rates to return to normal after strenuous exercise for students who exercise frequently and for students who exercise infrequently. Explain the difference.
24. How does exercise affect the relative levels of oxygen and carbon dioxide in the blood?
25. How can strenuous exercise over a period of time improve heart function?
26. How do you expect improved heart function to affect pulse rate?
27. How does hyperventilation upset the oxygen-carbon dioxide ratio in your blood?
28. Nicotine from smoking tobacco enters the bloodstream,

An inference may be more easily made if you base it on your own experience or on the experiences of friends or family.

constricting blood vessels and damaging bronchial linings. Based on your data, infer how nicotine would affect breathing and pulse rates.

29. The amount of blood pumped by the heart may be increased either by accelerating the heartbeat or by raising the volume of blood pumped during each beat. Which of these actions is represented by good physical conditioning? Why?

Further Investigations

1. Measure the amount of carbon dioxide you produce. Add 3 or 4 drops of phenolphthalein indicator to 100 mL water. Add 0.04% sodium hydroxide a drop at a time until the solution turns pale pink. Exhale through a straw into the water for one minute. CAUTION: Do not inhale the solution! The solution will turn clear. Add the 0.04% sodium hydroxide with a burette or graduated pipette until the pink color returns and remains pink for one minute. Multiply the milliliters of sodium hydroxide you used by 10 to get the number of **micromoles** of carbon dioxide you exhaled into the water. Repeat after strenuous exercise.

2. Determine your target heart rate zone for conditioning. Determine your maximum heart rate by subtracting your age from 220, and then multiply by 0.60 and 0.75. (Your target zone should be between 60 and 75% of your maximum heart rate.) Take your pulse rate after exercising for 10 minutes. Does this rate give you your target zone? Find out how conditioning can improve your heart and help you reach your target zone. (Make sure you are physically fit before doing this investigation.) The American Heart Association can give you more information.

Investigation

43.2 *Pulse and Breathing Rate*

1. _____

2. _____

3.–5. Enter your answers on the data chart.

Data

Trial	Time Breath Held (seconds)	Breathing Rate (breaths per minute)	Pulse Rate (beats per minute)	Time to Return to Resting Pulse (seconds)
At Rest (Steps A–C)				
Walking (Steps D–E)				
Running in Place (Steps F–G)				
Paper-Bag Breathing (Step H)				
Hyperventilating (Step I)				

6. _____

7. _____

_____

8. _____

9.–10. Enter your answers on the data chart.

11. _____

12. _____

13. Enter your answers on the data chart.

14. _____

15. Enter your answers on the data chart.

16. _____

17. _____

_____

18.–19. Enter your answers on the data chart.

20. _____

21. _____

_____

22. _____

23. _____

24. _____

25. _____

26. _____

_____

27. _____

28. _____

_____

29. _____

_____

44

Disease and Immunity Survey

Learning Objectives
- To design a survey.
- To compare different groups' understanding of infectious diseases and the immune system.

Process Objectives
- To analyze survey data.
- To communicate and compare your results with others.

Materials

For each student

Part I
- 2 Sheets of paper for survey data

Part II
- Sheet of graph paper
- 2 Colored Pencils

Strategy for Communicating
Read your six statements to a classmate. Have him or her say whether the statement is true or false. Discuss the correct answer for each statement. Were any answers surprising to your classmate? Were some statements more interesting than others? Were your statements clearly worded? You may decide that you need to reword some of your statements or that some are more interesting and challenging for the survey than others.

How well do people understand basic facts about infectious diseases and the immune system?

Introduction
If you get caught in the rain, you will catch a cold. True or false? Through our knowledge of **infectious diseases** and the immune system, we know that this statement is false. The chill from the rain does not cause a cold. Only infection with a cold virus will cause colds. Many people, however, base their knowledge about health issues on such common erroneous ideas and superstitions. In this investigation, you will design a survey to determine how prevalent these misconceptions are among different groups of people. You will analyze collected data to determine the extent to which these people understand basic facts about infectious diseases and the immune system.

Prelab Preparation
1. Based on information in Chapter 44, library research, and your own experience, make a list of six statements that might be used in a survey about infectious diseases and the immune system.

Three statements should be true and three statements should be false. Try to think of statements associated with common erroneous ideas or superstitions. Consider these examples:

- The AIDS virus (HIV III) can be transmitted through the air or water. (F)
- The common cold virus can be transmitted through the use of an infected person's handkerchief. (T)

Procedure

Part I
A. As a class, pool all of your statements and decide on 12 statements that will be used in this survey. The 12 statements should cover a range of topics. Be especially careful that the true statements are, indeed, true, and the false statements are, indeed, false. This is essential to the validity of your survey results. Your teacher will write the final form of the 12 statements on the chalkboard.

B. As a class, decide on two different types of groups to interview. Because you are determining your subjects' understanding of infectious diseases and the immune system, you may want to categorize each group according to your hypotheses about their level of knowledge on health-related issues. For example, you may want to interview younger students with less education, and older students who have more. Another possibility is to interview people from two generations. You may want to interview one group of people that reads the newspaper regularly and one group that does not. Whichever groups you choose, it is important that the whole class use the same criteria to identify members of each group. This will be important in the processes of collecting, compiling, and analyzing your data in Part II.

2. On two separate sheets of paper, prepare survey data sheets for each group as shown in the model below.

Model for Survey Sheet

Group Number

Statement	True	False	Total

C. Select five people who belong to each group. As you interview each person, read the statement aloud. Ask your subject whether he or she believes the statement to be true or false. NOTE: Do not indicate the answer in reading the statement. Do not show the data sheet to the subject.

3. Record answers in the true or false column of your survey sheet by making a check (√).
4. After you have completed interviews with each group, tabulate the answers recorded on your data sheets. Record the total number of people who responded to each statement. The sum of the total responses should be the same for all the survey statements.

Part II

Raw data, such as you have collected, do not serve a purpose until they are put into a useful form. The percentages you will obtain from the next tabulation will be graphed to provide a visual representation of the data collected by the class.

D. As a class, pool your results.

5. Using your survey data sheets and the class results, complete items 1–12 for Columns 1–4 in the chart provided.

E. Graphs are an effective way to communicate your results. The bar graph that you will construct from your data will chart the percentages of correct answers for each question obtained from each group. To obtain these percentages from the class data, follow this procedure:

6. At the bottom of your data chart, record the total number of correct answers for *all* statements for groups I and II from pooled class data (Columns 3 and 4).

For your own data, there are five subjects in each group. The *total* number of possible correct responses for *each* statement is 1×5, or 5 for each group.

7. What is the number of correct answers *possible* for *each* statement for the class data? Record this number in Columns 5 and 6 of the chart for each statement.

For your own data, there are a total of 12 questions and 5 subjects in each group. The *total* number of possible correct answers for *all* statements is 12×5, of 60 for each group.

8. What is the *total* number of possible correct answers for *all* statements for the pooled class data? Record this number at the bottom of the chart.

9. To obtain the percentage of correct responses for each question for each group from the pooled class data, divide the number of correct answers by the number of possible correct answers obtained in Question 7. Multiply this number by 100 to get the percentage of correct answers. Record each percentage in Columns 7 and 8.

10. To obtain the percentage of correct responses for *all* statements, divide the number of correct answers by the total number of possible correct answers obtained in Question 8 and multiply by 100. Record the percentage for each group at the bottom of the chart.

11. After you have calculated the percentages for each question (Columns 7 and 8), graph them on a bar graph (similar to the sample graph below). Use a separate sheet of graph paper. Use one pencil color for one group, another for the second group.

Interview Data Results — Sample Graph

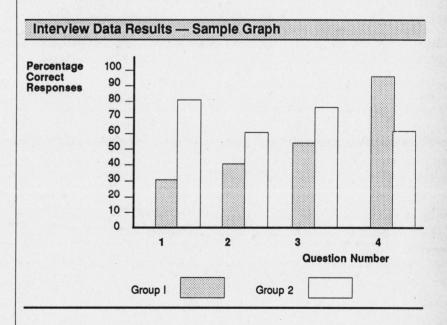

Strategy for Analyzing

To interpret your graph, note similarities and differences between the responses of the two groups. Think about the survey itself and the conditions under which the survey was conducted that might explain those similarities or differences.

Strategy for Communicating

For the statements identified in Question 14, write brief paragraphs explaining why each of these statements is true or false. Read them to a classmate. Are your ideas clearly stated? Is there additional information that is needed in order to understand the statement? Are there other ways by which you could communicate the information besides writing it? Discuss these questions with your classmate.

Postlab Analysis

12. Did you find a difference between the overall knowledge of your two groups?
13. If there was a difference between the two groups, how would you account for it?
14. On which statement(s) did the subjects indicate the greatest need for updated information?
15. If you had the opportunity to teach the people who answered the statement(s) identified in Question 14 incorrectly, how would you explain the biological information necessary for them to understand their mistakes?

Further Investigations

1. Determine where educational efforts should be placed to inform the general public about disease and immunity issues.
2. Based on the data that have been gathered and graphed, develop a public education campaign targeted to the groups in the survey. Design posters or billboards. Create sample ads for newspapers and magazines. Develop mock radio and television public service ads.
3. Design a questionnaire relating entirely to one disease that is a current issue in society and survey various groups to determine their knowledge regarding various aspects of the disease. You might look at another factor for grouping individuals, such as their perceptions about their knowledge of biology. For example: (a) Group I, my knowledge of biology is very good; (b) Group II, my knowledge of biology is average; (c) Group III, my knowledge of biology is below average.

Investigation

44 | *Disease and Immunity Survey*

1. _____

_____

_____

_____

_____

_____

_____

_____

_____

_____

_____

_____

_____

2.–4. Record your data on a separate sheet of paper.

5.–10. Enter your answers on the data chart on the next page.

11. Make a bar graph on a separate sheet of paper.

12. _____

_____

_____

13. _____

_____

_____

14. _____

_____

_____

_____

15. _____

_____

_____

Data Tabulation Form

Statement Number	Individual Number Correct Group #1	Group #2	Class Number Correct Group #1	Group #2	Total Number Correct Possible Group #1	Group #2	Percentage Correct Group #1	Group#2
1								
2								
3								
4								
5								
6								
7								
8								
9								
10								
11								
12								
Total Correct								
Total Number Correct Possible	60	60						
Percentage Correct								

45.1 *Nutrition*

Learning Objectives
- To review food consumption for one day.
- To evaluate the nutritional value of the foods you commonly eat.

Process Objectives
- To organize dietary data collected over a 24-hour period.
- To analyze how effectively your diet meets your nutritional needs.

Materials
- Lined paper
- Colored pencils
- Graph paper

How well do you meet your nutritional needs?

Introduction

It would seem that most Americans have enough food to eat. Yet several of the leading causes of deaths in the U.S. have been associated with **malnutrition**, an unbalanced diet. A malnourished person may be suffering from **undernutrition**: not enough of certain types of nutrients. For instance, a lack of calcium in a diet can result in breakage and deformities in the bones and teeth. A lack of vitamin C can cause poor healing, bleeding gums, and bones that fracture easily.

In some cases, the effects of undernutrition are only obvious after years of chronic deficiencies. We now understand that a lack of fiber in a diet is associated with the development of cancer of the colon. As a result, fiber is being added to processed foods to balance the loss of fiber during the refining process.

An inadequate diet can also be viewed in terms of **overnutrition**: eating too much of certain foods, often at the expense of more valuable nutrients. Additives that were once eaten only infrequently have, for many people, become large portions of the daily diet. The habits of adding sodium chloride to meals and snacking on salted foods can contribute to hypertension and heart disease. An abundance of refined carbohydrates in a diet (sweet desserts and candy snacks) can lead to tooth decay, weight problems, and diabetes.

As a young adult, it may be difficult to believe that a serious, diet-related disease could ever happen to you. Statistics show, for example, that approximately 10 million Americans have some type of diabetes and 60 million have hypertension. There are 145,000 new cases of cancer of the colon each year in this country. If you know of any instances of these diseases in your family, you should be especially careful of your eating habits.

Prelab Preparation

This lab will help you evaluate the foods you eat and analyze how balanced your diet is. It is essential to be honest as you record your present diet, especially the snacks between meals.

1. On a separate sheet of paper, keep a record of everything you eat for 24 hours. Mark down amounts as well—numbers of cookies or how many servings of each item.

The following are examples of "one serving": 1 slice of pizza, 8 oz of yogurt, 2 small cookies, a slice of bread, a bowl of soup or cereal, a milkshake, 1 doughnut, a hamburger patty, 1/2 cup of vegetables, fruit, or cooked grain.

Try to eat the foods that you usually eat. You will be referring to this list during the lab. Later, the list will provide a means of evaluating your personal diet.

Procedure

A. Use your 24-hour record of foods you ate.

2. Sort all of the foods you ate under the following categories: vegetables and fruits, cereals and breads, protein foods, and milk products. If you ate any foods that are high in sugar, such as hard candy, gum, or candy bars, list these in a separate category. Include the numbers of servings for each food item.

NOTE: Mixed foods may present problems. Divide up the combinations into parts of servings. For example, a cheese pizza may be counted as 1/2 serving protein food, 1/2 serving vegetable, 1 serving bread; one cup of stew may be considered as 1/2–1 serving protein, 1–2 servings vegetables, depending on the composition; presweetened cereal may be counted as 1/2 serving cereal, 1/2 serving sweet, etc.

3. Add the number of servings within each food category and then find the total number of servings from all categories consumed for the day.

4. Compare your results to the following recommendations: vegetables and fruits, 4 servings; cereals and breads, 4 servings; protein foods, 2 or more servings; milk products, 3 or 4 servings. More than 2 sweets or sweet snacks per day is not recommended.

5. Find what percentage of your day's diet comes from each category. Divide the total number of servings you ate in the day into the number of servings you ate of each category.

6. Organize the data from Question 5 into a bar graph on a separate sheet of graph paper. Distinguish the bar for each category by a different color or shading and label each with its name and percentage.

Strategy for Organizing Data

A bar graph can help you to quickly compare the differences in proportions. You can see relationships immediately. Make the contrasts clear by using bright colors.

B. To effectively evaluate your diet, you must first become familiar with the nutrient values of various foods. For each of the 6 types of nutrients in the chart that follows, foods are listed in groups that have a range of high, medium, or low proportions of that nutrient. For each food that you have eaten, find its placement with respect to each type of nutrient. Work with other members of your group to figure out where to place foods that are not specifically listed.

NOTE: Additives and means of preparation should be considered for each food. A baked potato, by itself, is low in calories, high in complex carbohydrates, and low in cholesterol. If you add butter or sour cream, these should be listed in the calories and cholesterol categories. The oil used to turn the potato into fries or chips should be listed, as should the sodium values.

Nutrients Chart

Calories

The amount of energy available in any type of food is measured by calories. Foods high in fat content are highest in calories (9 kilocalories per gram). Carbohydrates (starches and sugars) and proteins also can be broken down to release calories (4 kilocalories per gram).

Evaluate your foods as high or low in caloric value by considering the proportions of fats, butter, oils, sugars and starches.

Requirements: Calorie needs vary with the individual (genetics, body build, height) and with the amount of activity. In general, children, young adults, and pregnant women need more calories than older adults.

(continued)

High	Medium	Low
(240–420 kcal per serving)	(70–160 kcal per serving)	(40–50 kcal per serving)
roasted peanuts, Grape-nut cereal, chicken pie, hamburger on a bun, ice cream, chocolate candy bars.	whole and skim milk, yogurt, doughnuts, potato chips, mayonnaise, butter, bananas, baked potato, bread.	tomato, tangerine, lettuce, carrot, dill pickle, celery.

Carbohydrates

Complex carbohydrates, such as in whole grain products, are high in the proportion of starches to sugars. Sugars, such as monosaccharides and disaccharides, provide energy (calories) but few nutrients. These sweet foods are sometimes called empty calories. Per gram, the whole grain foods have fewer calories and many more vitamins and minerals than the refined sugar products.

Evaluate your foods by considering the proportions of starches (bread, pasta, grains, dough products) to sugars.

Requirements: 58 percent of your total diet should consist of complex carbohydrates. Refined sugars should come to only 10 percent.

High (greatest percentage starch) whole grain bread, brown rice, bulgar wheat dishes.	**Medium** (large percentage of refined starch) pasta, processed rice (white), white bread, pastries, cakes, cookies.	**Low** (more sugar than starch) candy bars, hard candies, flavored juices.

Proteins

Protein can be found in all meats, fish, milk products, and in certain combinations of grains and beans. Most fruits and candies are lacking in protein. Though protein can be converted to calories, its greatest value lies in the use of amino acids for the creation of tissues and for regulation of function, especially in brain tissue. Lack of protein in young children causes permanent brain damage.

Requirements: For people 15 to 18 years old, figure 0.39 gram of daily protein per pound of body weight. This would be fulfilled by eating 3 to 4 ounces of animal protein, more if using beans and grains.

High (24–26 gram per serving) turkey, flounder, tuna, beef, roasted chicken.	**Medium** (13–17 gram per serving) lamb, ham, cottage cheese, pork chop, white fish.	**Low** (3–10 gram per serving) split peas, sardines, cocoa (with whole milk), cream soups, scrambled eggs, hot dogs, pancakes, potatoes, corn on the cob.

Cholesterol

Fats and oils are important for a balanced diet. In small amounts, they help to regulate a number of body functions. The type of fat you eat is very important to consider. One type, saturated fat, is linked to high levels of cholesterol in the bloodstream of most people. Cholesterol is essential for the strengthening of cell membranes, protecting nerve fibers, and producing sex hormones, but too much of some types of cholesterol can literally clog up the blood vessels and lead to impairment or heart attack. Another type of fat, polyunsaturated oils, is less likely to cause cholesterol buildup on arterial walls. Saturated fats are usually solid at room temperature, while polyunsaturated fats are liquid oils.

Use the food list to choose a variety of fats.

Requirements: Cholesterol in foods should not exceed 300 mg per day (as averaged out over a week or two).

High (128–370 mg per serving) liver, eggs, shrimp.	**Medium** (50–85 mg per serving) lamb, veal, beef, pork, dark and light meat chicken, tuna, butter, milk, cheddar cheese, muffins, ice cream.	**Low** (3–35 mg per serving) lite cottage cheese, skim milk, most fruits and vegetables.

Sodium

The most common source is table salt, sodium chloride. Excessive sodium (including additives such as monosodium glutamate) can lead to hypertension.

You can evaluate most foods by remembering their taste. Many processed foods, however, are high in salt without tasting especially salty.

Requirements: Sodium is abundant in all foods as a natural component, so adding any at all as flavoring can exceed the recommended limit of 3–8 grams.

High (definite salty taste) corn and potato chips pickles, fast food cheeseburger, canned soups (chicken noodle).	**Medium** (salt is inconspicuous) processed American cheese, baked beans, canned bean salad, bologna, hot dogs, instant pudding.	**Low** (only natural salts present) fresh green beans, baked potato, fresh salad vegetables.

Fiber

We need fiber to keep the muscular contractions of our digestive system functioning smoothly. Peristalsis in the large intestine is more efficient when acting on bulky material. The complex fiber in vegetables and fruits provides not only bulk, but minerals, vitamins and trace nutrients; they may also promote absorption of certain nutrients. Insufficient fiber is linked to cancer of the colon and constipation.

Many of the foods listed in "complex carbohydrates" are also appropriate for the high fiber list. Evaluate for fiber by thinking about the amounts of natural vegetable material (high fiber) relative to the amount of processing (low fiber).

Requirements: Diets high in fresh produce, fruit, and complex carbohydrates seem to be healthier on the whole. Increase amounts of fiber slowly to allow the digestive system time to adjust.

High (high percentage of fiber) All-Bran cereal, shredded wheat, Grape-nuts, whole wheat bread, rolled oats.	**Medium** (some fiber, other carbohydrates) graham crackers, corn kernels, kidney beans, lentils, peas, potatoes, celery, brown rice, fruits.	**Low** (less fiber, more other materials) refined flour products, meats.

Strategy for Analyzing

If your analysis of your diet indicates that you have a potential health risk (if your habits continue as they are at present), remember that this does not mean that you are *certain* to get a disease, only that you are *more likely* to do so. Inherited characteristics and stress will greatly influence the outcome.

Postlab Analysis

8. Considering both the nutrient values and the amounts of the foods from your 24-hour record, how would you rate your diet relative to each type of nutrient?
9. Referring to the health problems described in the Introduction and in the table of Nutrient Values of Foods, what potential health risks do you infer may be related to an imbalance of nutrients in your diet?
10. What foods would you add to or subtract from your diet to make it better balanced?
11. What might be some of the obstacles to achieving a better-balanced diet?
12. How might you resolve those difficulties?

Further Investigations

1. Keep track of your diet for a week and then analyze your food list for nutritional value. Compare the diets of those eating in the school cafeteria with those bringing lunch prepared at home.
2. Record and evaluate the school lunch program for a week's time. Lunch programs follow general guidelines defined by a state health agency. Find out the guidelines for your school. Set up a procedure for comparing the intentions of the lunch program to what foods are actually eaten by the students who buy the lunch.
3. People of many cultures maintain their health on traditional diets that are low or lacking in animal proteins. By combining certain beans and grains, or by fermenting vegetable and dairy products, the protein needs are met. Research the concept and characteristics of complementary proteins; the book *Diet for a Small Planet*, by Frances Moore Lappe, is a good place to start.
4. Collect food packages. Using the information provided on the labels, compare the nutritional value of foods such as: a children's cereal (without milk), a frozen pizza, low-calorie yogurt, a canned pasta product, and a can of dog food. Which is the most nutritious per serving? What is the price per serving? What nonnutritive additives did you find on the labels? Research the purpose for each additive.

Investigation

45.1 *Nutrition*

1. Keep a record of your diet for a 24-hour period on a separate sheet of paper.

2. _____

3. _____

4. _____

5. _____

6. Make a bar graph on a separate sheet of paper.

7. Enter your answers on the data chart on the next page.

8. _____

9. _____

_____

_____

_____

10. _____

_____

_____

11. _____

12. _____

_____

Data Chart for Food Evaluation

Type of Nutrient	Foods Containing Nutrient		
	High Levels	Medium Levels	Low Levels
Calories			
Carbohydrates			
Protiens			
Cholesterol			
Sodium			
Fiber			

45.2 | *Starch Digestion*

Learning Objectives
- To understand the function of digestive enzymes in breaking down starch.
- To demonstrate how the breakdown of a molecule enables its transport across a barrier and absorption into the bloodstream.

Process Objectives
- To predict what will happen if the enzyme diastase digests starch inside a dialysis bag.
- To model starch digestion in the small intestines.

Materials
For Group of 2
- 4 pieces of presoaked dialysis tubing (2.5 cm flat width, 15 cm long)
- 4 100-mL Berzelius beakers, tall form
- 4 Rubber bands, cut once
- Clinistix or Testape strips
- 12 Test tubes
- Iodine solution
- Glass marking pencil
- Paper towels
- 4 50-mL Beakers
- 5 Medicine droppers
- Distilled water in wash bottle
- 80% Glucose solution (about 50 mL)
- 0.3% Starch suspension (about 50 mL)
- 1% Diastase solution (about 10 mL)
- 25-mL graduated cylinder

Strategy for Modeling
Draw a labeled diagram of the experimental set-up. Then, next to this, draw a labeled diagram of digestion in the small intestine. Connect those parts of each diagram that perform the same function.

What is the first step in starch digestion?

Introduction
Carbohydrates provide the most readily usable source of energy for the body. However, in its edible form, the molecules of starch are too large to cross semipermeable membranes and too insoluble to dissolve in the blood. Therefore, our digestive systems must break down starch molecules into simpler sugar molecules. Then these molecules, which are water soluble, can be absorbed into the bloodstream through the walls of the small intestine.

Starch digestion begins in the mouth. **Amylase**, an enzyme that is present in the saliva, breaks down starch into a complex sugar, a disaccharide called maltose. Amylase is also secreted by the pancreas into the small intestine to break down starch not yet digested by the salivary amylase.

Maltase, an enzyme produced by the small intestine, breaks down the maltose into simple sugars like glucose, a monosaccharide. Glucose and other monosaccharides are then absorbed into the bloodstream.

Prelab Preparation
Review Chapter 6 on diffusion and Lab 6 on modeling with dialysis bags.

You will test your solutions for the presence of starch with iodine. You will test for the presence of simple sugars using Clinistix or Testape strips. Instead of amylase, you will use diastase, a plant enzyme with equivalent enzymatic activity.

As you prepare the dialysis bags and beakers, be careful not to contaminate the solutions. False positive results may occur if contaminated. Make sure that the volume of water in the beaker is not so great that molecules diffusing out of the dialysis bags would be too dilute to detect.

The dialysis bag is a model of starch digestion in the small intestine.

1. What do the contents of the dialysis bag represent?
2. What does the dialysis tubing itself represent?
3. What represents the bloodstream in this model?

The model shows the passive diffusion of molecules from an area of greater to an area of lesser concentration. The model also demonstrates how a semipermeable membrane allows only certain molecules to pass through.

4. How might a membrane regulate the passage of molecules?

The small intestines, however, have an active transport membrane system that keeps molecules moving *from* the intestine *into* the bloodstream, even when the concentration of molecules is greater in the bloodstream.

5. How will your model work when there is greater concentration of molecules in the model bloodstream?

Procedure
A. Fill 2 test tubes each with 10 mL starch suspension and 2 test tubes each with 10 mL glucose solution. Be sure to label the tubes as to contents.

Add a drop of iodine to one starch tube and a drop of iodine to one sugar tube. Test the remaining 2 tubes with the Clinistix. Use the chart on the Clinistix package to judge color change. CAUTION: Use care when using iodine to avoid staining hands and clothing.

6. Describe the color changes for each kind of solution in your data chart for test standards.

B. Use these results as standards against which to check further results.

7. What test should you use when you want to determine if any starch has been broken down into simple sugars?

C. Obtain 4 presoaked pieces of dialysis tubing. Close off one end of each tube by wrapping a cut rubber band a few times around and tying securely. Any leakage from the tubing would confuse results. Place each tubing on its own marked paper towel for identification. Obtain about 15 mL of each solution listed in step D and place in a labeled 50-mL beaker.

D. Fill each tubing with about 10 mL of one of the following solutions. Make sure there is at least 5 cm of slack, but no air, at the top. Secure the top with a cut rubber band as in Step C. Rinse the bag with water to remove any spilled solution. Replace it on the marked paper towel to prevent confusion. The contents for each dialysis bag will be:
- 80% Glucose solution (Beaker I)
- 0.3% Starch suspension (Beaker II)
- 1:1 Glucose solution: Starch suspension (Beaker III)
- 1:1 Starch suspension: Diastase solution (Beaker IV)

E. Label the 4 Berzelius beakers to indicate the contents of the dialysis bags to be placed in each. Place each bag into its appropriately labeled beaker. Fill each beaker with a minimal amount of water, i.e., just enough to cover the dialysis bag. Allow 15 minutes for the reactions to occur.

8. Record the starting time.
9. What is the function and/or reaction of the contents of the dialysis bags in beakers I, II, III, and IV? What will they show?
10. Enter your predictions in your data chart for the results that will be detected in beakers I, II, III and IV after 15 minutes.

F. At the end of 15 minutes, remove the bag from each beaker and place on the labeled paper towel. From each beaker, take 2 samples of the water and place them in 2 clean test tubes. Test one tube with iodine and test the other tube with a Clinistix strip. Repeat the procedure for all 4 beakers so that 8 samples in all are tested.

11. Record these results in your data chart.
12. What did the results of the starch tests show?
13. What did the results of the sugar tests show?

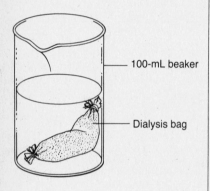

100-mL beaker

Dialysis bag

Strategy for Predicting

When you consider the results expected for each dialysis bag think about the relationship between the size of molecules and the movement of molecules across a membrane.

Postlab Analysis

14. How did your predictions compare with actual results? Do any discrepancies reflect experimental or conceptual errors?
15. According to your results, what is the direct action of diastase enzyme on starch? Why is this important?

Further Investigation

1. What are the optimal conditions for diastase? What would diastase activity be like in the stomach's acid environment or in the alkaline upper part of the small intestine? How do variations in temperature affect the activity of diastase? What is the effect of varying enzyme or substrate concentration? Design an experiment using information from Lab 4 on the enzyme catalase as well as your experience with dialysis bags.

Investigation

45.2 | *Starch Digestion*

1. _____

2. _____

3. _____

4. _____

5. _____

6. Enter your answers on the data chart.

Test Standards

	Iodine	Clinistix
Starch		
Sugar		

7. _____

8. _____

9. _____

10.–11. Enter your answers on the data chart.

Data Chart for Dialysis Bags

	Beaker I		Beaker II		Beaker III		Beaker IV	
	Sugar	Starch	Sugar	Starch	Sugar	Starch	Sugar	Starch
Prediction								
Result								

Key: + = positive test (present in beaker water)
       − = negative test (absent in beaker water)

12. _____

13. _____

14. _____

_____

_____

15. _____

_____

_____

46

Vision and Hearing

Learning Objectives

- To understand how your eyes and ears interpret light and sound waves.
- To distinguish amplitude and frequency characteristics in sound and visible light.

Process Objectives

- To observe your own responses and those of your lab partner to light and sound stimuli.
- To communicate with your classmates about your responses to light and sound.

Materials

For Group of 2

- Dark handkerchief or cloth
- Coin
- Pencil
- Cards and shapes cut from paper (colored with bright primary and secondary colors)
- 2 Blank sheets of heavy white paper
- Paper drinking straw
- Scissors
- 2 Tuning forks of the same pitch

What are the limits of what we hear and see?

Introduction

What we know about the world is determined by our senses. Look at the diagram of the electromagnetic spectrum and notice the relatively short range of wavelengths that results in visible light. What we see as a white flower may appear as a pattern of contrasting colors to a bee since bees see ultraviolet light. The keen senses of the hunting dog and the carrier pigeon have long been used as a means of extending our limited senses. Some migratory birds may even be sensitive to the earth's magnetism.

Electromagnetic Radiation

Wavelength

Ångstroms (Å)	microns (mm)	meters (m)		
10^{-4}				
10^{-3}			Gamma Rays	
10^{-2}				
10^{-1}			X-Rays	
1	10^{-4}			
10^{1}	10^{-3}			300
10^{2}	10^{-2}		Ultraviolet	400
10^{3}	10^{-1}		Visible Light	500
10^{4}	1			Purple / Blue / Green
10^{5}	10^{1}		Infrared	600 / Yellow / Orange
10^{6}	10^{2}			700 / Red
10^{7}	10^{3}			800
10^{8}	10^{4}		Microwaves	nm
10^{9}				
10^{10}		1	(Short waves)	
10^{11}		10^{1}	Radio waves	
10^{12}		10^{2}		
10^{13}		10^{3}	(Long waves)	

As our technology expands, we are increasing the limits to what we can sense with instruments. Your observations in this laboratory will give you a fuller understanding of the potentials and limits to better sight and hearing.

Prelab Preparation

Look at the diagram above and think about the wavelengths humans cannot perceive.

1. What other senses might have developed for humans, in-

stead of vision of reflected light and hearing of low frequency sounds? How might we perceive wavelengths outside the range of our senses?

Our ears are designed to convert the mechanical energy of sound waves into electrical impulses carried by neurons. Sound waves enter the outer ear and cause the ear drum to vibrate. This vibration is transmitted by 3 **ossicles** (small bones) to the inner ear. The inner ear contains the **cochlea**—a fluid-filled tube lined with hair cells (auditory receptors). Vibrations in the inner ear cause the hair cells to transmit neuronal impulses via the auditory nerve into the brain where they are perceived as sounds.

Hearing depends on the ability of the ear to interpret several aspects of sound waves. Loudness is determined by the **amplitude** of sound waves. **Pitch** is a measure of their frequency. **Timbre** results from the complexity of vibrational patterns. In addition, we are able to sense the direction from which a sound comes. The diagram shows the parts of an electromagnetic wave that we are referring to when we talk about frequency and amplitude.

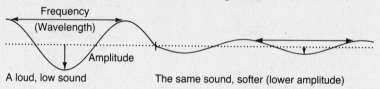

A. Low frequency electromagnetic wave

Frequency (Wavelength)

Amplitude

A loud, low sound   The same sound, softer (lower amplitude)

B. High frequency electromagnetic wave

As loud as A, but higher pitch   Same pitch but not as loud

Our eye is structured to focus light on the **retina**, the light-sensitive inner layer. The retina contains 2 types of photoreceptors—rods and cones. Nerves connect photoreceptors to the brain where impulses are organized. We sense the amplitude of light waves as brightness, the frequency as color, and the patterning as shapes. Without the eye's focusing ability, however, our visual world would consist of a fog of shifting light.

Procedure

Part I: Vision

A. **Dominance.** In the same way that you are right-handed or left-handed, one of your eyes is used more than the other. Find a small object at least 10 feet away, close one eye, and point to the object. Hold your hand steady and open both eyes. Are you still pointing to the object? Repeat the procedure with the other eye closed.

2. For which eye did your finger remain in line with the object? This is your dominant eye.
3. Are you right- or left-handed? What percentage of people in the class have dominance of the same eye and hand?

B. **Blind spot.** The place where the optic nerve enters the back of the eye has no rods or cones. Light that is focused here cannot be perceived. The approximate distance of focus for the blind spot in each eye can be determined as follows: Hold the following diagram at arm's length. Tilt your lab book slightly with the top of this page higher than the bottom. Closing your right eye, focus on the circle. Concentrate on looking at it only, and slowly bring the diagram closer.

4. Record the distance at which the cross disappears from view.

5. What happens as you continue to bring the diagram closer?

Repeat the procedure with your left eye closed; now focus on the cross.

6. Record the distance at which the circle disappears.

C. **Pupillary reflex.** Look into your partner's eyes, and note the color, shape, and size of the pupils and irises. The irises open and close by opposing muscles. Have your partner close both eyes and place a dark cloth over both of them. After 3 minutes, remove the cloth and observe the pupil and iris as the eye is opened. Repeat the exercise with your partner observing your response.

7. How does the pupillary reflex help the eye adjust to various amplitudes of light?

D. **Binocular vision.** The accommodation reflex helps us to judge distance. Observe the way your partner's eyeballs converge and how the size of the pupils adjusts as you bring a coin closer or farther away.

8. What is the relationship between convergence and pupil size in observing a near and far object?

Hold a pencil in your right hand with your elbow slightly bent. Focus on the tip of the pencil. Close your left eye; attempt to touch the point with the index finger of the left hand, starting with your arm at your side. Repeat with the right eye closed and touching with the left hand. Then switch the pencil to your left hand and try to touch the point with your right hand, first with your left eye closed and then with your right eye closed. As a control, try the experiment with both eyes open.

9. Is your dominant eye more accurate than the other?

10. If you are wearing glasses, repeat the experiment without them and record the results, including your inference on the effect of your glasses in adjusting your vision.

A person with normal vision has to integrate 2 views (one from each eye). Binocular vision involves convergence of the eyeballs, change in the size of the pupils, and the focusing of images by the lens.

E. **Afterimages.** Once a nerve pathway is strongly stimulated, chemicals involved in stimulus transmission are used up. Rods and cones continue sending an image to the brain, but the frequency or amplitude is the opposite of the initial stimulus. A bright image, such as a camera flash, will appear dark. A primary color will appear as its opposite color. Place a piece of red paper on a blank sheet of heavy white paper. Stare at the red paper for 30 seconds without shifting your gaze, then cover with another blank sheet of heavy white paper. Repeat for blue, green, and yellow.

11. What are the colors of the afterimages produced by red, blue, green, and yellow papers?

12. Can the afterimage relationship be reversed? For instance, red has a bright green afterimage. Does green have a red afterimage? Does this reversal work for the other pairs?

Part II: Hearing

F. **Frequency.** When we listen to a violin being played, the range of pitches (or frequency of sound) depends on the number of vibrations produced by the strings. The number of vibrations in a given time period is called frequency. A tight string will vibrate more quickly than a loose string and will sound higher in pitch. A short string will also sound higher than a long string. Pinch closed one end of a paper straw. Cut small, triangu-

Strategy for Observing

Test yourself for assumptions that might keep you from observing accurately. Do you already think you know what will happen? Check what you observe against your assumptions. Compare your results with those of your partner and check any discrepancies.

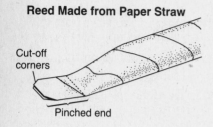

Reed Made from Paper Straw

Cut-off corners

Pinched end

lar shapes off both corners of the opening, so that the straw resembles the one shown. You have made a reed, like the mouthpiece of a clarinet or oboe. Carefully hold the reed section between your lips, and blow gently. Try various pressures until the reed vibrates. Have your partner cut sections off the outer end as you sound the reed.

13. What is the relationship between shortening the straw and the pitch of the sound? (Answer in terms of frequencies.)

G. **Amplitude.** If 2 objects are producing the same sounds and one is farther away, we perceive the closer object as louder, that is, of greater amplitude. The effect can be counteracted by making the more distant sound louder. Use 2 tuning forks of the same pitch to perform the following 2 experiments. Blindfold your partner. In experiment 1, rap the first tuning fork. After a few seconds (when the sound intensity starts to fall), rap the second fork and bring both forks to the same distance from one of your partner's ears. Then move the second tuning fork farther away. Ask your partner to tell you when the tuning forks sound equally distant from his or her ear. In experiment 2, rap both forks at the same time. Start both forks at the same distance from one of your partner's ear and then move one fork further away. Ask your partner to tell you when the forks sound at the same distance and when they sound apart.

14. How far apart were the forks in experiment 1 when they sounded as if they were at the same distance? How far apart were the forks in experiment 2 when they sounded as if they were at the same distance? Can you explain the difference?

Postlab Analysis

Compare your results with those of other students.

15. What aspect of vision is equivalent to the loudness of sound?
16. Accidental blindness is often accompanied by the perception of flashes of light for long periods after the complete loss of vision. The ringing in your ears (tinnitis) when everything is silent is a similar phenomenon. What do you think could explain these phenomena?
17. Imagine that you could rewire your optic and auditory nerves so that each arrived at the brain center for the other. How would your brain perceive colors? How would it perceive music?

Further Investigations

1. Vibrations emitted from a moving object have a different frequency coming toward you than going away. This phenomenon is called the Doppler effect. Listen to a car driving by you. The drop in pitch and amplitude is the Doppler effect in sound. The Doppler effect is used to measure the velocity of comets and planets. If a comet is moving toward us, the light it gives off is pushed to a shorter wavelength: this is called the "blue shift." What kind of a shift will occur if a planet is moving away from us? Go to the library to research the Doppler effect.
2. In aging, some of the senses become less sensitive than they were in youth. For the senses of vision and hearing, does the loss in sensitivity affect amplitude, frequency, patterning, or a combination of these factors? You may need to consult with an optometrist or audiologist to learn the answers.

Strategy for Communicating

Share descriptions of what you saw and heard in this lab with your classmates. Discuss why differences among your experiences may have occurred. How could individual variations in visual and auditory abilities have affected the results?

Investigation

46 | *Vision and Hearing*

1. _____

_____

_____

_____

2. _____

3. _____

4. _____

5. _____

6. _____

7. _____

_____

_____

_____

8. _____

_____

9. _____

10. _____

11. _____

_____

_____

12. _____

13. _____

14. _____

_____

_____

15. _____

16. _____

_____

17. _____

_____

_____

47

Embryonic Development

Learning Objectives
- To identify the stages of early animal development.
- To describe the changes that occur during early development.

Process Objectives
- To observe prepared slides of sea star embryonic development.
- To communicate microscopic observations by drawing.

Materials
For Group of 2
- Prepared slides of sea star development
- Compound microscope

What happens to a sea star egg after it is fertilized?

Introduction

Almost all animal life begins from a fertilized egg. After an egg is fertilized, it must divide and grow many times before it becomes a completely developed organism. In this lab, you will study the early stages of development of a sea star embryo, because the eggs are easy to obtain. However, the early stages of human development are in many ways similar.

For several reasons, the early stages of mammalian development are the most difficult to study. First, mammal eggs are the smallest of all animal eggs; tiny eggs are difficult to manipulate. Second, mammal eggs are not produced in great numbers; it is difficult to obtain enough eggs for studies. Third, mammal embryos develop inside an organism rather than outside; internal conditions can be difficult to duplicate in a lab.

Prelab Preparation

Before beginning this lab, review human embryonic development in Section 47.3 in your textbook. The fertilized egg (a **zygote**), divides by mitosis; it goes through a 2-cell stage, then a 4-cell stage, and then 8-, 16-, 32-, and 64-cell stages. These cell divisions are termed **cleavages**—a series of mitotic divisions that divide the large amount of cytoplasm in the egg into smaller cells.

The following chart compares the cleavage of the echinderm and the mammal. Despite the differences shown, we can still use the sea star to learn about developmental processes.

Echinoderm and Mammal Zygote Cleavage	
Echinoderm	**Mammal**
First cleavage is vertical and divides the egg into 2 approximately equal-sized cells.	First cleavage is vertical and divides the egg into 2 approximately equal-sized cells.
Second cleavage is also vertical and divides the egg into 4 approximately equal-sized cells.	Second cleavage, one cell divides vertically and the other cell divides horizontally, dividing the egg into 4 approximately equal-sized cells.
Third cleavage is horizontal and divides the egg into 8 approximately equal-sized cells. Successive cleavages alternate between horizontal and vertical. Cells are always similar in size.	Successive cleavages do not all occur at the same time; there is often an odd number of cells.
After the 128-cell stage, a hollow ball is formed (blastula) that has cilia and swims.	After third cleavage, cells compact, forming a solid ball of cells (morula). After the 32-cell stage, fluid creates a cavity (blastocoel) in the ball, which becomes hollow (blastocyst).
Cleavage is rapid; blastula formed by second day.	Cleavage slow (12–24 hours for each cleavage); blastocyst formed by eighth day.

After the formation of the **blastula,** the cells begin to form three distinct germ layers. The zygote is now called a gastrula. It is these layers that differentiate into complex tissues and organs.

The gastrula's innermost layer is **endoderm** and will form the digestive system and lungs. The outermost layer is **ectoderm** and will form the skin and nervous system. The middle layer is **mesoderm** and will form the bones, muscles, blood, and circulatory system.

1. Describe one way in which the cleavage of echinoderm and mammal eggs is identical.
2. Describe 2 ways in which the cleavage of echinoderm and mammal eggs are different.

Procedure

A. Obtain a set of prepared slides showing sea star eggs at different stages of development. Examine each slide under a compound microscope at low power. For each slide, focus on one good example of the stage. Then switch to high power.

3. Draw a diagram of each stage you examine. Label each diagram with the name of the stage it represents. Arrange your diagrams in order. Choose slides labeled unfertilized egg, zygote, 2-cell stage, 4-cell stage, 8-cell stage, 16-cell stage, 32-cell stage, 64-cell stage, blastula, early gastrula, middle gastrula, late gastrula, young sea star larva.
4. Compare the zygote size with the blastula size. At what stage does the embryo become bigger than the zygote?
5. At what stage do the cells not look exactly like each other?
6. How do cell shape and size change during successive stages of development?
7. Are the nuclei the same size, larger, or smaller as the stages progress?

Postlab Analysis

8. Compare the number of chromosomes in a fertilized egg with the number of chromosomes in one cell of each of the following: 2-cell stage, blastula, gastrula, and adult organism.
9. From your observations of changes in cellular organization why do you think the blastocoel is important during embryonic development?
10. Label the endoderm and ectoderm in your drawing of the late gastrula stage.
11. How is the symmetry of a sea star embryo and larva different from the symmetry of an adult sea star? Would you expect to see a similar difference in human development?
12. What must still happen to the sea star gastrula before it becomes a mature sea star?
13. In what ways may sea star embryos be used to study early human development?

Further Investigations

1. Examine prepared slides of developing chick embryos. Identify the different stages and describe the order of progressive development. How is chick development similar to and different from sea star development?
2. Make a chart of the average cell diameters of each of the stages.

47 | *Embryonic Development*

1. _____

_____

2. _____

_____

3.

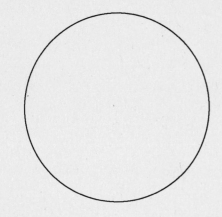

4. _____

_____

5. _____

6. _____

_____

7. _____

8. _____

_____

9. _____

10. Label your drawing of the late gastrula.

11. _____

_____

12. _____

_____

13. _____

48 *Reaction Times*

How fast can you react to a stimulus?

Learning Objectives
- To compare reaction times to a visual stimulus with those to an auditory stimulus.
- To compare reaction times to a stimulus without distractions with those to a stimulus with distractions.

Process Objectives
- To organize data using tables and graphs.
- To analyze and interpret team and class data.
- To make inferences about the effect of psychoactive drugs on reaction time based on the results of your experiment.

Materials
For Group of 2
- Meter stick
- *Modern Biology* textbook
- 2 Sheets of graph paper

Introduction

A dark blur against the ice marks the path of the hockey puck as it sails toward the goal at 80 km per hour. In a fraction of a second, the goalkeeper must plot the puck's course and move to block it.

The time required to sense a stimulus, analyze its meaning, and respond appropriately is called the reaction time. What factors affect reaction time? Do you respond to all stimuli with equal speed? Can your reaction time be improved? How does the use of alcohol, tobacco, or other drugs affect your ability to react?

Every person must be able to respond to stimuli in the environment. In many cases, the speed at which you react is not important. In other instances, reaction time makes the difference between success or failure and even life or death.

Prelab Preparation

1. Describe at least 3 experiences that have occurred during the last week when your reaction time has been important.
2. A motorist swerves the car to avoid a dog that has run into the road. Which parts of the nervous system are involved in the reaction and what is the function of each part?
3. How might tobacco, alcohol, and other drugs affect reaction time? Read the procedures for Steps A–G. The procedure described in Steps A–C will be used to obtain a measure of your reaction time.
4. Why is it necessary to work with a partner and not perform the experiment by yourself?

Procedure

Part I

Reaction Time to a Visual Stimulus

You must work with a partner to complete the following procedures. One of you will act as the recorder while testing your partner's reaction time. After completing Steps A through D, exchange roles and repeat the procedure.

A. Place your forearm on the surface of the table with your hand extended over the edge. Have your partner position the zero end of the meter stick between your thumb and forefinger, as shown on the next page.

B. Determine your reaction time by catching the meter stick between your fingers after your partner releases it without warning.

  5. In your data table, record the distance that the meter stick falls. Use the distance versus time table, to convert the distance of fall to reaction time. Use the distance on the table that is closest to your distance of fall value.

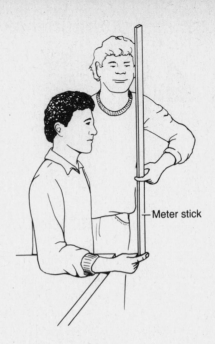

—Meter stick

C. Repeat the procedure for a total of 20 trials.

    6. Record your data after each trial in your data chart.

    7. Calculate the average distance of fall. Use this number, referring again to the distance versus time table, to determine the average reaction time. Record these figures in your data table.

    8. Converting the average distance of fall to the average reaction time gives a better measure than averaging the reaction times for the 20 trials. Why is this true?

Reaction Time to an Auditory Stimulus

D. Position the meter stick as described in Step A. Close your eyes. As the meter stick is released, your partner will say, "Go." React to this auditory stimulus as quickly as possible by closing your fingers on the meter stick.

    9. Record your data in your data table as in Step B.

    10. Repeat the process for a total of 20 trials, recording your results for each trial. Then find your average reaction time to an auditory stimulus as in Question 7.

Reaction Time with Distractions

    11. How might reaction times be affected when a person is distracted and less prepared to receive a stimulus?

E. Two teams should work together to test the effect of distractions on reaction time. While one person is tested, a second will act as the recorder, and the 2 others will act as distractors. Follow the procedure in Step D for testing reaction time to an auditory stimulus, except for the following changes. The distractors will select a previously studied chapter from your biology textbook. During the testing procedure, the distractors will ask the person being tested questions from the Review section at the end of the chapter. The person being tested is expected to respond to these questions while waiting for the "Go" signal from the recorder.

    12. Record your results for 20 trials in your data table and determine your average reaction time, as in Question 7.

F. Exchange roles and repeat the procedure to test all members of your team.

Distance v. Time Fall	
Distance of Fall (cm)	**Time of Fall** (sec)
0.2	.02
0.8	.04
1.1	.05
1.7	.06
2.4	.07
3.1	.08
3.9	.09
4.9	.10
7.0	.12
9.6	.14
12.5	.16
15.8	.18
19.5	.20
23.6	.22
28.9	.24
33.0	.26
38.2	.28
43.9	.30
48.8	.32
56.4	.34
63.5	.36
70.8	.38
78.4	.40
86.4	.42
94.8	.44
103.7	.46
112.9	.48
122.5	.50

13. Record the data after each trial.
14. You and your partner should each use separate sheets of
    graph paper to prepare graphs showing your own data for the
    3 types of stimulus. Label the horizontal axis of the graph
    "Trial Number" and the vertical axis "Reaction Time." Make
    sure that you and your partner use the same scales on your
    graphs so that you can compare data. Plot your data for each
    type of stimulus separately. Use different colored lines or dif-
    ferent symbols in your graph to distinguish between the types
    of stimuli.
15. In what way does this graph tell you more about your reaction
    time than the number calculated as "Average Reaction
    Time"?
16. Does your reaction time remain constant over 20 trials or is
    there any variation? Is there any evidence of a trend toward
    faster or slower reaction times over the 20 trials? What might
    account for such trends?
17. How does your reaction time to an auditory stimulus compare
    to that for a visual stimulus? What might account for this dif-
    ference?
18. How does your reaction time compare to that of your partner?
    What might account for this difference?

Part II

G. Pool the data for the entire class. Your teacher will fill in the figures on
   an enlarged copy of the table on the chalkboard.
H. Work as a class with your teacher to make a bar graph showing the variation among the average reaction time of all class members to the 3 forms
   of stimuli.

19. How does your reaction time compare to that of the class?
    What could account for any differences?

Model Bar Graph

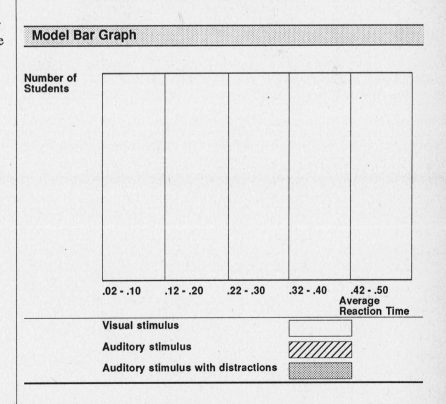

Number of Students

.02 - .10   .12 - .20   .22 - .30   .32 - .40   .42 - .50
Average Reaction Time

Visual stimulus
Auditory stimulus
Auditory stimulus with distractions

Postlab Analysis

20. Why might psychoactive drugs affect a person's reaction time? How does the data collected in Step E indirectly relate to the effect of psychoactive drugs on reaction time?
21. Briefly speculate on how each type of psychoactive drugs might affect reaction time.
22. What types of activities would be adversely affected because of the influence of psychoactive drugs on reaction time?

Further Investigations

1. Can your diet, fatigue, the use of caffeine, regular execise, or other factors affect your reaction time? Design an experiment that tests a hypothesis relating to this question. Have your teacher approve your design. Test your hypothesis by completing your experiment. Present a report of your work to the class that includes a statement of your hypothesis, your experimental design, a summary of your data, and your conclusion.
2. Conduct library research into the biological effects of a specific psychoactive drug. Based on your research, write a report that explains how the drug produces its effects, and speculate on its effect on reaction time.

Investigation

48 | *Reaction Times*

1. _____

_____

2. _____

_____

_____

3. _____

4. _____

5.–7. Enter your answers on the data table on the next page.

8. _____

9.–10. Enter your answers on the data table on the next page.

11. _____

12.–13. Enter your answers on the data table on the next page.

14. Make a graph on a separate sheet of paper.

15. _____

16. _____

_____

17. _____

_____

18. _____

_____

19. _____

_____

20. _____

_____

_____

21. _____

_____

22. _____

Student Data - Reaction Times

Visual Stimulus			Auditory Stimulus			With Distractions		
Trial	Distance of Fall (cm)	Reaction Time (sec)	Trial	Distance of Fall (cm)	Reaction Time (sec)	Trial	Distance of Fall (cm)	Reaction Time (sec)
1			1			1		
2			2			2		
3			3			3		
4			4			4		
5			5			5		
6			6			6		
7			7			7		
8			8			8		
9			9			9		
10			10			10		
11			11			11		
12			12			12		
13			13			13		
14			14			14		
15			15			15		
16			16			16		
17			17			17		
18			18			18		
19			19			19		
20			20			20		
Avg.			Avg.			Avg.		

49

Brine Shrimp

Learning Objectives
- To understand how a particular set of physical and chemical factors can affect the survival of a population.
- To understand the adaptations that allow brine shrimp to survive in specialized ecosystems.

Process Objectives
- To analyze the survival data of brine shrimp which were subjected to physical and chemical variables.
- To use compiled data in hypothesizing what the best environment is for brine shrimp.

Materials

For Class
- 2 Hatching trays (with covers)
- 2 Aeration pumps
- A flashlight or other bright source of light
- A fish feeding ring or substitute floating ring

For group of 2
- At least 400 brine shrimp eggs (*Artemia*)
- 8 Culture dishes (or cereal-sized bowls of glass, ceramic, or enamel)
- 5%, 10%, and 20% saline (sodium chloride) solutions
- Distilled water
- Dried yeast
- 1 Pipette or dropper
- 1 Saline rinse bottle (2%)
- 1 Aeration pump
- 1 Metric balance
- 1 Dissecting microscope
- 1 Grease pencil or marker
- 1 Compound microscope
- 2 Microscope slides
- 2 Coverslips
- Graph paper
- Colored pencils

What are the best physical and chemical conditions for brine shrimp survival?

Introduction

Every ecosystem on earth, from sandy beach and ocean bottom to prairie land and mountain top, can be distinguished by various combinations of water, light, temperature, and available mineral nutrients. These physical and chemical conditions, the **abiotic** factors, give underlying structure to the living, or **biotic**, factors.

Each type of environment is made up of a multitude of subunits, called **niches**. A niche defines the life of an organism in terms of its ability to grow and reproduce successfully. In turn, every organism shows some specializations in its form or behavior that adapt the organism to the its environment.

If the ecosystem is relatively stable with an abundance of light, moisture and nutrients, there are many niches, many varieties of organisms, and a complexity of relationships. Coral reefs and tropical rain forests are examples of rich ecosystems in which competition is the main selective pressure. Ecosystems support relatively fewer species of organisms if the life forms are stressed by extreme abiotic conditions.

Some organisms have adaptations that allow them to live in unusually stressful environments. The brine shrimp is an example of such an organism. It maintains a steady internal osmotic pressure by secreting excess salt through its gills. This way, the brine shrimp can accomodate to large changes in salt concentrations. The brine shrimp also produces many fertilized eggs. These eggs are capable of drying up completely while remaining viable for more than 12 years. Such adaptations reflect the characteristics of the brine shrimp's habitat.

Prelab Preparation

You will hatch brine shrimp eggs under controlled conditions to determine which conditions promote the hatching of the greatest number of larvae. You will then grow larvae under a variety of carefully controlled conditions. The Procedure will assign some variables, and you and your partner will choose others. The results obtained will help you to determine the optimum conditions for brine shrimp survival.

1. Shallow salt waters are the brine shrimp's natural habitat. High salinity and exposure to radiation and predation make survival difficult for most organisms in this habitat. List the adaptations that might help a population of organisms survive in such a place.
2. What does the lengthy survival time of dried eggs tell you about the brine shrimp's habitat?

Procedure

Part I

A. Add 1 g dried brine shrimp eggs to each of 2 hatching trays 2–3 days prior to doing Part II of the investigation. Fill one tray to half full with 5% saline solution and fill the other tray to half full with 10% saline solution. Place the eggs within each floating ring. Cover each tray, leaving one-fourth open to light. Maintain the trays at 20–25°C.

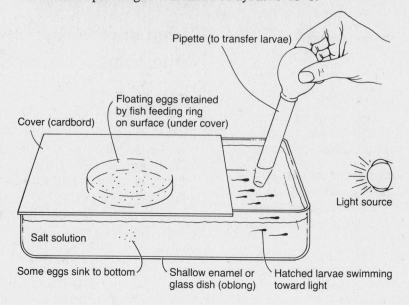

Pipette (to transfer larvae)

Floating eggs retained by fish feeding ring on surface (under cover)

Cover (cardbord)

Light source

Salt solution

Some eggs sink to bottom

Shallow enamel or glass dish (oblong)

Hatched larvae swimming toward light

B. Place 1 aeration pump tube in the 5% saline tray and another in the 10% saline tray.

   3. In which hatching tray do you expect the most larvae to hatch?

Part II

C. On the day of the investigation, shine a light on the uncovered section of the tray. Larvae will be attracted to the light. Try to keep any eggs that haven't yet hatched under the cover. Pipette a sample of the hatched brine shrimp from the 5% saline and place a drop on a slide. Cover with a cover slip and view under a compound microscope at 100x magnification. Count the number of brine shrimp in each of 10 fields. Then calculate the average for these fields. Repeat this procedure for the 10% saline hatching tray.

   4. Record your counts on a separate piece of paper and compare the results of the 5% and 10% solutions.
   5. Which sample showed the most hatching of brine shrimp eggs?
   6. Does increased salinity affect brine shrimp's hatching?
   7. What does the ability of brine shrimp to tolerate high levels of salinity tell you about the abiotic factors of the ecosystem to which brine shrimp are adapted?

D. Now set up 8 culture dishes in which to grow larvae from the 5% saline hatching tray. All dishes except number 5 should be well-lighted and maintained at 25°C. With a grease pencil, label 6 of the dishes as follows:

   1. fresh water
   2. 10% saline solution
   3. 20% saline solution

4. oxygenation of a 5% saline solution
5. darkness and a 5% saline solution
6. a control dish at 5% salinity

Additionally, select two more variables and label your own dishes 7 and 8.

E. After the 8 culture dishes are labeled, fill the dishes with equal amounts of their appropriate solutions. Place 50 brine shrimp larvae into each. If you have trouble extracting larvae, place several milliliters of solution from the hatching tray into a petri dish cover on top of a white paper to provide contrast. Pick out the larvae with a dropper. Make sure you rinse out your dropper so all the animals fall into the culture dishes.

F. Feed the larvae once every 3 days by dropping one granule of powdered yeast into each culture dish . Do not feed the larvae again until all signs of the previous feeding are gone, or the water is clear. Too many feedings will kill the larvae. Maintain the water levels in the dishes by adding fresh distilled water when needed.

   8. What do you think will be the best conditions for cultivating brine shrimp larvae? The worst conditions?
   9. Do you expect the larvae to grow best under the same physical conditions in which the eggs hatch? Explain.

Part III

G. Count the number of larvae in each of the 8 culture dishes every 3 days.

   10. Record the numbers in your data chart. Also record any other differences you notice, such as differences in size, activity level, or numbers of dead animals. Continue this procedure for 2 weeks.

Strategy for Analyzing

Data analysis will show the relationship of an independent variable and a dependent variable.

H. At the end of the 2 week period, make growth curve graphs by plotting the number of brine shrimp versus time. Then compare the growth curves for the populations in each dish to determine what conditions were most favorable to the growth and survival of the brine shrimp larvae.

   11. Make your graphs on separate graph paper. You can plot several graphs on one set of axes if you use a different color pencil for each culture dish and label each graph clearly.
   12. Can you see any trends among the 8 dishes?

Postlab Analysis

   13. Which concentration of salinity was best for the brine shrimp survival?
   14. How did physical conditions such as light and temperature affect the brine shrimp growth?

Strategy for Hypothesizing

All observations of brine shrimp response must be taken into account in the process of hypothesizing about the ideal brine shrimp habitat.

   15. Make a hypothesis about the optimal physical and chemical conditions for brine shrimp. Use evidence from this investigation to support your answers.
   16. Brine shrimp eggs retain their viabiliy for several years in a dried condition. If this is an adaptation to particular conditions, what might you infer about the brine shrimp's habitat?

Further Investigations

1. Hatch dried brine shrimp eggs under other physical and chemical conditions. One method is to freeze them; then heat them to 35°C. You might use iodized salt or double the number of eggs to produce crowded conditions. Think of other variations that would duplicate natural conditions. Always run a control. Record and evaluate the results of the hatching

experiments and explain what these results tell you about the brine shrimp's reproductive tolerance for these conditions.

2. Using the information you have about optimum conditions for brine shrimp, grow the larvae until they reach reproductive maturity. Examine the specimens and the tray bottom for new eggs. Separate the adult brine shrimp before any of the new eggs hatch, and put them in a separate tray. Use a dissecting microscope to watch for larvae that show any differences in traits. Does the length of the cycle of egg broods indicate any adaptations to season changes? Explain.

Investigation

49 | *Brine Shrimp*

1. _____
   _____
   _____
   _____

2. _____

3. _____
   _____

4. _____
   _____
   _____
   _____

5. _____

6. _____
   _____

7. _____
   _____

8. _____
   _____

9. _____
   _____
   _____

10. Enter your answers on the data chart on the next page.

11. Draw a graph on a separate sheet of paper.

12. _____
    _____
    _____
    _____
    _____

13. _____

14. _____
_____

15. _____
_____
_____

16. _____
_____
_____

Population Growth for Brine Shrimp

Number of Days After Hatching	Culture Dish Number							
	1	2	3	4	5	6	7	8

50

Ecosystem in a Jar

Learning Objectives
- To learn how different organisms affect one another in a closed ecosystem.
- To compare the relationships in a closed ecosystem to those you have observed in nature.

Process Objectives
- To form hypotheses about the relationships among several organisms in a closed ecosystem.
- To observe relationships in a closed ecosystem at regular intervals.
- To model these relationships and, on the basis of your model, form hypotheses to be tested in the future.

Materials

Part I
- Large glass jar with lid that seals
- Water (pond or dechlorinated tap)
- Gravel, or rocks and soil
- Plants for Ecosystem I, II, or III

Part II
- Animals for Ecosystem I, II, or III
- Graph paper
- 8 1/2" x 11" Acetate sheets
- Wax pencils (different colors)

Strategy for Modeling
Think of the type of ecosystem you wish to model and mentally construct a "box" within that ecosystem. Make a list of all the components in your box and try to include as many as possible in your model.

How do species interact in a closed ecosystem?

Introduction

As we study living organisms, we may tend to think of them one at a time, as if they were in cages in a zoo. While such categories are convenient, they are far from natural. In nature, organisms depend upon one another for food, places to live, and even the air they breathe. This investigation will allow you to observe some of these interactions as they occur in a closed ecosystem.

Prelab Preparation

Read Sections 50.1 and 50.2 in your textbook. Then answer Question 1.

1. Suggest 2 different ways that organisms obtain nutrients to sustain life.

Organisms in 3 hypothetical ecosystems are listed below. Read each list and then answer the questions.

ECOSYSTEM I	ECOSYSTEM II	ECOSYSTEM III
Pinch of grass seeds	*Anacharis* strand	Pinch of clover seeds
Pinch of clover seeds	*Fontinallis* strand	Pinch of grass seeds
10 Mung bean seeds	Foxtail strand	3 Pond snails
3 Earthworms	Duckweed	2 Anacharis strands
4–6 Isopods	*Chlamydomonas culture*	10 *Daphnia*
6 Mealworms	Black Ram's horn snail	3 Strands *Fontinallis*
6 Crickets	4 Guppies or platies	*Chlamydomonas*
		4 Guppies or platies

2. Which of these closed ecosystems would you most like to observe ?
3. For the ecosystem you chose in Question 2, model a possible **food chain.**
4. Write a hypothesis about the relationships among several organisms in the closed ecosystem.

Procedure
Part I
A. Form a group with classmates who chose the same ecosystem as you did in Question 2. As a group, prepare the environment in the jar with a chosen **substrate** (gravel and water, or rocks and soil). Ecosystem III indicates both land and water environments. You will need to set up the water environment in a small dish that fits inside the larger container.
B. Plant the seeds and/or add the algae to your ecosystem. Put a lid on the container and let the ecosystem sit in indirect sunlight for a week.

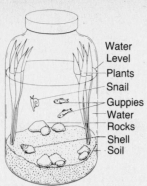

Water
Level
Plants
Snail
Guppies
Water
Rocks
Shell
Soil

Sample Jar for Ecosystem III

Strategy for Hypothesizing
As you think about the nature of the relationships among the organisms you have included in your jar, you can make hypotheses about how these organisms might affect population levels of other organisms.

5. Why should the ecosystem sit for a week before the animals are added?

Part II
C. Place the chosen animals into the jar and lightly close it.

6. What might you learn if more than one jar were set up in an identical manner? What might you learn from several jars if you varied one component in each jar?

D. Observe the jar for a few minutes every day.

7. Have any organisms increased in number? Decreased? Make a chart to record the number of each species.

8. What are some possible causes of the population changes you observed?

E. Make a graph for each species in your chart, plotting the number of organisms as a function of time. Place a clear acetate sheet over each graph.

9. Use a wax pencil to transfer the graph of the population levels onto the acetate sheet. Use a different color and a different acetate sheet for each organism.

F. Compare the acetate sheets of 2 organisms that you have hypothesized would interact—a predator and its prey, for instance. Hold one sheet on top of the other and analyze both graphs.

10. Did the population levels change in the same or in opposite directions?

11. What kind of relationship was there? Does it support or contradict the hypothesis you made in Question 4?

12. Did one population change more, or more quickly, than the other?

13. Was your model of the food chain in Question 3 accurate?

Postlab Analysis

14. Did you expect all of the organisms to reproduce or to die at the same rate? If not, how did you expect them to differ?

15. What effect might this have had on your ecosystem?

16. How would you modify the ecosystem if you were to do this investigation again?

Further Investigations

1. Continue to observe your ecosystem and add to your graphs. Look for trends or cycles.

2. We have concentrated on interactions among living organisms in this exercise. Abiotic factors such as temperature, light, and the nature of the substrate also play a role in ecosystems. Set up ecosystems that demonstrate what these roles might be.

3. Search news magazines for examples of environmental changes that have far-reaching effects on a variety of organisms.

4. Pretend you are a scientist and a chemical company asks you to determine the effect of its chemical (fertilizer or insecticide) on a crop plant, its pests, and the wildlife that uses the area. Instead of jars, you have fields to work with. Use your experience in this exercise to design a study that would answer the chemical company's question.

Investigation

50 | *Ecosystem in a Jar*

1. _____

_____

_____

_____

2. _____

_____

3. _____

_____

_____

4. _____

_____

_____

5. _____

_____

_____

6. _____

_____

_____

7. _____

_____

8. _____

9. Draw a graph on a separate sheet of paper.

10. _____

_____

11. _____

_____

_____

12. _____

13. _____

14. _____

_____

15. _____

_____

16. _____

_____

51

Owl Pellet

Learning Objectives
- To study the food consumed by owls.
- To learn how to use a field guide effectively.

Process Objectives
- To classify small animals whose skeletal remains you have dissected from owl pellets.
- To analyze data in order to determine characteristics of owl prey.

Materials

For Class
- Sheet of blank paper
- Identification guides to mammals, birds

For Group of 2
- Owl pellet
- Metric ruler
- 2 Petri dishes
- Needle probe
- Fine forceps
- Dissecting microscope

What prey do owls eat?

Introduction

Owls are frequently caricatured because of their large, round eyes and fluffy feathers. These features are adaptations that help the owl in its quest for food as a nocturnal predator. The large eyes enable the owl to see in dim light. The light, fluffy feathers allow the owl to swoop silently down on its prey; they also keep the owl warm on cold nights of foraging.

In this lab you will analyze the food consumed by an owl. From these data you will determine common characteristics of owl prey and make inferences about predator—prey interaction in an ecosystem.

Prelab Preparation

An owl catches a small animal and eats the entire thing. It cannot, however, digest the hard, tough parts such as bones, hair, and feathers. In fact, these materials are blocked from reaching the intestines by the **pyloric opening**. Eventually, after enough indigestible material has accumulated, the owl coughs it out as one solid lump. This lump is called an **owl pellet**.

The owl pellet you will use has been dried and fumigated. You can learn what the owl has eaten by carefully picking apart the pellet and dissecting out the skulls and bones of the prey animals. You will not need to analyze every tiny bone, but a careful count of the major bones in the body will tell you how many animals are in your pellet. You will identify the animals mainly by the skulls, mandibles, and teeth, so be especially careful when dissecting these.

In this lab you will be using field guides in order to identify the skulls found in the owl pellet. Get a field guide for mammals from the classroom collection or the library. Flip through it to familiarize yourself with its layout and contents.

1. What are some ways you can predict which species of animals you might find in an owl pellet?
2. In what way might the formation of owl pellets increase an owl's survival in an ecosystem?

Generalized Bird Skeleton

Generalized Mammal Skeleton

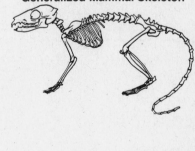

Strategy for Classifying

First sort bones by shape rather than by size. Sort by size after you determine the relationship among different kinds of bones. When identifying prey by size remember that some animals may actually be the young of large animals.

Shrew

House mouse

Meadow vole

Deer mouse

Mole

Rodent

Rabbit

Procedure

A. Measure the length and width of your owl pellet.

3. Record your data on the Chart of Major Bones. Place the pellet on the bottom half of a petri dish. Carefully tease apart the materials with your needle probe and forceps. Gently remove all the bones and place them in the petri dish cover. Place any other materials representative of invertebrate animals, such as an insect carapace, in another dish. Examine small, unidentifiable materials under the dissecting microscope to see whether they are parts of animals.

4. What different kinds of materials have you identified in the owl pellet?

5. How can you distinguish between vertebrate and invertebrate material?

B. Group all similar bones into piles. Try to fit bones from different piles together by matching each skull with the rest of the animal's skeleton.

6. Assign a number to each skull you find. If you find more than 4 skulls, continue the Chart of Major Bones on a separate sheet of paper. For skull #1, put a check mark next to the name of each bone you found that matches its skeleton. Do the same for the remaining skulls.

7. Write short descriptions of any invertebrate materials you found.

8. How many individual vertebrates do you think your collection of bones represents? (Keep in mind that some animal skulls, such as small birds and lizards, might be too fragile to stay intact within the pellet.)

9. Do you think the skulls you have found are from different species? Why?

10. How many different species of animals do you think the remains represent?

Dental Charts of Some Common Small Mammals

Mammal	Number of Teeth Per Side				Total
	Incisors	Canines	Premolars	Molars	
Shrews					
Sorex, Microsorex, Blairina	U 3 / L 1	1 / 1	3 / 1	3 / 3	32
Notiosorex	U 3 / L 1	1 / 1	1 / 1	3 / 3	28
Cryptotis	U 3 / L 1	1 / 1	2 / 1	3 / 3	30
Moles					
Scapanus, Condylura, Parascalops	U 3 / L 3	1 / 1	4 / 4	3 / 3	44
Scalopus	U 3 / L 2	1 / 0	3 / 3	3 / 3	36
Mice and Rats					
Peromyscus, Microtus, Clethrionomys, Mus, Rattus	U 1 / L 1	0 / 0	0 / 0	3 / 3	16

U = Upper   L = Lower

C. With the help of a field guide containing pictures of mammal skulls and dental charts, try to identify the animals in your sample. In the Dental Charts of Some Common Small Mammals on the previous page, the U stands for upper teeth, L stands for lower teeth. The first entry under the heading Incisors is U 3. This entry means that all of the kinds of shrews listed have in their upper jaw 3 incisors on the left side and 3 incisors on the right side.

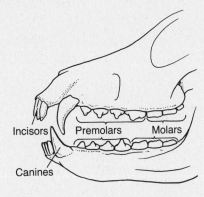

D. Your classification of prey species should include the habitats of the potential prey animals. In a field guide, look up the species that you have identified.

   11. On a separate piece of paper, list each different species and the normal habitats of each.
   12. Are the habitats, sizes, and behaviors of the animals you identified the same as those of animals that might be the prey of owls?

E. Pool data for your class on a class chart with the numbers and species of the animals identified in the pellets.

   13. Fill in a class data chart on the chalkboard.

F. Calculate percentage present. Add the number of each species found, and then divide by the total of all the animals found in all of the pellets for the class.

$$\text{Percent (P) for one species} = \frac{\text{Total of one species}}{\text{Total of all animals}}$$

   14. Record the percentages on the chalkboard.

Postlab Analysis

   15. What animals are represented most often in the diets of the owls your class studied?
   16. What are the common characteristics of these animals?
   17. What type of biotic relationship do the owl pellets provide evidence for?
   18. If all the owls whose pellets your class studied lived in the same ecosystem, what generalization could you make about the population size of the most common prey?

Further Investigations

   1. Assemble into a complete skeleton the bones of one of the small mammals you have dissected from the owl pellet.
   2. Design a model ecosystem in which the owl whose pellet you dissected probably lived. Include characteristic plant and animal species. Diagram

Strategy for Analyzing
To find percentage, divide the total number of animals found into the total number of each species found. This will give you a decimal number. Multiply this number by 100 to find the actual percentage.

your ecosystem, and explain the relationships among the different organisms. Do research to determine the diet of the animals that are represented most often. Are the prey animals primary or secondary consumers? Explain their position in the food chain.

3. Predators such as owls are subjected to poisons that have been concentrated through the food chain. Some species of owls are considered endangered, threatened by pesticides and heavy metals that contaminate the foods eaten by their prey. Do library research and write a report on this topic. Explain how a grain crop contaminated with lead can affect the owl population.

4. From your understanding of predator-prey relationships, explain why prey animals are not eliminated from an ecosystem. How can the existence of predatory animals actually strenghten prey populations?

Investigation

51 | *Owl Pellet*

1. _____

_____

_____

2. _____

_____

3. Enter your answers on the data chart.

Chart of Major Bones

Owl Pellet	Length: (cm)	Width: (cm)		
Bones	Skull #1	Skull #2	Skull #3	Skull #4
Mandible				
Scapula				
Clavicle (furcula—bird)				
Humerus				
Radius				
Ulna				
Femur				
Tibia				
Fibula				
Sternum				
Pelvis				
Tarsometatarsus (bird)				

4. _____

5. _____

6. Enter your answers on the data chart.

7. _____

_____

_____

_____

8. _____

9. _____

10. _____

11. Make a list on a separate sheet of paper.

12. _____

13. Record your answers in the class data chart.

14. Record your answers in the class data chart.

15. _____

_____

16. _____

_____

17. _____

18. _____

_____

52

Population Sampling

Learning Objectives
- To investigate populations within a sector of a natural ecosystem.
- To understand the technique of random sampling.

Process Objectives
- To organize the information you collect on populations.
- To classify populations of different organisms within the ecosystem you study.

Materials
For Class
Part I
- Meterstick or tape measure
- 4 Stakes
- Hammer
- 45 Meters of string

For Group of 2
Part I
- Meterstick or tape measure
- 4 Stakes
- Right triangle
- 5 Meters of string
- Field guides to trees, shrubs, insects, birds, and mammals
- Straight spade or shovel
- 2 Sealable polyethylene bags
- 10 mL 34% Formaldehyde
- Medicine dropper
- Lined paper

Part II
- Fine mesh screen
- White enamel specimen pan
- Fine forceps
- Hand lens or dissecting microscope (optional)
- Petri dish

How do different populations live together in an ecosystem?

Introduction

A population is a group of individuals of the same species living in the same place. Do you know the number of people in your community? This "number of people" may have subcategories, such as the numbers of men, women, and children. Is the population of your classroom representative of the population of your community? For example, imagine that you are a poll-taker asking people's views. Would you expect to get the same results from your class as from your community?

Ecologists face similar problems trying to determine populations of different organisms in an ecosystem. An ecosystem is a community of interacting organisms. If ecologists study too narrow a representation of that ecosystem, they cannot be sure their results are valid. Also, for sampling to be valid, it must be done in a random manner. A random sampling procedure requires that all portions of the area or group being studied are equally represented in the samples taken.

In addition to making actual counts of organisms, ecologists must also consider the behavior of organisms. For example, migratory birds only live within a specific ecosystem for part of the year. Ecologists must also consider environmental conditions that affect each population within an ecosystem. If the sampling is completed during a drought, the data will not show some organisms because they will have died.

Prelab Preparation

In Part I, your class will visually select a representative section of the ecosystem you will investigate. The area should include most of the different kinds of plants contained within the ecosystem. Within that area, each group will select smaller areas, or **quadrats,** to concentrate their population study. Quadrats are plots used for ecological or population studies. Be sure to read through the entire procedure carefully before you go to the site.

1. Think about areas that may be suitable for your study. What kinds of bias do you think you might have introduced in choosing locations to study ecosystems?
2. Review Chapters 50–52 of your textbook. Use the chapters to define the terms *dominant organism, herbaceous,* and *deciduous.*
3. Look through the library or your teacher's field guides if they are available. Discuss with your teacher which types of guides will be most helpful during a sampling of populations in the ecosystem you will study. Record the titles of the guides you use.

Procedure

Part I: Field Study of an Ecosystem

A. Select a class site. Volunteers should stake out a square that is 10 meters on each side. To make the square, place one stake at each corner of the site. Then, loop string around one stake and continue to the next stake until the boundaries have been formed. With your classmates, survey this section by walking through the section and noting the dominant species of trees and shrubs.

4. List these plants by type, relative heights, and layer in a data chart on a separate sheet of lined paper. Use the information in the table on Categories of Plants. If you cannot identify a dominant tree or shrub, ask your teacher to collect one small branch that you may look up in reference books later.

Categories of Plants

Plant types	Relative Heights	Layer
Trees	Tallest	Canopy
Trees	Shorter than tallest	Subcanopy
Trees—not fully grown	About 1.3 m tall	Saplings
Woody plants	0.5–3 m tall	Shrubs
Very young trees	Less than 30 cm tall	Seedlings
Non-wood plants (dies down in winter)	Near ground	Herbaceous vegetation (grasses, weeds)
Ground cover like lichens, mosses	Close to the surface	Flat plants

5. Would you typify your ecosystem as containing mostly trees, shrubs, herbaceous plants, or grasses?

6. What trees and shrubs (or other plants) are dominant in the section of the ecosystem that you are investigating?

7. Are the plants mostly deciduous or evergreen?

B. Each group of 2 will select an area within the 10-meter square to count vegetation. CAUTION: Watch for nettles, stinging insects, and poisonous plants and animals. Groups should spread out and choose areas with different types of growth so that the total counts will be representative of the whole section. Make a quadrat by staking corners of a 1-m square. Use a right triangle to make corners. Loop string around the stakes to outline your quadrat. Each member of your group will measure a particular type of plant. For example, one person will count trees, saplings and shrubs; another will count herbaceous plants. If grasses are dominant in your quadrat, count the grasses in 10 squares, 10 cm long on each side. Then, average your counts and multiply by 100 to get the total for your quadrat.

8. In the data chart, under Plants—Species in Quadrats, list all the types of plants in your quadrat.

Tabulate your counts and identify members of dominant species.

9. What are the dominant plants in your quadrat?

C. As you study and count the plants, observe any insects that live on the plants. Look for flying insects and birds. Note any bird and mammal tracks and fecal material. Look for small mammal burrows. While you are making a quantitative record of plants in your quadrat, make qualitative notes on mobile animals. Be aware, however, that mobile animals cannot be accurately counted in an ecosystem in a short period of time.

For identifying specimens, refer to a field guide or to text Chapters 29–41 when you return to the classroom.

10. List all of the animals and indications of animals each group member has observed during the field study. Identify species if you can.

11. Do you think the animals you see or find traces of actually live in the quadrat or just come and go?

12. What animal species do you think are dominant in this ecosystem?

D. You will collect soil samples in order to examine the animals that burrow in or live on **detritus** (di-TRITE-us), decaying materials at the soil surface. In order to pick a random sample, throw a rock or beanbag into your quadrat. You and your partner should each do this. Dig into the soil to remove a sample from the place where your rock or beanbag landed. Observe the texture and color of the soil. Observe the kinds of materials that form the detritus.

13. Characterize the soil in your sample. Is it rich in decaying materials, clay, or sand? Cut a chunk of soil 10 cm deep and 10 cm on each side. Remove it and place it into a sealable plastic bag. Fill in the holes with extra soil and detritus.

E. In the classroom, add to the soil sample enough water to moisten and a dropperful of formaldehyde. Reseal and mix well. The formaldehyde will preserve your soil sample until you are ready to analyze it. CAUTION: Use formaldehyde only in a well-ventilated area. Do not inhale its fumes or get any on your skin.

14. What kinds of invertebrates do you expect to find in your type of soil?

Strategy for Organizing Data
Once you have organized your raw data into a chart and have calculated the percent for each species of plants, you will be able to compare different plants.

F. In class, complete your data chart of the plants in each quadrat. For each plant species present in your quadrat, calculate its percent among all the plants in your quadrant. Divide the total number of each plant species by the total number of plants.

$$\text{Percent } P = \frac{\text{total plants of one species}}{\text{total number of plants in area}}$$

15. Calculate the percent present of the dominant plant in your quadrat. Does your number agree with the percent calculated for that kind of plant in all the quadrats studied by the other teams? Why or why not?

Part II: Analyzing Soil Samples

G. You will use a fine mesh screen to filter the soil away from the organisms in your sample. Place the sample on the screen over a sink or dish. Carefully pour water over the soil sample, repeatedly rinsing until small soil particles are gone. Turn the screen over into a white enamel specimen pan. Use fine forceps to pick through the materials left from filtering. Look for macrofauna (animals larger than 1 cm), mesofauna (animals 0.2 cm to 1 cm) and microfauna (animals less than 2 mm). Sort the organisms by class. Use a hand lens or dissecting microscope if you need to. Each team member will carry out this procedure with his or her own soil sample.

16. Is one population of invertebrates clearly dominant?

Typical Soil and Detritus Organisms

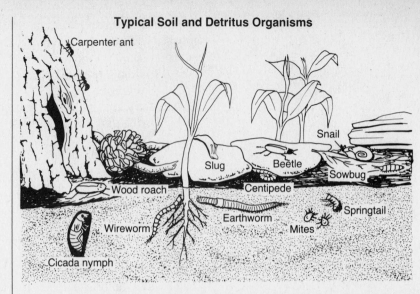

17. Prepare a class chart of soil fauna as you did for vegetation. Count the organisms according to class. Enter the numbers of each population from each student's sample under that quadrat number. Calculate percent present for each population.
18. Then, multiply the numbers of each population for your 10 cm x 10 cm x 10 cm sample by 100 to calculate the numbers of each population.

Postlab Analysis

19. Which organisms in your ecosystem are the producers? Which are the primary consumers? Which are the secondary consumers?
20. List any predators and their prey that you found in your ecosystem. Are any of the prey organisms especially limited in quantity? How would you expect predators in the ecosystem to be affected by this?

Further Investigations

1. Use the data you collected in this lab to construct a numbers pyramid to show the relationship of the different trophic levels in the ecosystem.
2. Research a different ecosystem from the one studied in this investigation. Use library resources to determine weather patterns, populations, and how environmental conditions affect the ecosystem.
3. Find out how your state fish and wildlife agency determines populations to (a) stock lakes and ponds with fish and (b) assess kill limits during hunting season. Find out how environmental conditions affect their decisions from year to year.

Investigation

52 | *Population Sampling*

1. _____

2. _____

   _____

   _____

   _____

3. _____

4. List your data on a chart on a separate sheet of paper.

5. _____

6. _____

7. _____

8. Enter your data on the chart.

Plants—Species in Quadrats

Quadrats	Deciduous Trees	Evergreen Trees	Saplings	Shrubs	Herbs	Grasses	Low Plants
Total							
Average							
Percent Present							

9. _____

10. _____

   _____

   _____

   _____

11. _____

12. _____

13. _____

14. _____

15. _____

16. _____

17.–18. Enter your data on the chart.

Chart of Soil Fauna

Quadrat #	Macrofauna			Mesofauna			Microfauna		Other
	Earth-worms	Large Arthropods	Molluscs	Mites	Spring-tails	Pot worms	Nematodes	Small Arthropods	
Total									
Average									
Percent Present									

19. _____

_____

20. _____

_____

_____

53

Effects of Acid Rain

Learning Objectives

- To learn about the effects of acid rain on seeds from 3 different species of plants.
- To develop an understanding of the consequences of acid rain in the environment.

Process Objectives

- To predict the effects of acid solutions on the germination of seeds.
- To organize class data into a graph that shows the rate of seed growth as a function of pH.

Materials

For Group of 2

- 20 Seeds, bean, pea or corn
- 2 Metric rulers
- 250-mL Beaker
- Mold inhibitor
- 2 Labels
- 2 Plastic bags
- Paper towels
- Water solutions, 2 different pH values
- Sheet of graph paper

How would acid rain affect the germination of seeds?

Introduction

Perhaps you live in an area where smog alerts and air pollutant reports are regular features in the news. Such pollutants include nitrogen oxide and sulfur dioxide. They may irritate your lungs and eyes, but they pose an even greater threat to the environment as a component of acid rain. Acid rain forms when nitrogen oxide and sulfur dioxide combine with water vapor in the air to produce nitric acid and sulfuric acid. Acid rain is now widespread over large areas of western Europe, the eastern United States, and southeastern Canada. While the damage to the fish populations of lakes has been extensively publicized, the effects of acid rain on plants are not as well understood.

Prelab Preparation

Most of the plants we eat come from seeds that germinate only when proper conditions are present. During germination, the seed coat softens, starch within the seed changes to usable sugars, and the embryo grows. This investigation simulates one consequence of acid rain: the effect on seed germination. Review seed germination in Chapter 27. Note the differences mentioned in your textbook between corn seeds and bean or pea seeds.

1. What are the proper conditions for seed germination? How will you control these conditions during your investigation?
2. How would you determine whether precipitation is acidic?
3. Develop a hypothesis to explain the possible effects of acids on the germination of your seeds.

Procedure

CAUTION: Mold inhibitor and acid solutions can damage skin and clothes.

Part I

A. For this procedure, you will work with one water solution, while your partner works with a different water solution. Both you and your partner will work with the *same kind of seed.*

4. Use the metric ruler to measure the length in millimeters of each of your 20 seeds. Determine the average length. Record the seed type, your solution's pH, and the average seed size on your data table.
5. Based on the hypothesis you developed in Question 3, predict what percentage of your seeds will germinate at each of the pH values being used.

B. Place the 20 seeds into a beaker, 10 seeds for you and 10 seeds for your partner. Slowly add mold inhibitor until the seeds are covered. Soak the

Strategy for Predicting

To make a prediction, consider reasons for and against each possible prediction.

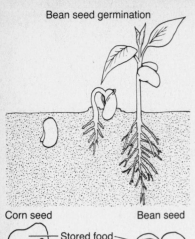

Bean seed germination

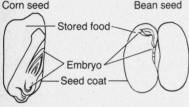

Corn seed   Bean seed

- Stored food
- Embryo
- Seed coat

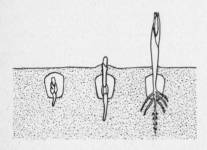

Corn seed germination

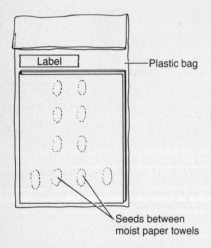

Label

Plastic bag

Seeds between
moist paper towels

Strategy for Organizing

When you graph the class data, plot all the points for one type of seed at a time.

---

seeds for 10 minutes. While the seeds soak, label a plastic bag with your name and the pH of the water solution assigned to you. Moisten 3 layers of paper towels with your assigned water solution. Your partner will label his or her own plastic bag and use a different assigned solution. Fold the moistened towels in half and put them aside. Drain the mold inhibitor from the beaker and gently rinse the seeds with water. Place the seeds on clean paper towels. Blot the seeds dry.

C. Count 10 seeds and place them between the layers of the moistened paper towels. Slide the towels and seeds into your plastic bag. Fold the bag closed. Keep the plastic bag flat while you place it in a warm location. Your partner should do the same with the other 10 seeds.

Part II

D. After 2 days, examine your seeds. Note the shapes and colors. Squeeze gently to see if the seeds have softened. Look for split seed coats. A seed has germinated if you can see a small embryo poking through the seed coat.

6. What overall changes in appearance do you observe?
7. Check for changes in length. Record the average seed size.
8. Count the germinated seeds and record your data.
9. Add your data to the class chart on the chalkboard.
10. Compare your results with those of your classmates. How does the pH value appear to affect seed germination?
11. Look at the class data and predict what changes will occur in the seeds during the next 24 hours.

E. Moisten the paper towels with your assigned solution and return the towels and seeds to the plastic bag.

F. Observe your seeds again after 2 to 3 days. Be sure to check the moisture level of the paper towels and add more of your solution when necessary.

12. Record the average length and number of germinated seeds.
13. What changes have occurred since your last observations?

G. Make your final observations and measurements 7 to 10 days after you began this investigation.

14. What overall changes in appearance do you observe?
15. Repeat Questions 7–9.
16. Subtract the Day 0 average seed size from the last observation average seed size. This is the overall average seed growth. Record this on your data table and the class chart.
17. Create one graph to organize the class data on the effects of pH on the germination and growth of each type of seed. Use a different color to graph data for each type of seed. Label the vertical axis with the number of millimeters of overall average seed growth. Along the horizontal axis show the range of pH values for the entire class.

Postlab Analysis

18. Study the class data. Did seeds from all species of plants studied in this investigation need the same pH range? What appears to be the optimum pH for the germination of seeds?
19. In which solution did each type of seed germinate first?
20. What pH appears to be best both for successful germination and continued seed growth? Which is the least beneficial?

Further Investigation

1. Do you think different stages in the life cycle of plants are more susceptible to acid rain? Design an experiment to investigate this question.

Investigation

53 | *Effects of Acid Rain*

1. _____

   _____

   _____

   _____

   _____

2. _____

3. _____

   _____

4. Enter your answers on the data chart.

Team Data Table

	You	Your Partner
Seed type		
pH of solution		
Average Seed Size Day 0		
First Observation		
Second Observation		
Last Observation		
Overall Average Seed Growth (Last Observation)		
Number of Seeds Used		
Number of Seeds That Germinated First Observation		
Second Observation		
Last Observation		

5. _____

   _____

6. _____

   _____

7.–8. Record your data in the chart above.

9. Record your data on the class chart.

10. _____

11. _____

12. Record your data on the chart on the previous page.

13. _____

_____

14. _____

15.–16. Record your data on the appropriate charts.

17. Make a graph on a separate sheet of paper.

18. _____

_____

19. _____

20. _____

_____